Biomathematics

Volume 22

Managing Editor
S. A. Levin

Editorial Board
C. DeLisi M. Feldman J. Keller R. M. May J. D. Murray
A. Perelson L. A. Segel

Yuri I. Lyubich

Mathematical Structures in Population Genetics

Translated by D. Vulis and A. Karpov
Edited by E. Akin

Springer-Verlag
Berlin Heidelberg New York
London Paris Tokyo
Hong Kong Barcelona
Budapest

Yuri I. Lyubich
State University of New York
at Stony Brook
Department of Mathematics
Stony Brook, NY 11794, USA

Ethan Akin
Department of Mathematics
The City College of
The City University of New York
New York, NY 10031, USA

Original Russian edition, entitled
„Matematicheskie struktury v populyatsionnoi genetike"
published by Naukova Dumka, Kiev, 1983

With 17 Figures

ISBN-13:978-3-642-76213-0   e-ISBN-13:978-3-642-76211-6
DOI: 10.1007/978-3-642-76211-6

Library of Congress Cataloging-in-Publication Data
Lyubich, Yu.I. (Iurii Il'ich) Mathematical structures in population genetics/Yuri I. Lyubich:
translated by D. Vulis and A. Karpov: edited by E. Akin.   p.   cm.
(Biomathematics; v. 22)
Translated from Russian.
Includes bibliographical references and index.
ISBN-13:978-3-642-76213-0
1. Population genetics – Mathematical models. I. Akin, Ethan, 1946 –.   II. Title.   III. Series.
QH455.L58   1992   575.1'5'0151–dc20   91-29941   CIP

© Springer-Verlag Berlin Heidelberg 1992
Softcover reprint of the hardcover 1st edition 1992

Typesetting: Camera-ready by author
41/3140 - 543210   Printed on acid-free paper

Preface

Mathematical methods have been applied successfully to population genetics for a long time. Even the quite elementary ideas used initially proved amazingly effective. For example, the famous Hardy-Weinberg Law (1908) is basic to many calculations in population genetics. The mathematics in the classical works of Fisher, Haldane and Wright was also not very complicated but was of great help for the theoretical understanding of evolutionary processes. More recently, the methods of mathematical genetics have become more sophisticated. In use are probability theory, stochastic processes, nonlinear differential and difference equations and nonassociative algebras. First contacts with topology have been established. Now in addition to the traditional movement of mathematics *for* genetics, inspiration is flowing in the opposite direction, yielding mathematics *from* genetics. The present monograph reflects to some degree both patterns but especially the latter one. A pioneer of this synthesis was S. N. Bernstein. He raised—and partially solved- -the problem of characterizing all stationary evolutionary operators, and this work was continued by the author in a series of papers (1971-1979). This problem has not been completely solved, but it appears that only certain operators devoid of any biological significance remain to be addressed. The results of these studies appear in chapters 4 and 5. The necessary algebraic preliminaries are described in chapter 3 after some elementary models in chapter 2. The later chapters concern the dynamics of populations (chapters 6, 7 and 9) and their formal analogue (chapter 8). Here when selection is absent a very effective algebraic approach was introduced by Reiersöl (1961). This approach was extended by the author to describe explicit solutions of the general evolution equation (1971). These results are in chapter 6 and some more abstract algebraic- dynamical theory, discovered by Etherington (1939) and his followers, is described in chapter 7.

The dynamics of selection (chapter 9) require different methods. Here the leading role goes to Fisher's "Fundamental Theorem" concerning the increase of mean fitness of a population due to natural selection. An approach based on this theory uses a relatively complicated "relaxation" technique to establish the global convergence to equilibrium under conditions more general than have been previously achieved. Due to limitations of length we have omitted several topics which could be studied using these methods: polyploidy, overlapping generations, migration etc. Also not discussed in the book are stochastic considerations such as the effect of finite population size. These important topics deserve a separate presentation.

References and comments are collected in a special concluding Appendix. We assume as background the standard material of the linear algebra, mathematical analysis and probability theory which. For more special information see the books (SchR66; JacN79; GalD60). For the reader unfamiliar with classical genetics we have included a brief sketch in section 1 of chapter 1.

I am grateful to Prof. Akin for his initiative in preparing the translation of the book and for the work done by him.

Contents

9 Selection Dynamics   307

List of Figures

List of Tables

Chapter 1

Introduction to Population Dynamics

1.1 Elements of Classical Genetics

The fundamental concept of classical genetics, the *gene*, was introduced by Mendel under the name "constant character" to explain the observed statistics of inheritance. According to the modern concept, a gene is the distinct portion of a linear molecule of nucleic acid (DNA or RNA) which codes for a certain amino-acid sequence. We view the gene as the elementary unit of inheritance.

Each cell carries in its nucleus an enormous number of genes. The set of these genes is the *genotype* of the cell. The genotypes of all the cells of an organism—excepting the sex cells, or *gametes*, and a few other specialized cell types—is identical to the genotype of the *zygote*—the fertilized egg from which the organism develops. This uniformity is secured by the mechanism of *mitosis*—a process of cell division in which each gene is duplicated and then the two copies depart to the two daughter cells. This description and those below are greatly simplified; however, they are sufficient for the purposes of this book. They can be considered as a preliminary verbal model that will be subject to further mathematical formalization.

The carriers of the genes in the nucleus are thread-like structures called *chromosomes*. Every chromosome belongs to a class of *homologous* chromosomes that are very much alike. This class is called a *linkage group*. The number of linkage groups and their chromosome set is fixed for each species. In species with bisexual differentiation two linkage groups are associated with bisexuality and are called *sex linkage* groups. The chromosomes in these groups are called *sex chromosomes.* These linkage groups are usually labeled X and Y (in humans the Y-chromosome determines the male sex). The remaining linkage groups (and chromosomes) are called *autosomal* (*autosomes*).

The position at which a gene occurs on a chromosome is called the *locus* (of this gene). A chromosome can be conventionally represented as a segment of a straight line cut into consecutive loci. In homologous chromosomes there is a one-to-one correspondence between the loci. The genes occurring at corresponding loci (i.e. identical places) in homologous chromosomes are called *allelic* or *alleles.* Alleles can spontaneously (with certain probabilities) turn into each other. These transformations are called *mutations.* Mutations may also result in occurrence of new alleles.

Considering the corresponding loci as identical, let us say that allelic genes belong to the same locus or are alleles of this locus. In this sense the locus belongs not to a given chromosome but to a given linkage group. Two loci that belong to the same linkage group are called *linked.* The latter can be called *autosomal* or X- or Y-linked, depending on the linkage group of the locus. X- and Y-linked loci are called sex- linked. Some loci should be considered simultaneously as X- and Y-linked. They are called XY-*linked.* In this case X- and Y-chromosomes are considered to be homologous.

Each locus contains not less than two alleles. In the case of exactly two alleles the locus is called *diallelic* and in the general case it is called *polyallelic.* The set of N homologous chromosomes is called N-ploid (number N is the *ploidy* of the set). For $N = 1$ it is *haploid,* for $N = 2$ it is *diploid,* for $N = 3$ it is *triploid* and so on. For $N > 2$ the set is called *polyploid.*

In each cell there is not less than one autosome from each linkage group and not less than one sex chromosome. The *ploidy* of the cell in the general case is described by ploidies $N_1, \ldots, N_r$ and N_X, N_Y of autosomal and sex groups respectively. If $N_1 = 1, \ldots, N_r = 1$, $N_X + N_Y = 1$, then the cell is called *haploid,* if $N_1 = 2, \ldots, N_r = 2$, $N_X + N_Y = 2$ it is called *diploid.* In a diploid cell there are either two X-chromosomes, i.e. $N_X = 2$, $N_Y = 0$ or one X- and one Y-chromosome, i.e. $N_X = 1$, $N_Y = 1$. The combination of two Y-chromosomes ($N_X = 0$, $N_Y = 2$) doesn't occur. In a haploid cell there is either a single X- or a single Y-chromosome. In the simplest case that we will consider further on, the gametes are haploid and the zygotes are diploid. If the zygote is diploid then the same or different genes may occur at a locus belonging to two homologous chromosomes. In the first case the cell is called a *homozygote* and in the second case it is called a *heterozygote.*

It often proves useful to consider not the whole of the genotype but a part, that is, a singled out and fixed system of loci. The definitions given above apply to such incomplete genotypes and the word "incomplete" is not used only for the sake of brevity. For example, in the simplest case in Mendel's work there is just one diallelic autosomal locus with A- and a-alleles. Haploidic gametes had A- and a-genotypes, while the diploid zygotes had AA, aa and Aa-genotypes (the alleles at a locus are unordered). The AA and aa-genotypes are homozygotes. The Aa is a heterozygote.

Each organism of a species with sexual reproduction is the direct offspring of two parents, "father" and "mother". We will only consider species of this kind. That doesn't exclude *hermaphrodism*, i.e. bisexuality of a zygotic organism with a sexual differentiation of gametes. For instance, this is the case with monoecious plants. These plants can exhibit self- fertilization.

Let's consider the formation of the offspring genotype from the parents' genotypes. This process is carried out in two stages— meiosis and fertilization—according to the two levels of existence of nuclei-zygotic and gametic. This reproduction in two stages is the formal criterion of its sexual character. Asexual reproduction is carried out by mitosis.

I. *Meiosis* is a process of forming of gametes from a parental zygote by the independent separation and departure of homologous chromosomes to two germ-cells. Here an enormous number of possibilities (2^{23} in humans) occurs. In fact, the number is many orders greater due to recombination.

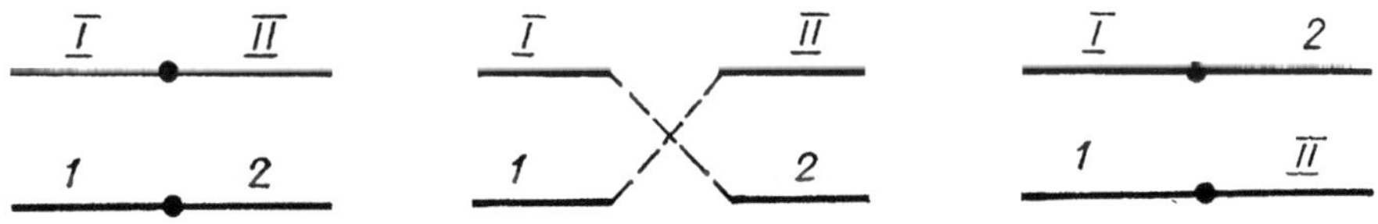

Fig. 1.1

During *crossing-over* the homologous chromosomes are broken in the same place with exchange of the corresponding parts. This exchange of parts (*recombination*) is shown in Fig. 1.1. This is a *single* crossing-over. During one act of meiosis, there can simultaneously occur breaks in several places and therefore *multiple* crossings-over are possible (*double, triple* and so on). Formally speaking, the homologous chromosomes can be altogether decomposed to genes that will be arbitrarily, though with loci intact, distributed over two new homologous chromosomes. Actually, such total decomposition or even most of the partial ones are physically impossible.

Recombination between two loci occurs if and only if a crossing- over of odd multiplicity takes place. Let the probability of i- multiple crossing-over equal $p_i(i = 0, 1, 2, \ldots,)$. Thus if $p_0 \geq p_1 \geq p_2 \geq \ldots$, then $r \leq \frac{1}{2}$. However $r > \frac{1}{2}$ is possible. That only happens when the probability of multiple crossings-over doesn't depend monotonically on the multiplicity.

II. *Fertilization.* During fertilization two haploid gametes picked at random from the gamete pool of their parents fuse to form a zygote that is diploid. (In some species, for example bees, there can occur a process of parthenogenesis, that is reproduction through the development of unfertilized gamete by reduplication of the chromosomal set and subsequent mitosis). The zygotic genotype is produced by pooling the genotypes of parental gametes. The male gamete is called the *sperm cell* and the female cell is called the *egg cell* or *ovum.* With parental genotypes given, different genotypes of the offspring may occur with certain probabilities.

Let us consider the rather intriguing question of influence of the genotype on the phenotype.

The phenotype of an individual is usually defined as a set of all its experimentally or visually distinctive hereditarily determined *characters* such as sex, color of the body parts, for example the eyes, number of bristles, size, for example the length of the body or the height, characteristic behaviors and so on. Obviously, the concept of character should be considered as primary, from the formal point of view. Every character is an element of a certain set of "the same type" characters. The examples of the same type characters are male and female sex, different eye color, different values of height. Thus, there is a system of sets of the same type characters for every species. These sets are called *kinds of characters* and to which are applied such terms as "sex", "eye-color", "height" and so on. The kind of characters is called *qualitative* if it comprises a finite number of possibilities (for example, sex) and *quantitative* in the opposite case (for example, height, theoretically, at least).

The *principle of genotype control* plays the fundamental role in genetics. It states that the phenotype of an individual is uniquely determined by its genotype under strictly fixed environment conditions. Formally speaking, the genotypic control is a certain mapping of the set of genotypes on the set of phenotypes. Biochemically, the genotypic control is realized by the system of protein synthesis operating in an organism, the program for which is coded in its DNA. As a rule, the character of a certain kind is determined not by the complete genotype but by an associated part. That allows one to speak of a locus or a system of loci controlling the character. For instance, sex is determined by the following rule (in human): the zygote carrying two X-chromosomes is female, the zygote carrying an X- and a Y-chromosome is male. The gamete carrying an X-chromosome is called *female* and it is called *male* if it carries a Y-chromosome. The egg cells are always female, the sperm cells can be male and female. A mother passes on one or another of her X-chromosomes to each of her children, and a father passes over his X-chromosome to his daughter and his Y-chromosome to his son. In this sense the sex of a child only depends on the father. Theoretically, parents with many children should have daughters and sons approximately in a 1:1 ratio. However, statistically significant deviations are observed that have not been thoroughly explained.

The color of peas in Mendel's experiments was controlled by an autosomal diallelic locus via the following rule: yellow in the presence of A gene, i.e. homozygotes AA and heterozygotes Aa have yellow color, and green in the absence of gene A, i.e. the homozygotes aa have green color. For such interrelations of genes we use the following terms: gene A is *dominant* or A is dominant over a, while gene a is *recessive*. Mendel crossed "pure lines", i.e. AA and aa ("pure" yellow with "pure" green) and only got the yellow "hybrids" Aa in the first generation of the offspring. Crossing in this generation led to $\frac{1}{4}AA + \frac{1}{4}aa + \frac{1}{2}Aa$ in the next generation, i.e. $\frac{3}{4}$ yellow ones and $\frac{1}{4}$ green ones phenotypically. In fact, the ratio 3:1 was approximate as fluctuations are unavoidable in finite samples. Let us cite Mendel: "Test 2. Color of seed. 258 plants gave 8023 seeds, 6022 yellow ones and 2001 greens ones so the first ones are related to the second ones by a ratio 3.01:1". Mendel got the analogous results for other kinds of characters, such as seed coat, length of stem, color of flower, etc.

It should be mentioned that this disappearance of the character in the intermediate generation gives evidence of its control by a recessive gene.

One of blood group systems in human is autosomal. The locus has two alleles M and N. Here the genotypes are phenotypically distinct, MM, NN, MN. The main blood group system in human is controlled by an autosomal locus with A, B, O alleles. Each of the A and B genes is dominant over O and they "don't interact" in AB combination. So four phenotypes are possible: $\mathbf{O} = \{OO\}$, $\mathbf{A} = \{AA, AO\}$, $\mathbf{B} = \{BB, BO\}$, $\mathbf{AB} = \{AB\}$. If the father and mother of a child have **O** blood group then the child can only have the same group. If the father has **AB** blood group the child cannot have **O** group, no matter what the mother's group is, and so forth. Applying these rules resulting from genotypic control of blood groups, fatherhood can be legally ruled out in some cases and sometimes the fact of the fatherhood can be legally admitted though the latter can be only established by "the logic of probabilities".

The so-called rhesus-factor Rh of blood in humans is under control of an autosomal locus, that carries more than ten alleles, eight of which are clearly phenotypically distinguishable.

The kind of characters may be expanded with the development of experimental techniques. That is exactly what happened to the kind Rh. Originally, it contained only two elements Rh^+ and Rh^- that were serologically distinct, i.e. were distinguishable by serum reaction. The inheritance of the Rh- factor was satisfactorily explained by an autosomal locus with R- and r-alleles where R is dominant over r. In the presence of R the zygote is rhesus-positive (Rh^+) and in its absence it is rhesus-negative (Rh^-). A more subtle serological analysis using three kinds of sera, i.e. eight combinations $+++,++-,\ldots,---$, let to distinguishing 8 alleles: $R^0, R^1, R^2, R^z, r, r', r'', r^y$. The first four of these were formerly pooled as R, the other four as r. Still in the examined samples of people about 0.09% lacked any of these eight characters. This indicated the existence of some other alleles at the locus. The discovery of new sera may reveal more alleles.

The blood disease hemophilia is caused by the gene h and is a recessive allele of the "normal" gene H. These genes occur on the X-chromosome. If a father isn't afflicted with the disease then his X-chromosome carries H. A mother who is heterozygone while not afflicted is still a carrier, i.e. one of her X-chromosomes carries H and the other carries h. None of her daughters would be subject to hemophilia, however, they will be carriers with $\frac{1}{2}$ probability. A son will be hemophiliac with a probability $\frac{1}{2}$. So hemophilia is practically a male disease (the probability of female genotype hh is too small) but it is passed "in the female line" from mother to son.

Color vision in humans is controlled by five alleles of an X- chromosomal locus.

Y-linked gene can only be passed "in the male line" from father to son. However, the only known case of such inheritance in humans is hypertrichosis of the ear-lobe. So the main role of Y-chromosome is sex-determination. It follows that the partial sex-linkage is not found in humans. It was found in some species of fish (a partially sex-linked locus controls the color pattern on the body).

The characters determined by one locus are called *single* and the characters determined by more than one locus are called *multiple*.

The action of genes is manifested statistically in sufficiently large communities of mating individuals (belonging to the same species). These communities are called *populations*. Open and closed populations should be distinguished. In an *open* population *migrations* are possible, i.e. there may occur an inflow of individuals from other populations of the same species. This results in gene exchange with other populations. A *closed* population or an *isolate* is not a subject to migration either due to its *geographical* isolation from the other population of the same species or due to specific factors of environment (*ecological* isolation). In the closed populations new genes occur only by the way of mutations.

Actually, not just the fact of migration but the balance of inflow and outflow of migrants is what matters, i.e. we can speak of *effective* isolation even for an open population. Further on only effectively closed populations will be under consideration. The territory that is occupied by a population is called its *areal* and the total of its factors of environment (typical food, climate, characteristics of habitat, circadian rhythm etc.) is called its *ecological niche*.

The population exists not only in space but also in time, i.e. it has its own life cycle. The basis for this phenomenon is reproduction by *mating*. Mating in a population can be free (*panmixia*) or subject to certain restrictions. For example, in human populations marriages between parents and children, between brothers and sisters (siblings) and sometimes between cousins are usually forbidden because of their probable undesirable hereditary consequences.

It is often possible to think that the whole population in space and time comprises *discrete generations* $F_0, F_1, F_2, \ldots$. The generation F_{t+1} is the set of individuals whose parents belong to the F_t generation. The generation F_0 is a priori fixed. That is the situation when the generations are said not to *overlap*. In the general case only those individuals belong to F_{t+1} whose parents belong to the generations $F_i, F_j (i, j \leq t)$. In the present monograph only the situation of discrete non-overlapping generations is considered. The number of generation is identified with the moment of time.

In each generation F_t the population is characterized by its size, S_t. If the numbers are great then the population must be characterized in terms of statistics. Only the limiting case $S_t \equiv \infty$ will be studied using the state of the population as a means of description. A *state* of a population is a distribution of probabilities of the different *types* of organisms in every generation. Type-partition is called *differentiation.* Speaking of a population we will consider it to be already differentiated (in this sense the concept of "type" is primary). Further on the set of types will be considered finite.

The simplest example is sex differentiation. In bisexual population any kind of differentiation must agree with the sex differentiation, i.e. all the organisms of one type must belong to the same sex. Thus, it is possible to speak of male and female types. That doesn't exclude the identification of heterosexual types under certain conditions. A bisexual population in which the types are sex-independent is called *autosomal.* It often occurs that the kind of characters is controlled only by autosomal loci (see examples cited above). In this case sex differentiation ought to be introduced as partition of a population into two parts of which the set of genotypes, the kind of characters and the genotypic control are the same.

In evolutionary theory it is the genotypical differentiations that play the main role. *Genotypical* differentiation is the partition of a population into genotypes for a given locus or system of loci. Along with genotypical differentiation one may consider *phenotypical* differentiation though in a certain sense this kind of differentiation is not hereditary. We shall go into more detail in the next section.

The *evolution* (or *dynamics*) of a population comprises a determined change of state in the generations as a result of reproduction and selection. The evolution of a population can be also considered as a random process. The present book doesn't study the latter point of view.

Selection occurs through a process of organism environment interactions and internal processes affecting the probabilities of survival before the reproductive period.

1.2 General Evolutionary Equations

We now move to the formal description of a population and its evolution. Assuming the population is bisexual we suppose that the set of females can be partitioned into finitely many different types indexed by $\{1, \ldots, n\}$ and, similarly, that the male types are indexed by $\{1, \ldots, \nu\}$. The number $n + \nu$ is called the *dimension* of the population. The population is described by its *state vector* (x, y) in $\Delta^{n-1} \times \Delta^{\nu-1}$, the product of the unit simplices in $\mathbf{R}^n$ and $\mathbf{R}^\nu$ respectively. x and y are the probability distributions of the females and males over the possible types:

$$x_i \geq 0, \ \sum_{i=1}^{n} x_i = 1; \ y_k \geq 0, \ \sum_{k=1}^{\nu} y_k = 1.$$

Thus, the *state space* $S = \Delta^{n-1} \times \Delta^{\nu-1}$ is a compact, convex subset of $\mathbf{R}^{n+\nu}$.

We call the partition into types *hereditary* if for each possible state $z = (x, y)$ in S describing the current generation, the state $z' = (x', y')$ is uniquely defined describing the next generation obtained by reproduction and selection. This means that the association $z \mapsto z'$ defines a map $V : S \to S$ called the *evolutionary operator* and

$$z' = V(z) \qquad (z \in S) \qquad (1.2.1)$$

is called the *evolutionary equation* (or *equation of evolution*) for the system.

Thus for a hereditary partition the distribution of types in the parental generation provides sufficient information to describe the distribution of offspring types. Other characteristics of the parental generation or the environment either have no effect or these effects can be averaged away. Also we assume the system is time-independent meaning that the functional relation between the generations given by (1.2.1) remains the same over time. This means we can describe the evolution of the population distributions as a dynamical system on S by iterating the map V. That is, we apply the equation (1.2.1) recursively defining:

$$z^{(t+1)} = V(z^{(t)}). \qquad (t = 0, 1, 2, ...) \qquad (1.2.2)$$

for every initial state $z^{(0)}$ in S. Obviously, $z^{(t)} = V^t(z^{(0)})$.

We assume that the operator V is continuous and so the general theory of discrete time dynamical systems applies to (1.2.2). In particular, because S is a compact, convex set, the Brouwer Fixed-Point Theorem implies that the map V has at least one fixed point in S. z^* is a fixed point when $z^* = V(z^*)$ and so z^* is an *equilibrium* for the system. The set of fixed points is the subset on which the map V and the identity id_S are equal. So the set is a closed, nonempty subset of S.

All points are fixed precisely when V is the identity map $,V = \mathrm{id}_S$.

For any point $z^{(0)}$ in S the associated sequence $z^{(t)} : t = 0, 1, \ldots$ defined by (1.2.2) is called the *trajectory* of $z^{(0)}$. If $z^{(0)}$ is an equilibrium then the trajectory remains at $z^{(0)}$. More generally, if the trajectory sequence converges and $z^{(\infty)} = \lim_{t \to \infty} z^{(t)}$ then $z^{(\infty)}$ is an equilibrium by (1.2.2) and continuity of the operator V.

In general, the set $\omega(z)$ of limit points of the trajectory $\{z^{(t)}\}$ where $z^{(0)} = z$, is a nonempty, closed subset of the compact space S. $\tilde{z} \in \omega(z)$ if some subsequence $\{z^{(t_k)}\}$ converges to $\tilde{z}$ in which case $\{z^{(t_k+1)}\}$ converges to $V(\tilde{z})$ by continuity of V. Consequently, $\omega(z)$ is V invariant. Moreover, $V(\omega(z)) = \omega(z)$ by compactness of S. The trajectory is convergent if and only if $\omega(z)$ consists of a single point. In the majority of population genetics examples all trajectories do converge. The following Lemma provides a useful test for convergence.

Lemma 1.2.1 *The set $\omega(z)$ consists entirely of equilibrium points if and only if*

$$\lim_{t \to \infty} d(z^{(t+1)}, z^{(t)}) = 0. \tag{1.2.3}$$

In that case, $\omega(z)$ is a connected set.

Proof If (1.2.3) holds and $\{z^{(t_k)}\}$ converges to $\tilde{z}$ then $0 = \lim_{k \to \infty} d(z^{(t_k+1)}, z^{(t_k)}) = d(V(\tilde{z}), \tilde{z})$ and so $V(\tilde{z}) = \tilde{z}$.

In general, if U is any open set containing the entire set of limit points $\omega(z)$ then the sequence $z^{(t)}$ lies in U when t is sufficiently large. If $\omega(z)$ consists entirely of equilibria then the real valued function $\delta(z) = d(V(z), z)$ vanishes on $\omega(z)$. Then for every $\epsilon > 0$ the set $U_\epsilon = \{z : \delta(z) < \epsilon\}$ is an open set containing $\omega(z)$ and so $z^{(t)} \in U_\epsilon$ for t sufficiently large. This implies (1.2.3).

Now suppose $\omega(z)$ consists of equilibria and is the disjoint union of closed sets A_1 and A_2. We prove that A_1 or A_2 is empty. Choose $\epsilon > 0$ so that the distance between A_1 and A_2 is at least ϵ. Then choose T so that $t \geq T$ implies $z^{(t)}$ lies in the open set $W = \{x : d(x, \omega(z)) < \frac{\epsilon}{3}\}$ and in $U_{\frac{\epsilon}{3}}$ from the previous paragraph. Notice that W is the disjoint unions of $W_i = \{x : d(x, A_i) < \frac{\epsilon}{3}\}$ for $i = 1, 2$. If $z^{(T)} \in W_1$ then $d(z^{(T+1)}, z^{(T)}) = \delta(z^{(T)}) < \frac{\epsilon}{3}$ and so $z^{(T+1)} \notin W_2$. Hence, $z^{(T+1)} \in W_1$. Repeating the argument inductively we have $z^{(t)} \in W_1$ for all $t \geq T$. So the entire set of limit points $\omega(z)$ is contained in the closure of W_1 which is disjoint from W_2. But $A_2 \subset W_2$ and A_2 is a subset of $\omega(z)$. Hence, A_2 is empty. It follows that $\omega(z)$ is a connected set. $\square$

Corollary 1.2.2 *If $\omega(z)$ is a finite set of equilibria then the trajectory is convergent.*

Proof A connected, nonempty, finite set has one element. $\square$

The form of the evolutionary operator depends on assumptions about the reproductive and selection processes. The simplest and at the same time most important system of mating is *random mating* or *panmixic*. This means that mating partners are randomly selected from the population with no dependence on type.

Formally, if (x, y) is the population state then the probability that a female of type i mates a male of type k is given by the product $x_i y_k$. Although we will not consider them, departures from panmixia do occur in natural populations. The most important are: *assortative mating* (preference for similar type), *inbreeding* (preference for relatives) and *self-sterility* (incompatibility of like types). We will restrict attention to panmixic populations.

(1) *Inheritance coefficients.* $p_{ik,j}^{(f)}$ and $p_{ik,l}^{(m)}$ are defined as the probability that a female offspring is type j and, respectively, that a male offspring is of type l, when the parental pair is ik ($i, j = 1, \ldots, n$ and $k, l = 1, \ldots \nu$).

$$p_{ik,j}^{(f)} \geq 0, \qquad \sum_{j=1}^{n} p_{ik,j}^{(f)} = 1$$

$$p_{ik,l}^{(m)} \geq 0, \qquad \sum_{l=1}^{\nu} p_{ik,l}^{(m)} = 1.$$

The inheritance coefficients include the effects of recombination, mutation, gamete selection, variation of fertility.

(2) *Survival coefficients.* $\lambda_j^{(f)}$ and $\lambda_l^{(m)}$ are defined as the probability that a female individual of type j and that a male individual of type l will survive to reproductive maturity. The complementary probabilities $1 - \lambda_j^{(f)}$ and $1 - \lambda_l^{(m)}$ are the *mortality* coefficients for each type.

We will assume that these coefficients are constants, independent of time and the current state of the population.

Now let (x, y) be the state of the parental generation. To be precise x and y are the probability distributions of the female and male types at the mature reproductive stage.

Let $(\overline{x}, \overline{y})$ denote the state of the offspring population at the birth stage. This is obtained from the inheritance coefficients by using the assumption of panmixia:

$$\overline{x}_j = \sum_{i,k=1}^{n,\nu} p_{ik,j}^{(f)} x_i y_k; \quad \overline{y}_l = \sum_{i,k=1}^{n,\nu} p_{ik,l}^{(m)} x_i y_k. \tag{1.2.4}$$

From the birth stage to the reproductive stage of the offspring generation the survival coefficients describe the effects of selection due to differential viability. We have by formulae of Bayes:

$$x'_j = \frac{\lambda_j^{(f)}\overline{x}_j}{\sum_{r=1}^n \lambda_r^{(f)}\overline{x}_r}; \ y'_l = \frac{\lambda_l^{(m)}\overline{y}_l}{\sum_{\rho=1}^\nu \lambda_\rho^{(m)}\overline{y}_\rho}. \tag{1.2.5}$$

Combining (1.2.4) and (1.2.5) we get the evolutionary equations:

$$x'_j = \frac{1}{W^{(f)}(x,y)} \sum_{i,k=1}^{n,\nu} w_{ik,j}^{(f)} x_i y_k$$

$$y'_l = \frac{1}{W^{(m)}(x,y)} \sum_{i,k=1}^{n,\nu} w_{ik,l}^{(m)} x_i y_k \tag{1.2.6}$$

where $w_{ik,j}^{(f)} = p_{ik,j}^{(f)}\lambda_j^{(f)}$ and $w_{ik,l}^{(m)} = p_{ik,l}^{(m)}\lambda_l^{(m)}$. These are *fitness coefficients* (corresponding to the male and female portions of the population) with mean values:

$$W^{(f)}(x,y) = \sum_{i,k,j=1}^{n,\nu,n} w_{ik,j}^{(f)} x_i y_k$$

$$W^{(m)}(x,y) = \sum_{i,k,l=1}^{n,\nu,\nu} w_{ik,l}^{(m)} x_i y_k. \tag{1.2.7}$$

These bilinear forms are called the *mean fitnesses* for the male and female genders when the population state is (x,y).

If a population is *autosomal* then the male and female types are identical and, in particular, $n = \nu$, the inheritance and survival coefficients are the same for male and female offspring, defining

$$p_{ik,j} \equiv p_{ik,j}^{(f)} = p_{ik,j}^{(m)}; \ \lambda_j \equiv \lambda_j^{(f)} = \lambda_j^{(m)}.$$

$$w_{ik,j} \equiv p_{ik,j}\lambda_j = w_{ik,j}^{(f)} = w_{ik,j}^{(m)}. \tag{1.2.8}$$

Observe that with the assumptions of (1.2.8) the offspring distributions for males and females are the same, both at the birth stage, $\overline{x} = \overline{y}$ from (1.2.4), and at the adult stage, $x' = y'$ from (1.2.5) and (1.2.6). Thus, even if $x \neq y$ at the initial point (x,y) describing the parental state, after one generation the state of the population moves into the diagonal subset

$D = \{(x, x) : x \in \Delta^{n-1}\} \subset S$. In other words, the evolutionary operator V on S has image contained in D. Hence, all evolutionary trajectories $\{z^{(t)} : t = 0, 1, 2, \ldots\}$ lie in D for all $t \geq 1$. In particular, every equilibrium lies in the invariant set D.

We can identify Δ^{n-1} with the diagonal D by the map $\delta : \Delta^{n-1} \to S$ given by $\delta(x) = (x, x)$. So we obtain an operator $\hat{V}$ on Δ^{n-1} by restricting V to the invariant set D and then using the identification δ. So the diagram

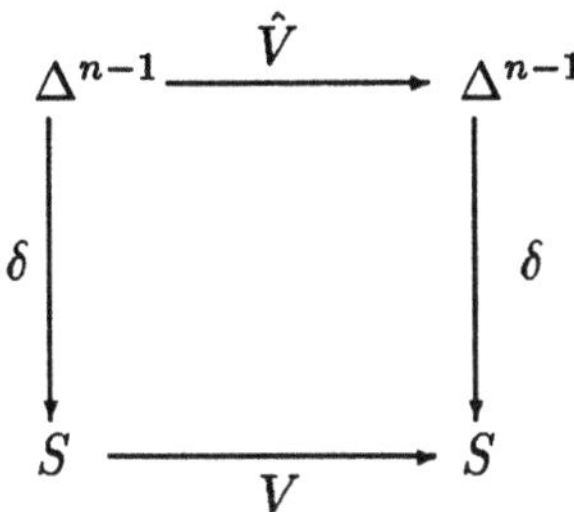

is commutative. We refer to $\hat{V}$ as the *reduced evolutionary operator* in the autosomal case (this case is defined by $n = \nu$ and assumptions (1.2.8)). Since the image $V(S)$ is contained in $D = \delta(\Delta^{n-1})$ every trajectory of the original system can be identified with a trajectory of the reduced system after at most one iteration of the map.

Further we will write V instead of $\hat{V}$ for the reduced map on the *autosomal state space* which is the reduced space state Δ^{n-1}. We call n the *dimension* of the reduced population. Now $x' = V(x)$ is defined by:

$$x'_j = \frac{1}{W(x)} \sum_{i,k=1}^{n} w_{ik,j} x_i x_k, \qquad (l \leq j \leq n)$$

$$W(x) = \sum_{i,k,j=1}^{n} w_{ik,j} x_i x_k. \qquad (1.2.9)$$

This quadratic form is called the *mean fitness* of the autosomal population. Because we are only looking at the restricted operator we can impose the symmetry assumptions:

$$p_{ik,j} = p_{ki,j}; \ w_{ik,j} = w_{ki,j} \qquad (1 \leq i, k, j \leq n). \qquad (1.2.10)$$

For even if the male and female parents do not have symmetric effect on inheritance, i.e. so that (1.2.10) does not hold, we can replace our original inheritance coefficients by their symmetrizations $\frac{1}{2}(p_{ik,j} + p_{ki,j})$ which similarly symmetrizes $w_{ik,j}$ as well. Because the evolutionary equations (1.2.9) use only the diagonal case $x = y$ the function V on Δ^{n-1} is unaffected by this change. So we may assume (1.2.10) without loss of generality.

A population is called *selectively neutral* if the survival coefficients are type independent, i.e. $\lambda_j^{(f)}$ is a constant independent of j and $\lambda_k^{(m)}$ is independent of k. From (1.2.5) it is clear that $x' = \overline{x}$ and $y' = \overline{y}$. That is, the offspring distribution is the same at maturity as it was at the initial birth stage. So for a selectively neutral population the evolutionary equations are given by (1.2.4) with (x', y') replacing $(\overline{x}, \overline{y})$. An alternative route to this result begins by observing that the mean fitnesses $W^{(f)}(x, y)$ and $W^{(m)}(x, y)$ are, in the selectively neutral case, independent of x and y and are, in fact, just the common constant values $\lambda^{(f)}$ and $\lambda^{(m)}$, respectively. In (1.2.6) these constants are common factors in $w_{ik,j}^{(f)}$ and $w_{ik,j}^{(m)}$. When pulled out and cancelled they transform (1.2.6) to (1.2.4).

An autosomal, selectively neutral population is called *free*. The reduced evolutionary equations of a free population of dimensions n are characterized by the inheritance coefficients $\{p_{ik,j} : 1 \leq i, k, j \leq n\}$ satisfying:

$$p_{ik,j} \geq 0, \quad \sum_{j=1}^{n} p_{ik,j} = 1$$

$$p_{ik,j} = p_{ki,j}. \tag{1.2.11}$$

The reduced operator $V : \Delta^{n-1} \to \Delta^{n-1}$ is defined by

$$x'_j = \sum_{i,k=1}^{n} p_{ik,j} x_i x_k \qquad (1 \leq j \leq n). \tag{1.2.12}$$

The transformation of (1.2.9) to (1.2.12) in the reduced case is just like that of (1.2.6) to (1.2.4).

We see from (1.2.12) that for a free population the evolutionary operator is a quadratic mapping of the simplex into itself. This map is closely related to an algebra structure on the vector space $\mathbf{R}^n$ containing the simplex Δ^{n-1}. Recall that any vector x in $\mathbf{R}^n$ can be written as $x = \sum_{i=1}^{n} x_i e_i$ with $\{e_1, \ldots, e_n\}$ the canonical basis on $\mathbf{R}^n$ with $e_1 = (1, 0, \ldots, 0)$, $e_2 = (0, 1, 0, \ldots), \ldots$ i.e. in coordinate form $(e_k)_i = \delta_{ki}$, the Kronecker delta given by

$$\delta_{ki} = \begin{cases} 1 & k = i \\ 0 & k \neq i \end{cases} \quad (1 \leq i, \ k \leq n). \tag{1.2.13}$$

Now introduce on $\mathbf{R}^n$ a multiplication defined by

$$e_i e_k = \sum_{j=1}^{n} p_{ik,j} e_j. \tag{1.2.14}$$

Thus we identify the coefficients of inheritance as the structure constants of an algebra, i.e. a bilinear mapping $\mathbf{R}^n \times \mathbf{R}^n$ to $\mathbf{R}^n$ $,x \times y \mapsto xy$. The algebra is commutative because $p_{ik,j} = p_{ki,j}$, but is not in general associative. The general formula for multiplication is the extension of (1.2.14) by bilinearity for multiplication:

$$xy = \sum_{i,k,j=1}^{n} (p_{ik,j} x_i y_k) e_j. \tag{1.2.15}$$

In particular, the evolutionary operator defined in coordinate form by (1.2.12) can be written:

$$x' = V(x) = x^2 \qquad (x \in \Delta^{n-1}). \tag{1.2.16}$$

This algebraic interpretation is very useful. For example, a population state x is an equilibrium precisely when x is an idempotent element of the unit simplex Δ^{n-1}.

If we write $x^{[t]}$ for the power $(\ldots (x^2)^2 \ldots)$ (t times) with $x^{[0]} \equiv x$ then we see that the trajectory with initial state x is $V^t(x) = x^{[t]}$.

We note that the commutative multiplication $x \times y$ on the space $\mathbf{R}^n$ satisfying $Vx = x \times x$ for a given evolutionary operator $V : \Delta^{n-1} \to \Delta^{n-1}$ is unique. Indeed, the identity $x \times x = x^2$ for $x \in \Delta^{n-1}$ implies the same identity for all $x \geq 0$ (i.e. for $x_1 \geq 0, \ldots, x_n \geq 0$) by the substitution $x \mapsto \lambda x$ ($\lambda > 0$). Then $x \times x = x^2$ for all $x \in \mathbf{R}^n$ because if a quadratic form is zero for $x \geq 0$ then it is zero identically. Now we have

$$x \times y = \frac{1}{4}((x+y) \times (x+y) - (x-y) \times (x-y)) = \frac{1}{4}((x+y)^2 - (x-y)^2) = xy.$$

The commutative algebra $\mathcal{A}_V$ generated by evolutionary operator V is called the *evolutionary algebra*.

We have three closely connected objects, namely, the free population, its evolutionary operator and the corresponding evolutionary algebra. All notions for one of them are automatically applicable to the others. For example, the linear form

$$s(x) = \sum_{i=1}^{n} x_i$$

is called the *weight* of the evolutionary algebra $\mathcal{A}_V$, because it is the fixed algebra homomorphism $\mathcal{A}_V \to \mathbf{R}$, i.e. $s(xy) = s(x)s(y)$. It is called the weight of the operator V as well as of the population. Obviously, $s(V(x)) = s^2(x)$.

A systematic presentation of the algebraic foundations see in chapter 3.

Let us note in conclusion that in the process of deriving the evolutionary equations it was not supposed that differentiation is genotypical (though this is the main case in applications) but only the existence of completely definite inheritance and survival coefficients were assumed which implied the hereditary nature of differentiation.

The phenotypic differentiation in the presence of dominance is an example of a nonhereditary differentiation. If, for example, a gene A dominates allele a, then there exist two phenotypes: **A** corresponding to the genotypes AA and Aa, and **a** corresponding to the genotype aa. If for some generation the distribution of probabilities of the phenotypes is given, then the distribution of probabilities of the phenotypes for the next generation is not uniquely determined, as it depends on a proportion of genotypes AA and Aa. If, for example, all individuals in some generation are of the genotype AA then the same is true for the next generation, in contrast to the situation where the heterozygote Aa is present. Hence, the distribution of probabilities of phenotypes cannot be used as a state of population as it is impossible to construct a dynamic theory upon this foundation (but this distribution may be considered as an "output" of a system).

The genotypical differentiation, in contrast to the phenotypical one, is hereditary. It was with the discovery of this differentiation and its hereditary nature by Gregor Mendel in 1865 that genetics had its start.

1.3 Two-level Populations

From the point of view of genetics a population has two levels of organization, namely the zygote and gamete levels, tied together by the processes of meiosis and fertilization. Usually, we consider a population as consisting of zygotes or, more precisely, of organisms which are genetically identified with zygotes. As a result, the types of zygotes are considered. Instead of "type of zygote" one briefly speaks of "zygote". In parallel with the population of zygotes there exists and evolves the "population" of gametes, or the *gamete pool* of a given population. The differentiation of the gamete pool may not be directly given, but may be only indirectly dictated by the observable properties of the population. Therefore, a priori, it is not, generally speaking, clear what the state of the gamete pool is. Suppose that a space S_γ of states of the gamete pool is a subset of the space $\mathbf{R}^r$ for some r. Every state of the gamete pool should be uniquely determined by some state of a population, i.e. a mapping should exist (the *meiosis operator*) $\mu : S_\zeta \to S_\gamma$, where $S_\zeta \equiv S$ is the space of states; $S_\gamma = \mathrm{Im}\,\mu$, i.e. we assume that μ is surjective. The natural assumption (fulfilled in all known concrete situations) is that the mapping μ is affine, i.e.

$$\mu(\sum_i \alpha_i z_i) = \sum_i \alpha_i \mu(z_i)$$

for all convex combinations $(\alpha_i \geq 0, \sum_i \alpha_i = 1)$ of any points $z_i \in S_\zeta$. We shall always suppose this condition to be fulfilled. Note that it implies continuity of the mapping μ as μ may be extended to an affine mapping of the affine hull of the set S_ζ onto the affine hull of the set S_γ. Recall that the affine hull of a set M is the set

$$\mathrm{Aff}(M) = \{u : u = \sum_k \lambda_k u_k \text{ where } u_k \in M, \sum_u \lambda_k = 1\}.$$

As S_ζ is a compact convex polyhedron (the simplex or the Cartesian product of two simplices), its affine image S_γ, the space of states of the gamete pool, is also a compact convex polyhedron.

In the following it is supposed that
1) $0 \notin \mathrm{Aff}(S_\gamma)$; 2) $\dim \mathrm{Aff}(S_\gamma) < r$.

The state of the population in the next generation is then described by a map $\phi : S_\gamma \to S_\zeta$ which summarizes the process of fertilization and selection. Under selective neutrality, we call ϕ the *fertilization operator*.

It follows that the zygote version of the evolutionary operator is the composition $V = V_\zeta \equiv \phi\mu : S_\zeta \to S_\zeta$. The gamete state in the next generation is given by $V_\gamma \equiv \mu\phi : S_\gamma \to S_\gamma$. The situation can be summarized by the following commutative diagram.

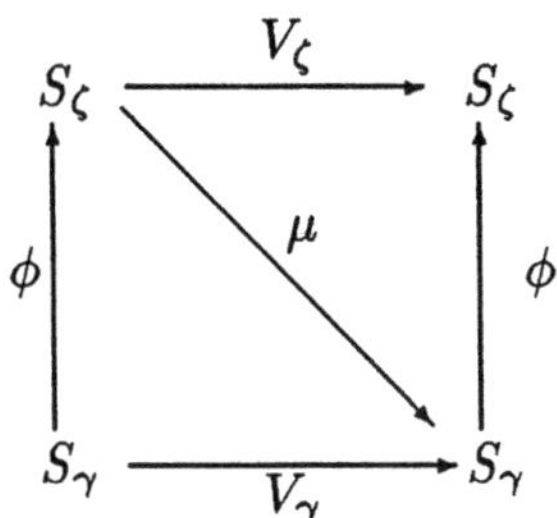

This diagram with μ affine and onto and V_ζ continuous describes formally what we will call a *two-level population*. To study the dynamics we use:

Lemma 1.3.1 *The meiosis operator* $\mu : S_\zeta \to S_\gamma$ *admits a continuous right inverse. That is, there is a continuous map* $\overline{\mu} : S_\gamma \to S_\zeta$ *such that* $\mu\overline{\mu} = \mathrm{id}$ *(the identity on* S_γ*).*

Proof As a compact, convex set, S_γ can be triangulated, that is, written as the union of a finite collection of simplices the intersection of any two of which is either empty or a common face. Let $\{g_1, \ldots, g_N\} \subset S_\gamma$ list the vertices of the triangulation. For each $i = 1, \ldots, N$ choose z_i in S_ζ so that $\mu(z_i) = g_i$. First, define $\overline{\mu}(g_i) = z_i$ for $i = 1, \ldots, N$. Now on each simplex of the triangulation the definition of $\overline{\mu}$ is extended linearly from its value on the vertices of the simplex. As this extension is uniquely defined the definitions of $\overline{\mu}$ on overlapping simplices agree on the common face. In particular, $\overline{\mu}$ defines a continuous function from S_γ to S_ζ. $\mu\overline{\mu}$ is affine on each simplex of the triangulation and is the identity on the vertices. So it is the identity map. $\square$

Remark The inverse map $\overline{\mu}$ is piecewise linear but need not be affine on all of S_γ unless S_γ was itself a simplex. For example, when S_ζ is the unit simplex in $\mathbf{R}^4$ and μ maps the vertices to the corners of a square $= S_\gamma$ then there is no inverse map $\overline{\mu} : S_\gamma \to S_\zeta$ which is affine on the entire square.

Corollary 1.3.2 *The fertilization-selection operator* ϕ *is determined by* μ *and the evolutionary operator* $V = V_\zeta$*. The operator* ϕ *is continuous.*

Proof If $\overline{\mu}$ is a continuous right inverse to μ then $\phi = V\overline{\mu}$ $\square$

Corollary 1.3.3 *The evolutionary operator on the gamete level,* V_γ*, is determined by* μ *and the evolutionary operator* $V = V_\zeta$*.* V_γ *is continuous.*

Proof $V_\gamma = \mu\phi = \mu V\overline{\mu}$. $\square$

Remember that we call a selectively neutral autosomal population free and in that case the evolutionary operator $V_\zeta = V$ is given as a quadratic map by (1.2.12). We analyze the meiosis and fertilization operators in this case:

Theorem 1.3.4 *For a free, two-level population the meiosis operator* $\mu : S_\zeta \to S_\gamma$ *is the restriction of a linear map* $\mu : \mathbf{R}^n \to \mathbf{R}^r$ *with* $p = \mu(x)$ *satisfying:*

$$p_i = \sum_{j=1}^{n} \pi_{ij} x_j \qquad (1 \le i \le r). \qquad (1.3.1)$$

The fertilization operator $\phi : S_\gamma \to S_\zeta$ is quadratic with $x = \phi(p)$ satisfying:

$$x_j = \sum_{i,k=1}^{r} c_{ik,j} p_i p_j \qquad (1 \le j \le n). \qquad (1.3.2)$$

with $c_{ik,j} = c_{ki,j}$. The evolutionary operator V_γ on the gamete level is quadratic.

Proof We observed above that the affine assumption is equivalent to the assumption that μ is the restriction of a linear map from $\mathbf{R}^n$ to $\mathbf{R}^r$ where $\mathbf{R}^n$ is the space with canonical basis the vertices of the simplex S_ζ. With (π_{ij}) the matrix of the linear map μ we get (1.3.1). Note in particular that the coefficients are uniquely defined by the operator μ. The subspace W spanned by the convex set S_γ might be a proper subspace of $\mathbf{R}^r$. So the linear map $\mu : \mathbf{R}^n \to \mathbf{R}^r$ is only onto the subspace W. By our assumptions there exists a linear form f on the space $\mathbf{R}^r$ such that $f(p) = 1$ for $p \in S_\gamma$. Then $f(\mu(x)) = 1$ for $x \in S_\zeta$ and thus $\mu^* f \equiv f \circ \mu = s$ where s is the weight of evolutionary operator V_ζ. Therefore, $\ker \mu \subset \ker s$, i.e.

$$s = \sum_{i=1}^{r} \lambda_i p_i.$$

Now construct a linear map $\tilde{\mu} : \mathbf{R}^r \to \mathbf{R}^n$ in two steps. First, let $\tilde{\mu} : W \to \mathbf{R}^n$ be a linear right inverse for μ. For example, suppose after reordering that $\{g_1 \equiv \mu(e_1), \ldots, g_{r'} \equiv \mu(e_{r'})\}$ is a basis for W. Then define $\tilde{\mu}(g_i) = e_i$ for $1 \le i \le r'$. Observe that $r' = \dim W \le r$. By extending $\{g_1, \ldots, g_{r'}\}$ to a basis for all of $\mathbf{R}^r$ and extending the definition of $\tilde{\mu}$ arbitrarily on the new basis vectors, we get the linear map $\tilde{\mu}$ we want. $\mu\tilde{\mu}(p) = p$ if p is in W and, in particular, if $p \in S_\gamma$. However, it will usually not be true that $\tilde{\mu}$ maps S_γ back into S_ζ. Let $(\tilde{\pi}_{\alpha i})$ be the matrix for $\tilde{\mu}$. We prove that for $p \in S_\gamma$,

$$\phi(p) = (\tilde{\mu}(p))^2. \qquad (1.3.3)$$

(1.3.2) then follows with

$$c_{ik,j} = \sum_{\alpha,\beta=1}^{n} p_{\alpha\beta,j} \tilde{\pi}_{\alpha i} \tilde{\pi}_{\beta k}.$$

Because μ maps S_ζ onto S_γ there exists $x \in S_\zeta$ such that $\mu(x) = p$ and for any such x equation (1.2.16) implies:

$$\phi(p) = (\phi\mu)(x) = V(x) = x^2. \qquad (1.3.4)$$

By continuity, it suffices to prove (1.3.3) only for p in the image under μ of the interior of the simplex S_ζ, interior with respect to the affine subspace $H = \{x \in \mathbf{R}^n : s(x) = 1\}$.

So assume $x_0 \in \mathrm{Int} S_\zeta$ and $\mu(x_0) = p$. Let $u = \tilde{\mu}(p) - x_0$ and let $x_\epsilon = x_0 + \epsilon u$. But $\mu(u) = (\mu\tilde{\mu})(p) - \mu(x_0) = 0$ i.e $u \in \ker \mu$. Thus we have $s(u) = 0$ and so the line x_ϵ beginning at x_0 remains in H. By bilinearity:

$$x_\epsilon^2 = x_0^2 + 2\epsilon x_0 u + \epsilon^2 u^2. \tag{1.3.5}$$

For all ϵ, $\mu(x_\epsilon) = p$ and for $|\epsilon|$ small enough x_ϵ is still in S_ζ. Applying (1.3.4) with $x = x_\epsilon$ for small ϵ we see that x_ϵ^2 remains constant at $\phi(p)$ for ϵ small. This means that the coefficients of 2ϵ and ϵ^2 are 0 and so $x_\epsilon^2 = \phi(p)$ for all ϵ. In particular, when $\epsilon = 1$, $x_1 = \tilde{\mu}(p)$ implies (1.3.3). The last part of the Theorem follows immediately as $V_\gamma = \mu\phi$ where μ is linear and ϕ is quadratic. $\square$

Remark The quadratic representation (1.3.2) of the fertilization operator ϕ is not unique if the linear forms p_i are linear dependent on the space W. This fact causes some difficulties in the further theory .

In the natural biological interpretation the variables x_j are the probabilities of zygotes $1,\ldots,n$ and the variables p_i are the probabilities of gametes $1,\ldots,r$. Then π_{ij} is the probability that gamete i arises from zygote j in meiosis. The coefficient $c_{ik,j}$ is the probability of zygote j in the next generation arises from the parental gametes i,k by fertilization. Certainly, for such interpretation it is necessary to have $\pi_{ij} \geq 0$, $c_{ik,j} \geq 0$ (see chapter 4).

We now compare the dynamics of the two systems which have the evolutionary operators V_ζ, V_γ respectively. They are intertwining by the operators μ and ϕ, i.e.

$$\mu V_\zeta = V_\gamma \mu, \quad V_\zeta \phi = \phi V_\gamma. \tag{1.3.6}$$

Theorem 1.3.5 *If $\{z^{(t)}\}$ is a trajectory for V_ζ in S_ζ with limit point set ω_ζ, then $\{\mu(z^{(t)})\}$ is a trajectory for V_γ in S_γ with limit point set the image $\mu(\omega_\zeta)$. If $\{g^{(t)}\}$ is a trajectory for V_γ in S_γ with limit point set ω_γ then $\{\phi(g^{(t)})\}$ is a trajectory for V_ζ in S_ζ with limit point set $\phi(\omega_\gamma)$.*

Proof Apply μ to $z^{(t+1)} = V_\zeta(z^{(t)})$ and get from (1.3.6) $\mu(z^{(t+1)}) = V_\gamma(\mu(z^{(t)}))$. So if $\lim_{k\to\infty} z^{(t_k)} = z^*$ in ω_ζ then (by continuity of μ) $\lim_{k\to\infty} \mu(z^{(t_k)}) = \mu(z^*)$ and this one lies in the limit set for $\{\mu(z^{(t)})\}$. Conversely, if $\lim_{k\to\infty} g^{(t_k)} = g^*$ and $z^{(t)} = \phi(g^{(t-1)})$ we can assume that $\{z^{(t_k)}\}$ convergent to a point z^* and $\mu(z^*) = g^*$. The proof for ϕ is similar. $\square$

In addition to (1.3.6) we have $V_\zeta = \phi\mu$ and $V_\gamma = \mu\phi$ and so for the trajectories of the Theorem:

$$\phi(\mu(z^{(t)})) = z^{(t+1)}, \quad \mu(\phi(g^{(t)})) = g^{(t+1)}. \tag{1.3.7}$$

In other words, applying first μ and then ϕ merely shifts the original V_ζ trajectory by one time step. In particular, we can recover the original trajectory from its μ image, except for the initial position. Hence

Corollary 1.3.6 *Two V_ζ-trajectories which map to the same V_γ- trajectory under μ agree except possibly for the initial state. Two V_γ-trajectories which map to the same V_ζ-trajectory under ϕ agree except possibly for the initial state.*

Corollary 1.3.7 *A V_ζ-trajectory is convergent if and only if its μ- image is so. A V_γ-trajectory is convergent if and only if its ϕ-image is so.*

Proof If ω_ζ consists of a single point then $\mu(\omega_\zeta)$, the limit set of $\{\mu(z^{(t)})\}$ is a single point and so the latter is convergent. Similarly, if $\{\mu(z^{(t)})\}$ is convergent then its image $\{\phi(\mu(z^{(t)}))\}$ is convergent but this is just the shift of $\{z^{(t)}\}$ and so the latter is convergent as well. $\square$

Corollary 1.3.8 *Let $E_\zeta \subset S_\zeta$ be the set of equilibria for V_ζ and $E_\gamma \subset S_\gamma$ be the set of equilibria for V_γ. $\mu(E_\zeta) = E_\gamma$, $\phi(E_\gamma) = E_\zeta$ and the maps $\mu : E_\zeta \to E_\gamma$ and $\phi : E_\gamma \to E_\zeta$ are inverses of one another providing inverse homeomophisms between these two closed subsets.*

Proof $E_\zeta = \{z : V_\zeta(z) = z\}$ and so is a closed subset of S_ζ. Similarly, $V_\gamma(g) = g$ precisely on the closed subset E_γ of S_γ. That $\mu(E_\zeta) \subset E_\gamma$ and $\phi(E_\gamma) \subset E_\zeta$ follows from the fact that $z^{(t)} \equiv z \in E_\zeta$ and $g^{(t)} \equiv g \in E_j$ are trajectories. In addition, $\phi(\mu(z)) = V_\zeta(z) = z$ if $z \in E_\zeta$ and similarly, $\mu(\phi(g)) = g$ for $g \in E_\gamma$. Thus, on the equilibrium sets μ and ϕ are inverse of one another. $\square$

Corollary 1.3.9 *All the limit points of a V_γ -trajectory are equilibrium if and only if the same is true for the correspondent V_ζ -trajectory.*

Proof If $z \in \omega_\zeta$ then $z = \phi(g)$ where $g \in \omega_\gamma$ and $V_\gamma(g) = g$ by assumption. Therefore,

$$V_\zeta(z) = (V_\zeta\phi)(g) = (\phi V_\gamma)(g) = \phi(g) = z.$$

The proof for a given $g \in \omega_\gamma$ is similar. $\square$

Chapter 2

Elementary Models

2.1 The Free Single Locus Population

With two alleles A and a at a locus the zygotes are AA, aa, Aa. Table 2.1 presents multiplication in the corresponding evolutionary algebra. It reflects the fact that during meiosis

$$AA \mapsto A, \ aa \mapsto a, \ Aa \mapsto \frac{1}{2}A + \frac{1}{2}a \qquad (2.1.1)$$

(the coefficients at the gametes are equal to the probabilities of their formation from a given zygote), and during fertilization father's and mother's gametes are chosen at random and independently. For example, it follows formally that

$$AA \times Aa = A(\frac{1}{2}A + \frac{1}{2}a) = \frac{1}{2}AA + \frac{1}{2}Aa$$

$$(2.1.2)$$

$$Aa \times Aa = (\frac{1}{2}A + \frac{1}{2}a)^2 = \frac{1}{4}AA + \frac{1}{4}aa + \frac{1}{2}Aa$$

The remaining cases are similar. Note that here and in what follows the absence of mutations and selection of gametes is assumed.

m	f		
	AA	aa	Aa
AA	AA	Aa	$\dfrac{1}{2}\,AA + \dfrac{1}{2}\,Aa$
aa	Aa	aa	$\dfrac{1}{2}\,aa + \dfrac{1}{2}\,Aa$
Aa	$\dfrac{1}{2}\,AA + \dfrac{1}{2}\,Aa$	$\dfrac{1}{2}\,aa + \dfrac{1}{2}\,Aa$	$\dfrac{1}{4}\,AA + \dfrac{1}{4}\,aa + \dfrac{1}{2}\,Aa$

Table 2.1

The adduced multiplication table yields the *Mendel First Law.* Therefore, it (and the corresponding algebra) is called the *Mendel diallelic zygote algebra.*

According to (1.2.16) if in some generation the state is $x = x_1 AA + x_2 aa + x_3 Aa$, then in the next generation it will be $x' = x^2 = (x_1 AA + x_2 aa + x_3 Aa)^2$. Using the multiplication table, one derives

$$x_1' = x_1^2 + x_1 x_3 + \frac{1}{4}x_3^2,$$

$$x_2' = x_2^2 + x_2 x_3 + \frac{1}{4}x_3^2, \tag{2.1.3}$$

$$x_3' = 2x_1 x_2 + x_1 x_3 + x_2 x_3 + \frac{1}{2}x_3^2.$$

This is *evolutionary operator of the population.*

It is clear, essentially, that *the population under consideration is two-level* in view of (2.1.1), (2.1.2). But these formulas apply to the gamete pool of individuals, namely of father and mother. Amalgamation of individual gamete pools in the gamete pool of the whole population creates the population of gametes. In this case formulas (2.1.1) are averaged in correspondence with the current state of the population of zygotes.

The two-level character follows formally from (2.1.3):

$$x_1' = p^2, \ x_2' = q^2, \ x_3' = 2pq \tag{2.1.4}$$

where

$$p = x_1 + \frac{1}{2}x_3, \; q = x_2 + \frac{1}{2}x_3. \tag{2.1.5}$$

The map $\mu : \Delta^2 \to \Delta^1$ defined by (2.1.5) is the *meiosis operator* and the map $\phi : \Delta^1 \to \Delta^2$ defined by (2.1.4) is the *fertilization operator* (that $\operatorname{Im}\mu = \Delta^1$ is obvious: $p + q = 1$, $p \geq 0$, $q \geq 0$ and any point $(p,q) \in \Delta^1$ enters $\operatorname{Im} \mu$ being the image of $(p,q,0) \in \Delta^2$).

Formulas (2.1.5) correspond to averaging of (2.1.1) with respect to a state x with p and q the probabilities of gametes A, a in the gamete pool of the entire population in a state x. The formulas (2.1.4) correspond to the independent combination of parent gametes chosen at random from the gamete pool in the state (p,q). The coefficient 2 in the last formula is connected with distinction between father's and mother's gametes, as there are two ways of generating the heterozygote Aa, that is, by uniting the father's A with the mother's a, and vice versa.

Let us compute the *evolutionary operator of the gamete pool*. It is equal to $V_\gamma = \mu\phi$, i.e.

$$p' = p^2 + \frac{1}{2} \cdot 2pq = p(p+q) = p$$

$$\tag{2.1.6}$$

$$q' = q^2 + \frac{1}{2} \cdot 2pq = q(p+q) = q$$

The operator turns out to be the identity, i.e. *all states of the gamete pool are in equilibrium*. To interpret this remarkable fact let us identify the gametes A and a with the genes of the same name and, corresponding, let us call the gamete pool as the *gene pool*. So in the process of evolution of a given population the probabilities of the genes remain constant. This is quite natural in the absence of mutation and selection, as in this case the genes neither arise nor disappear—they are only passed on to the offspring. Thus, (2.1.6) is the *Gene Conservation Law* . Let us see where it leads on the level of zygotes.

Let $V = V_\zeta$ be the evolutionary operator of the population. Then $V = \phi\mu$, $V^2 = \phi(\mu\phi)\mu = \phi\mu = V$, since $V_\gamma = \mu\phi = \text{id}$. Thus we obtain due to the Gene Conservation Law,

$$V^2 = V \tag{2.1.7}$$

i.e.

$$x_1'' = x_1', \; x_2'' = x_2', \; x_3'' = x_3'. \tag{2.1.8}$$

The latter means that *all the states of population in the first generation of offspring*, i.e. in F_1, are equilibria. Therefore, all the trajectories are convergent even at the first step.

Distribution (2.1.4) is called the *Hardy-Weinberg Law* and is of great importance in population genetics. The evolutionary operator (2.1.3) is called the *Hardy-Weinberg operator* . The latter, usually written as a combination of (2.1.4) with (2.1.5) is the population equivalent of the Mendel mechanism of heredity in one diallelic locus.

If an evolutionary operator V of some free population is such that $V^2 = V$, then the population is called *stationary* or *Bernstein* to honor S.N. Bernstein who was the first to investigate such populations from a general viewpoint. The condition $V^2 = V$ is called the *Stationarity Principle*. Further on, we shall see that among classical models of population genetics only a free one-locus population with any number of alleles satisfies the Stationarity Principle (with some exotic exceptions, for example, Y-linkage). Thus, it is reasonable to accept the Stationarity Principle as an axiom to characterize elementary mechanisms of heredity. So, the problem of the explicit description of all stationary evolutionary operators posed by S.N. Bernstein, looks to be very interesting, though, apparently, the question is too wide for genetics, where one need only describe the two-level populations with identically equilibrious gamete (gene) pool. In the latter case we say that the described operator has *stationary gene structure.*This structure will be investigated in chapters 4, and in chapter 5 we will consider the stationary populations without the Gene Conservation Law.

Example There is the next extension of the Hardy-Weinberg Law:

$$x_1' = p^2 + 2a_{12}pq, \ x_2' = q^2 + 2a_{21}pq, \ x_3' = 2bpq,$$

where

$$p = x_1 + c_1 x_3, \ q = x_2 + c_2 x_3$$

and the parameters c_1, c_2, b are positive, a_{12}, a_{21} are nonnegative and

$$c_1 + c_2 = 1, \ a_{12} + c_1 b = \frac{1}{2}, \ a_{21} + c_2 b = \frac{1}{2}.$$

It's easy to see that the Stationarity Principle for the corresponding evolutionary operator V is fulfilled, i.e. $V^2 = V$. This follows from the Gene Conservation Law ($p' = p$, $q' = q$) again.

Now we consider the situation of m alleles, namely $A_1, \ldots, A_m$. Here $\frac{1}{2}m(m+1)$ zygotes are present: m homozygotes $A_i A_i$ ($1 \le i \le m$) and

$\frac{1}{2}m(m-1)$ heterozygotes $A_iA_k(1 \le i < k \le m)$,, with $A_kA_i = A_iA_k(i < k)$. In meiosis

$$A_iA_k \mapsto \frac{1}{2}A_i + \frac{1}{2}A_k. \qquad (2.1.9)$$

Thus, in the corresponding evolutionary algebra

$$A_iA_k \times A_jA_l = (\frac{1}{2}A_i + \frac{1}{2}A_k)(\frac{1}{2}A_j + \frac{1}{2}A_l)$$

$$= \frac{1}{4}A_iA_j + \frac{1}{4}A_iA_l + \frac{1}{4}A_kA_j + \frac{1}{4}A_kA_l. \qquad (2.1.10)$$

This is the multiplication table in *the Mendel multiallelic algebra of zygotes*.

Let x_{ik} ($\equiv x_{ki}$) be the probability of a genotype A_iA_k and let us consider the state

$$x = \sum_{1 \le i \le k \le m} x_{ik}A_iA_k.$$

Put $\xi_{ii} = x_{ii}$, $\xi_{ik} = \frac{1}{2}x_{ik}(i \ne k)$. Then $x = \sum_{i,k=1}^m \xi_{ik}A_iA_k$. It follows from (2.1.10)

$$x^2 = \frac{1}{4}\sum_{i,k,l,j=1}^m \xi_{ik}\xi_{jl}(A_iA_j + A_iA_l + A_kA_j + A_kA_l)$$

$$= \sum_{i,j=1}^m p_ip_jA_iA_j = \sum_{i=1}^m p_i^2A_iA_i + 2\sum_{1 \le i < j \le m} p_ip_jA_iA_j$$

where

$$p_i = \sum_{k=1}^m \xi_{ik} = x_{ii} + \frac{1}{2}\sum_{k \ne i} x_{ik}. \qquad (2.1.11)$$

Thus

$$x_i' = p_i^2 \; (1 \le i \le m), \; x_{ij}' = 2p_ip_j \; (1 \le i < j \le m). \qquad (2.1.12)$$

This is *the Hardy-Weinberg Law for a multiallelic locus*. Combining (2.1.12) with (2.1.11) one obtains *the Hardy-Weinberg operator*.

The population under consideration is two-level. Formulas (2.1.11) define the meiosis operator $\mu : \Delta^{\frac{1}{2}m(m+1)-1} \to \Delta^{m-1}$ while formulas (2.1.12) define the fertilization operator $\phi : \Delta^{m-1} \to \Delta^{\frac{1}{2}m(m+1)-1}$. The interpretation given for $m = 2$ remains unchanged. As before, the *Gene Conservation Law* is valid:

$$p_i' = p_i^2 + \frac{1}{2}\sum_{k \ne 1}^2 p_ip_k = p_i\sum_{k=1}^m p_k = p_i,$$

i.e. *the gamete (gene) pool is identically equilibrious* (the evolutionary operator of the gamete pool is the identity). Therefore, the Stationarity Principle is valid *the given population at the level of zygotes is stationary (Bernstein).*

Let us describe the set E_ξ of the equilibrious states of zygotes. According to Corollary 1.3.8, E_ξ coincides with Im ϕ as at the gamete level all states are equilibrious. Hence, E_ξ is determined by the parametric equations

$$x_{ij} = p_i^2, \ x_{ij} = 2p_ip_j \ (i < j) \qquad (2.1.13)$$

where a parametric point $p = (p_1,\ldots,p_m)$ runs over the simplex Δ^{m-1}. Topologically E_ξ is an $(m-1)$- dimensional simplex embedded "curvilinearly". Figure 2.1 shows the simplest case $m = 2$. Here E_ξ is an arc of the parabola $x_1 = p^2$, $x_2 = q^2$, $x_3 = 2pq$ $(0 \le p \le 1, \ q = 1 - p)$.

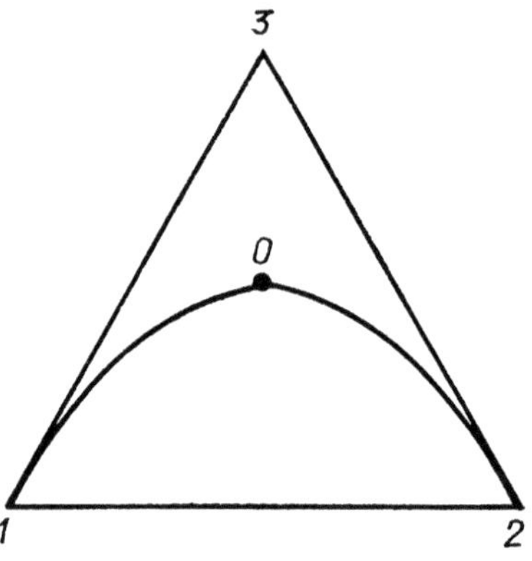

Fig. 2.1

The vertex of the parabola is at the point $O(\frac{1}{4},\frac{1}{4},\frac{1}{2})$ which corresponds to the equilibrium observed in Mendel's experiments. If A dominates over a, the equilibrious ratio of phenotypes at this point is 3:1, and at an arbitrary point (p,q) it equals $(1 - q^2) : q^2$.

2.2   The Free Two Locus, Two Allele Population

Let A and a be the alleles at one and B and b the alleles at another locus. Assume the loci are linked. Then the population contains zygote genotypes 1)

ABAB 2) *AbAb* 3) *aBaB* 4) *abab (homozygotes)* 5) *ABAb* 6) *aBab* 7) *ABaB*
8) *Abab (simple heterozygotes)*, 9) *ABab* 10) *AbaB (double heterozygotes)*.
These notations express that the two former genes are on one chromosome
and the two latter genes are on the other chromosome homologous to the
first one. Hence, the zygote *ABab* carries the chromosomes *AB*, *ab*, and the
zygote *ABaB* carries the chromosomes *AB*, *aB*.

Let us consider meiosis. For the homozygotes $ABAB \mapsto AB$ etc., for the
simple heterozygotes $ABAb \mapsto \frac{1}{2}AB + \frac{1}{2}Ab$ etc. For the double heterozygotes
with crossing-over probabilities r, ρ one has

$$ABab \mapsto \frac{1-r}{2}AB + \frac{r}{2}Ab + \frac{r}{2}aB + \frac{1-r}{2}ab,$$

$$AbaB \mapsto \frac{\rho}{2}AB + \frac{1-\rho}{2}Ab + \frac{1-\rho}{2}aB + \frac{\rho}{2}ab.$$

The probabilities r and ρ are, generally speaking, different, i.e. we take into
account the influence of gene localization upon cross-over probabilities. In
the *normal crossing-over* $r = \rho$. We will suppose that $r > 0$, $\rho > 0$, i.e. we
will rule out the *rigid linkage* to avoid going into unnecessary details.

The multiplication table for the evolutionary algebra in question is of
10×10 size. The way of compiling it is obvious: we use the formal product
of the right-hand terms of meiosis schemes. For example,

$$ABab \times aBab = (\frac{1-r}{2}AB + \frac{r}{2}Ab + \frac{r}{2}aB + \frac{1-r}{2}ab)(\frac{1}{2}aB + \frac{1}{2}ab)$$

$$= \frac{r}{4}aBaB + \frac{1-r}{4}abab + \frac{1}{4}aBab + \frac{1-r}{4}ABaB + \frac{r}{4}Abab + \frac{1-r}{4}ABab + \frac{r}{4}AbaB.$$

The probability that a zygote i is a state x is denoted by x_i ($1 \leq i \leq 10$).
The evolutionary equations corresponding to the described table are the
following:

$$x_1' = p_1^2, \ x_2' = p_2^2, \ x_3' = p_3^2, \ x_4' = p_4^2$$

$$x_5' = 2p_1p_2, ; x_6' = 2p_3p_4, \ x_7' = 2p_1p_3, \ x_8' = 2p_2p_4 \qquad (2.2.1)$$

$$x_9' = 2p_1p_4, \ x_{10}' = 2p_2p_3$$

where

$$p_1 = x_1 + \frac{1}{2}x_5 + \frac{1}{2}x_7 + \frac{1-r}{2}x_9 + \frac{\rho}{2}x_{10}$$

$$p_2 = x_2 + \frac{1}{2}x_5 + \frac{1}{2}x_8 + \frac{r}{2}x_9 + \frac{1-\rho}{2}x_{10}$$

$$\qquad (2.2.2)$$

$$p_3 = x_3 + \frac{1}{2}x_6 + \frac{1}{2}x_7 + \frac{r}{2}x_9 + \frac{1-\rho}{2}x_{10}$$

$$p_4 = x_4 + \frac{1}{2}x_6 + \frac{1}{2}x_8 + \frac{1-r}{2}x_9 + \frac{\rho}{2}x_{10}.$$

The population is two-level as in the one-locus case. Formulas (2.2.2) define *the meiosis operator* $\mu : \Delta^9 \to \Delta^3$; p_1, p_2, p_3, p_4 are interpreted as the *probabilities of gametes AB, Ab, aB, ab* in the gamete pool of population in a state x. Formulas (2.2.1) define *the fertilization operator* $\phi : \Delta^3 \to \Delta^9$ and correspond to an independent combination of randomly chosen parental gametes. More expressive notation would be as follows:

$$x'(ABAB) = p^2(AB), \ldots$$

$$x'(ABAb) = 2p(AB)p(Ab), \ldots$$

$$x'(ABab) = 2p(AB)p(ab),\, x'(AbaB) = 2p(Ab)p(aB).$$

The evolutionary operator of the gamete pool is of the form

$$p_1' = p_1^2 + p_1p_2 + p_1p_3 + (1-r)p_1p_4 + \rho p_2p_3$$

$$p_2' = p_2^2 + p_1p_2 + p_2p_4 + rp_1p_4 + (1-\rho)p_2p_3$$

$$p_2' = p_3^2 + p_3p_4 + p_1p_3 + rp_1p_4 + (1-\rho)p_2p_3 \tag{2.2.3}$$

$$p_4' = p_4^2 + p_3p_4 + p_2p_4 + (1-r)p_1p_4 + \rho p_2p_3.$$

This can be simplified since $p_1^2 + p_1p_2 + p_1p_3 = p_1(p_1 + p_2 + p_3) = p_1(1 - p_4) = p_1 - p_1p_4$. Using this and three similar identities for p_2, p_3, p_4 one obtains

$$p_1' = p_1 - \delta,\ p_2' = p_2 + \delta,\ p_3' = p_3 + \delta,\ p_4' = p_4 - \delta \tag{2.2.4}$$

where

$$\delta = rp_1p_4 - \rho p_2p_3. \tag{2.2.5}$$

It is natural to call the value δ *the measure of disequilibrium* of a state as, by (2.2.4), for equilibrium it is necessary and sufficient that

$$\delta = 0. \tag{2.2.6}$$

Note that now not all states of the gamete pool are equilibrious: equation (2.2.6) singles out the two-dimensional submanifold (with boundary) of the equilibrious states in the state space Δ^3. It can be conveniently parametrized by the probabilities of the genes A and B:

$$p_A = p_1 + p_2, \ p_B = p_1 + p_3 \tag{2.2.7}$$

(using also $p_a = p_3 + p_4 = 1 - p_A$, $p_b = p_2 + p_4 = 1 - p_B$). It looks especially simple in the normal crossing-over $(r = \rho)$ case, when (2.2.6) is reduced to the equation without r:

$$p_1 p_4 = p_2 p_3. \tag{2.2.8}$$

Then

$$p_1 = p_1^2 + p_1 p_2 + p_1 p_3 + p_1 p_4 = p_1^2 + p_1 p_2 + p_1 p_3 + p_2 p_3$$

$$= (p_1 + p_2)(p_1 + p_3) = p_A p_B.$$

Or

$$p(AB) = p_A p_B. \tag{2.2.9}$$

Similarly,

$$p(Ab) = p_A p_b, \ p(aB) = p_a p_B, \ p(ab) = p_a p_b. \tag{2.2.10}$$

Conversely, (2.2.8) follows from (2.2.9) and (2.2.10), i.e. p_A and p_B are the global parameters of the equilibrium, $0 \le p_A \le 1$, $0 \le p_B \le 1$. Incidentally, a remarkable circumstance follows *in the normal crossing-over case, for equilibrium it is necessary and sufficient for the probabilities of gametes to be the product of probabilities of the comprising genes, i.e. the genes are statistically independent.*

In a general case $(r \ne \rho)$, it is necessary to substitute the expressions

$$p_2 = p_A - p_1, \ p_3 = p_B - p_1, \ p_4 = p_a - p_B + p_1. \tag{2.2.11}$$

in (2.2.6), i.e. in the equilibrium equation $r p_1 p_4 - \rho p_2 p_3 = 0$. This leads to the quadratic equation in p_1:

$$(r - \rho)p_1^2 + \{(\rho - r)(p_A + p_B) + r\}p_1 - \rho p_A p_B = 0. \tag{2.2.12}$$

The unknown p_1 must lie in the segment

$$I : \max(0, p_A + p_B - 1) \le p_1 \le \min(p_A, p_B). \tag{2.2.13}$$

This is necessary and sufficient to define $p_i \ge 0$ $(2 \le i \le 4)$ by p_1 via (2.2.11). The equation (2.2.12) has a unique solution on the segment I, as its left-hand term is $-\rho p_A p_B \le 0$ for $p_1 = 0$, $-\rho p_a p_b \le 0$ for $p_1 = p_A + p_B - 1$ and

$rp_A p_B \geq 0$ for $p_1 = \min(p_A, p_B)$, if, to be definite, one takes $p_A \leq p_B$. It is this solution that, given the probabilities p_A and p_B of genes A and B, defines the unique equilibrious state.

By Corollary 1.3.8 the equilibrious states at the zygote level are defined from (2.2.1) using the equilibrious states (p_1, p_2, p_3, p_4) at the gamete level:

$$x_1 = p_1^2, \ldots; \; x_5 = 2p_1 p_2, \ldots; \; x_9 = 2p_1 p_4, \; x_{10} = 2p_2 p_3. \qquad (2.2.14)$$

This is an *analogue of the Hardy-Weinberg Law* for two diallelic autosomal loci. In the case of normal cross-over, (2.2.14) can be represented as $x(ABAB) = p_A^2 p_B^2, \ldots; \; x(ABAb) = 2p_A^2 p_B p_b, \ldots; \; x(ABab) = x(AbaB) = 2p_A p_B p_a p_b$ in accordance with the above description of the equilibrium at the gamete level.

Now let us consider the problem of convergence to equilibrium. By virtue of Corollary 1.3.7, it is sufficient to study it at the gamete level. Let us note that the equation (2.2.4) yields *the Gene Conservation Law*

$$p'_A = p'_1 + p'_2 = (p_1 - \delta) + (p_2 + \delta) = p_1 + p_2 = p_A$$

$$(2.2.15)$$

$$p'_B = p'_1 + p'_3 = (p_1 - \delta) + (p_3 + \delta) = p_1 + p_3 = p_B$$

and automatically $p'_a = p_a$, $p'_b = p_b$). Hence, by (2.2.11), it is sufficient to trace out an evolution of the probability p_1 (inside the segment I). But according to (2.2.5) and (2.2.11),

$$p'_1 = p_1 - \delta = f(p_1) \equiv (\rho - r)p_1^2 + \{(\rho - r)(p_A + p_B) - r\}p_1 + \rho p_A p_B \quad (2.2.16)$$

(cf. (2.2.12)). Using (2.2.11) it is possible to include any $p_1 \in I$ in the state $(p_1, p_2, p_3, p_4) \in \Delta^3$ with the given probabilities p_A, p_B and to apply afterwards the evolutionary operator (2.2.3). The value p'_1 obtained in such a way will be $f(p_1)$, and the probabilities p_A, p_B will not change. Hence, $f(p_1) \in I$ i.e. f maps the segment I into itself. Let us estimate the derivative $df/dp_1 = 2(\rho - r)p_1 + 1 - r + (\rho - r)(p_A + p_B)$. To this end let us represent it in the form

$$\frac{df}{dp_1} = 1 - r(p_1 + p_4) - \rho(p_2 + p_3). \qquad (2.2.17)$$

It follows that

$$\min(1 - r, 1 - \rho) \leq \frac{df}{dp_1} \leq \max(1 - r, 1 - \rho). \qquad (2.2.18)$$

Thus, the map $f : I \to I$ is contracting with a coefficient $\overline{\kappa} = \max(1 - r, 1 - \rho)$. Its iterations converge with an exponential rate $O(\overline{\kappa}^{(t)})$. Therefore, *the convergence of every trajectory to an equilibrium is proved. The limit equilibrium is defined by equation (2.2.12) with the constraints (2.2.13) where the probabilities of genes A, B are defined by the initial state.*

Since, in a neighborhood of the limit point

$$f(p_1) = p_1^{(\infty)} + \kappa(p_1 - p_1^{(\infty)}) + O((p_1 - p_1^{(\infty)})^2)$$

where $\kappa = (df/dp_1)_{p_1^{(\infty)}} \in (0,1)$, the asymptotic formula is valid (cf. Appendix to the present chapter)

$$p_1^{(t)} - p_1^{(\infty)} \sim c\kappa^t. \tag{2.2.19}$$

Here c is a positive constant depending on the initial state. Thus, *the equilibrium is achieved with an exponential rate.*

By (2.2.18)

$$\underline{\kappa} \leq \kappa \leq \overline{\kappa} \tag{2.2.20}$$

where $\underline{\kappa} = \min(1 - r, 1 - \rho)$. In the normal crossing-over $\underline{\kappa} = \overline{\kappa} = 1 - r$ and consequently

$$\kappa = 1 - r. \tag{2.2.21}$$

In this case the asymptotic formula (2.2.19) becomes an identity. Indeed, if $r = \rho$, then $\delta = r\epsilon$, $\epsilon = p_1 p_4 - p_2 p_3$ and it follows from (2.2.4) that

$$\epsilon' = p_1' p_4' - p_2' p_3' = (p_1 - \delta)(p_4 - \delta) - (p_2 + \delta)(p_3 + \delta)$$

$$= (p_1 p_4 - p_2 p_3) - \delta(p_1 + p_2 + p_3 + p_4) = \epsilon - \delta - (1 - r)\epsilon.$$

From this in the t-th generation

$$\epsilon^{(t)} = (1 - r)^t \epsilon^{(0)}. \tag{2.2.22}$$

But $p_1^{(t)} - p_1^{(t+1)} = r\epsilon^{(t)}$, which yields

$$p_1^{(t)} - p_1^{(\infty)} = r \sum_{k=t}^{\infty} \epsilon^{(k)} = (1 - r)^t \epsilon^{(0)} \tag{2.2.23}$$

as stated above. The latter reasoning does not depend on the preceding analysis of the general case, and directly indicates the convergence to the equilibrium in the normal crossing-over. Besides, it gives the explicit form of the limit state of a population in terms of the initial one: for $t = 0$ it follows from (2.2.23) that

$$p_1^{(\infty)} = p_0^{(0)} - \epsilon^{(0)}. \tag{2.2.24}$$

Similarly,

$$p_2^{(\infty)} = p_1^{(0)} + \epsilon^{(0)}, \ p_3^{(\infty)} = p_3^{(0)} + \epsilon^{(0)}, \ p_4^{(\infty)} = p_4^{(0)} - \epsilon^{(0)}. \tag{2.2.25}$$

Thus for $i = 1, 2, 3, 4$

$$p_i^{(t)} = p_i^{(0)} \pm (1 - (1 - r)^t)\epsilon^{(0)} \tag{2.2.26}$$

where the sign $(+)$ obtains for $i = 2, 3$ and $(-)$ for $i = 1, 4$. These are the explicit expressions for the state $p^{(t)}$ in the t-th generation in terms of the initial state $p^{(0)}$. There are no explicit formulas for $r \neq \rho$.

Note also that for $r = \rho = 1$ (2.2.22) implies $\epsilon^{(t)} = 0$ $(t \geq 1)$, i.e. the trajectories converge on the gamete level in the first generation. The gamete population is stationary (Bernstein), and on the zygote level two generations are needed for the convergence to occur, i.e. in this case the evolutionary operator V on the zygote level satisfies $V^3 = V^2$.

We will give the explicit formula for κ which, according to (2.2.19), characterizes the exact rate of convergence in an exponential scale. We have

$$\kappa = 1 - \sqrt{((r - \rho)(p_A + p_B) - r)^2 - 4\rho(r - \rho)p_A p_B}. \tag{2.2.27}$$

This follows from the equations (2.2.12) and

$$\kappa = 1 + 2(\rho - r)p_1 + (r - \rho)(p_A + p_B) - r$$

after eliminating p_1. For $r = \rho$ (2.2.27) is transformed in (2.2.21).

Evidently, κ depends on an initial state determining the gene probabilities. But $\overline{\kappa}$ and $\underline{\kappa}$, which are the upper and lower bounds of κ, do not depend on an initial state: they are determined by the parameters r and ρ of the population itself. Note that these bounds are achieved, namely, $\kappa = 1 - r$ for $p_A = p_B = 0$ and, $\kappa = 1 - \rho$ for $p_a = 1, p_B = 0$.

We have assumed that the loci are linked. With unlinked loci the primary novelty is the identification of the double zygotes $AbaB \equiv ABab$. In this case it is expedient to indicate the genes of the first locus first, and afterwards the genes of the second one, e.g. $AABB$, $ABab$, etc. By *Mendel's Second Law* a pair of homologous chromosomes carrying one locus splits into gametes independently from the pair carrying the other locus. This means that the aforesaid schema of meiosis are preserved, except the double heterozygote $AaBb \mapsto \frac{1}{4}AB + \frac{1}{4}Ab + \frac{1}{4}aB + \frac{1}{4}ab$, which corresponds to the linked loci with the normal cross-over: $r = \rho = \frac{1}{2}$. But a more general scheme

$$AaBb \to \frac{1-r}{2}AB + \frac{r}{2}Ab + \frac{r}{2}aB + \frac{1-r}{2}ab \qquad (2.2.28)$$

is not ruled out, despite the absence of a linkage. This *pseudolinkage* may be interpreted as if a chromosome, carrying the gene A, prefers to go to the same gamete as a chromosome, carrying the gene B, if $r < \frac{1}{2}$, and as a chromosome, carrying the gene b, if $r > \frac{1}{2}$. In the former case a simultaneously prefers to be associated with b, and in the latter case with B, the measure of this preference being the same. Formally, scheme (2.2.28) corresponds to the linked loci with $\rho = 1 - r$ due to the identification of double heterozygotes. So it is possible to extend the previous results, i.e. the convergence to an equilibrium, bounds of a rate of convergence, parametrization of the manifold of equilibrious states on the gamete level, explicit formulas (in the normal case) to the Mendelian or even more general situation of pseudolinkage.

Thus, for the Mendelian situation the convergence to an equilibrium occurs with the rate $O(2^{-t})$. Moreover, the explicit formulas can be written down (cf. (2.2.26))

$$p_i^{(t)} = p_i^{(0)} \pm (1 - 2^{-t})\epsilon^{(0)},.$$

(and, consequently, (2.2.24) and (2.2.25) are valid too). And equilibrium is characterized by the statistical independence of loci (which is in no way implied by their "physical" independence).

For a pseudolinkage the convergence to an equilibrium occurs with the rate $O(\bar{\kappa}^t)$, where $\bar{\kappa} = \max(r, 1 - r)$. Obviously, $\bar{\kappa} \geq \frac{1}{2}$. The exact rate κ is $1 - r$ for $p_A = p_B = 0$ and is r for $p_A = 1$, $p_B = 0$. Thus, for pseudolinkage, as compared to the Mendelian segregation of chromosomes, the convergence rate depends on the initial state.

Note finally that the fertilization operator for the unlinked loci differs from that for the linked loci by the fact that the equations for x_9' and x_{10}' are merged in the equation

$$x_9' = 2(p_1 p_4 + p_2 p_3).$$

Correspondingly, x_9 and x_{10} are merged in the meiosis operator:

$$p_1 = x_1 + \frac{1}{2}x_5 + \frac{1}{2}x_7 + \frac{1-r}{2}x_9$$

$$p_2 = x_2 + \frac{1}{2}x_5 + \frac{1}{2}x_8 + \frac{r}{2}x_9$$

$$p_3 = x_3 + \frac{1}{2}x_6 + \frac{1}{2}x_7 + \frac{r}{2}x_9$$

$$p_4 = x_4 + \frac{1}{2}x_6 + \frac{1}{2}x_8 + \frac{1-r}{2}x_9.$$

The evolutionary operator of the gamete pool remains, surely, the same, but with a constraint $\rho = 1 - r$. The analogue of the Hardy-Weinberg Law is modified at the point dealing with the double zygotes: $x(AaBb) = 4p_A p_B p_a p_b$.

2.3   The Dynamic Effects of Sex Linkage

We consider a bisexual, selectively neutral population whose differentiation is determined by two alleles A, a of some sex-linked locus. There are three possibilities to consider.

I. *Y-linkage.* Such a locus is on the Y-chromosome and hence of the male only. A male offspring inherits the father's Y-chromosome. Thus, any state $pA + qa$ is equilibrious: $p' = p$, $q' = q$.

II. *X-linkage.* In this case there are two male genotypes A, a, and three female ones: AA, aa, Aa. The meiosis of females is governed by (2.1.1), and that of males is $A \mapsto A$, $a \mapsto a$ for X-gametes. So the formation of offspring is described by Table 2.2. The father's genotype does not influence that of a male offspring, since the latter receives his mother's X-chromosome only.

The formation of a female offspring is described by Table 2.3.

m	f		
	AA	aa	Aa
A	A	a	$\frac{1}{2}A + \frac{1}{2}a$
a	A	a	$\frac{1}{2}A + \frac{1}{2}a$

Table 2.2

m	f		
	AA	aa	Aa
A	AA	Aa	$\frac{1}{2}AA + \frac{1}{2}Aa$
a	Aa	aa	$\frac{1}{2}aa + \frac{1}{2}Aa$

Table 2.3

The state of the population is determined by the distributions $x = x_1 AA + x_2 aa + x_3 Aa$, $y = y_1 A + y_2 a$. The state space is not a simplex now, but the Cartesian product of two simplices $\Delta^2 \times \Delta^1$. Following Tables 2 and 3 we have the *evolutionary equations*

$$x_1' = x_1 y_1 + \frac{1}{2} x_3 y_1$$

$$x_2' = x_2 y_2 + \frac{1}{2} x_3 y_2$$

$$x_3' = x_1 y_2 + x_2 y_1 + \frac{1}{2} x_3 y_1 + \frac{1}{2} x_3 y_2 \qquad (2.3.1)$$

$$y_1' = x_1 y_1 + x_1 y_2 + \frac{1}{2} x_3 y_1 + \frac{1}{2} x_3 y_2$$

$$y_2' = x_2 y_1 + x_2 y_2 + \frac{1}{2} x_3 y_1 + \frac{1}{2} x_3 y_2.$$

Put

$$p_X = x_1 + \frac{1}{2} x_3, \ q_X = x_2 + \frac{1}{2} x_3$$

$$p_Y = y_1, \ q_Y = y_2. \qquad (2.3.2)$$

Then

$$x_1' = p_X p_Y, \ x_2' = q_X q_Y,, \ x_3' = p_X q_Y + p_Y q_X$$

$$y_1' = p_X, \ y_2' = q_X \qquad (2.3.3)$$

(since $y_1 + y_2 = 1$). This is *the fertilization operator* and (2.3.2) is *the meiosis operator*. The space of states of the gamete (gene) pool is $\Delta^1 \times \Delta^1$. The gamete pool has two parts, namely, female and male parts; p_X, q_X being the probabilities of gametes (genes) A, a in the female part, p_Y, q_Y in the male part. Thus, now it is necessary to make distinction between gametes not only by their genotype but by their origin. Equations (2.3.3) corresponds to the independent combination of parental X- gametes in the formation of female offspring and of transference of a mother's X-chromosome to a male offspring. Let us calculate *the evolutionary operator of the gamete pool*: $p_X' = x_1' + \frac{1}{2} x_3' = p_X p_Y + \frac{1}{2}(p_X q_Y + p_Y q_X) = \frac{1}{2} p_X (p_Y + q_Y) + \frac{1}{2} p_Y (p_X + q_X) = \frac{1}{2}(p_X + p_Y)$. As it should be, half of X-gametes of the females have maternal origin and half are paternal.

Similarly, we find q_X', p_Y', q_Y'. Finally, we obtain

$$p_X' = \frac{1}{2}(p_X + p_Y), \ q_X' = \frac{1}{2}(q_X + q_Y)$$

$$p_Y' = p_X, \ q_Y' = q_X \qquad (2.3.4)$$

The equilibrium conditions $p_X = \frac{1}{2}(p_X + p_Y)$, $q_X = \frac{1}{2}(q_X + q_Y)$, $p_Y = p_X$, $q_Y = q_X$ are reduced to the two latter relations which express a balance of genes between sexes. Thus, *in a state of equilibrium* (and in such a state only) *the probability of every gene in the male sex is equal to its probability in the female sex.* Let us note that this condition for a gene a follows from the same condition for A since $q_X = 1 - p_X$, $q_Y = 1 - p_Y$. By Corollary 1.3.8 the set of equilibrious states on the level of zygotes is defined by the equations

$$x_1 = p^2, \; x_2 = q^2, \; x_3 = 2pq$$

$$y_1 = p, y_2 = q$$

Thus, *the Hardy-Weinberg Law is true for the female sex in an equilibrium state of the population*, p, $q = 1 - p$ being the probabilities of gene A, a for any sex.

Let us study the convergence of trajectories. As before, it is possible and expedient to carry it out on the gamete level. Let us introduce the probability vector of the genes of the sexes

$$P = \begin{pmatrix} p_X \\ p_Y \end{pmatrix}$$

and the "transition" matrix

$$S = \begin{pmatrix} \frac{1}{2} & \frac{1}{2} \\ 1 & 0 \end{pmatrix}.$$

Then (2.3.4) yields $P' = SP$. It follows that in the t- th generation $P^{(t)} = S^t P^{(0)}$. To compute S^t write out the characteristic equation

$$\begin{vmatrix} \frac{1}{2} - \lambda & \frac{1}{2} \\ 1 & -\lambda \end{vmatrix} \equiv \lambda^2 - \frac{1}{2}\lambda - \frac{1}{2} = 0.$$

Its roots are $\lambda_1 = 1$, $\lambda_2 = -\frac{1}{2}$. Therefore, every element of the matrix S^t is of a form $\alpha + \beta(-\frac{1}{2})^t$ (α, β being constant). Using the initial data $S^0 = I$ (the unit matrix), $S^1 = S$ one obtains

$$S^t = \begin{pmatrix} \frac{2}{3} + \frac{1}{3}(-\frac{1}{2})^t & \frac{1}{3} - \frac{1}{3}(-\frac{1}{2})^t \\ \frac{2}{3} + \frac{1}{3}(-\frac{1}{2})^{t-1} & \frac{1}{3} - \frac{1}{3}(-\frac{1}{2})^{t-1} \end{pmatrix}.$$

Hence, in the t-th generation

$$p_X^{(t)} = \frac{2p_X^{(0)} + p_Y^{(0)}}{3} + \frac{1}{3}(-\frac{1}{2})^t(p_X^{(0)} - p_Y^{(0)})$$

$$p_Y^{(t)} = \frac{2p_X^{(0)} + p_Y^{(0)}}{3} + \frac{1}{3}(-\frac{1}{2})^{t-1}(p_X^{(0)} - p_Y^{(0)}).$$

$$(2.3.5)$$

For $t \to \infty$ one finds the limit distribution

$$p_X^{(\infty)} = p_Y^{\infty} = \frac{2p_X^{(0)} + p_Y^{(0)}}{3}. \tag{2.3.6}$$

The similar formulas are valid for the gene a. Thus, *all the trajectories converge* (on the level of gametes and hence, on the level of zygotes). The exact order of the rate of convergence is $O(2^{-t})$, but the approach is of oscillatory.

Since the condition $p_X = p_Y$ is necessary and sufficient for an equilibrium and difference $\epsilon = p_X - p_Y$ may be regarded as *the measure of disequilibrium* of a state of the population. It follows from (2.3.5)

$$\epsilon^{(t)} = (-\frac{1}{2})^t \epsilon^{(0)}. \tag{2.3.7}$$

Thus, the modulus of the measure of disequilibrium is halved for one generation, and its sign alternates, i.e. an excess of genes is pumped from one sex to another.

It follows from (2.3.7) that if $\epsilon^{(0)} \neq 0$, then $\epsilon^{(t)} \neq 0$ for all t, i.e. an equilibrium is never reached, unless it existed from the very beginning. Therefore, the population under consideration is non-stationary.

Note additionally that the *Gene Conservation Law* in this case takes on the form

$$\frac{2}{3}p_X + \frac{1}{3}p_Y = const, \; \frac{2}{3}q_X + \frac{1}{3}q_Y = const.$$

These follow from (2.3.4). The coefficients $\frac{2}{3}, \frac{1}{3}$ correspond to the ratio 2:1 of X-chromosomes in female and male zygotes.

To study the trajectories one could employ the Gene Conservation Law. Putting $p_A = \frac{2}{3}p_X + \frac{1}{3}p_Y$ and eliminating p_Y from (2.3.4), we find $p'_X = -\frac{1}{2}p_X + \frac{2}{3}p_A$ which yields formulas equivalent to (2.3.5).

III. *XY-linkage.* In this case a meiosis of males may be accompanied by a recombination of alleles between X- and Y- chromosomes. Here it's necessary to distinguish between the male genotype Aa and aA, writing, for the sake of definiteness, the gene from X-chromosome first. Let us denote by ρ_1 the probability of the recombination $Aa \mapsto aA$, by ρ_2 the probability of the inverse recombination $aA \mapsto Aa$. Let $0 < \rho_i < 1$. If before meiosis A is on the X- chromosome and its allele a is on the Y-chromosome, then they change places with the probability ρ_1, and, as a result, the male sperm cell A and female sperm cell a arise; at the same time they remain in place with the probability $1 - \rho_1$, and this results in female sperm cell A and male sperm cell a. Similarly when before meiosis A is in Y-chromosomes, a is in X- chromosome with corresponding probabilities equal to ρ_2 and $1 - \rho_2$. Thus, the male meiosis follows the schema

$$Aa \mapsto (\overbrace{(1 - \rho_1)A + \rho_1 a}^{X}, \overbrace{\rho_1 A + (1 - \rho_1)a}^{Y}), \; AA \mapsto (\overset{X}{A}, \overset{Y}{A})$$

$$aA \mapsto (\rho_2 A + (1 - \rho_2)a, (1 - \rho_2)A + \rho_2 a), \; aa \mapsto (a, a)$$

(2.1.1) is valid for female meiosis as before. Production of male and female offspring is described by Tables 2.4 and 2.5.

m	f		
	AA	aa	Aa
AA	AA	aA	$\frac{1}{2}AA + \frac{1}{2}aA$
aa	Aa	aa	$\frac{1}{2}aa + \frac{1}{2}Aa$
Aa	$\rho_1 AA + (1 - \rho_1)\,Aa$	$(1 - \rho_1)\,aa + \rho_1 aA$	$\frac{\rho_1}{2}AA + \frac{1-\rho_1}{2}aa + \frac{1-\rho_1}{2}Aa + \frac{\rho_1}{2}Aa$
aA	$(1 - \rho_2)\,AA + \rho_2 Aa$	$\rho_2 aa + (1 - \rho_2)\,aA$	$\frac{1-\rho_2}{2}AA + \frac{\rho_2}{2}aa + \frac{\rho_2}{2}Aa + \frac{1-\rho_2}{2}aA$

Table 2.4

m	f		
	AA	aa	Aa
AA	AA	Aa	$\frac{1}{2}AA + \frac{1}{2}Aa$
aa	Aa	aa	$\frac{1}{2}aa + \frac{1}{2}Aa$
Aa	$(1 - \rho_1)\,AA + \rho_1 Aa$	$\rho_1 aa + (1 - \rho_1)\,Aa$	$\frac{1-\rho_1}{2}AA + \frac{\rho_1}{2}aa + \frac{1}{2}Aa$
aA	$\rho_2 AA + (1 - \rho_2)\,Aa$	$(1 - \rho_2)\,aa + \rho_2 Aa$	$\frac{\rho_2}{2}AA + \frac{1-\rho_2}{2}aa + \frac{1}{2}Aa$

Table 2.5

The state of a population is determined by a pair of distributions $x = x_1 AA + x_2 aa + x_3 Aa$, $y = y_1 AA + y_2 aa + y_3 Aa + y_4 aA$. The state space is $\Delta^2 \times \Delta^3$. The evolutionary equations based upon Tables 2.4 and 2.5 are as follows:

$$
\begin{aligned}
x_1' &= x_1 y_1 + (1 - \rho_1) x_1 y_3 + \rho_2 x_1 y_4 \\
&\quad + \frac{1}{2} x_3 y_1 + \frac{1 - \rho_1}{2} x_3 y_3 + \frac{\rho_2}{2} x_3 y_4 \\
x_2' &= x_2 y_2 + \rho_1 x_2 y_3 + (1 - \rho_2) x_2 y_4 + \frac{1}{2} x_3 y_2 \\
&\quad + \frac{\rho_1}{2} x_3 y_3 + \frac{1 - \rho_2}{2} x_3 y_4 \\
x_3' &= x_1 y_2 + \rho_1 x_1 y_3 + (1 - \rho_2) x_1 y_4 + x_2 y_1 + (1 - \rho_1) x_2 y_3 \\
&\quad + \rho_2 x_2 y_4 + \frac{1}{2} x_3 y_1 + \frac{1}{2} x_3 y_2 + \frac{1}{2} x_3 y_3 + \frac{1}{2} x_3 y_4 \\
y_1' &= x_1 y_1 + \rho_1 x_1 y_3 + (1 - \rho_2) x_1 y_4 \\
&\quad + \frac{1}{2} x_3 y_1 + \frac{\rho_1}{2} x_3 y_3 + \frac{1 - \rho_2}{2} x_3 y_4 \\
y_2' &= x_2 y_2 + (1 - \rho_1) x_2 y_3 + \rho_2 x_2 y_4 \\
&\quad + \frac{1}{2} x_3 y_2 + \frac{1 - \rho_1}{2} x_3 y_3 + \frac{\rho_2}{2} x_3 y_4 \\
y_3' &= x_1 y_2 + (1 - \rho_1) x_1 y_3 + \rho_2 x_1 y_4 \\
&\quad + \frac{1}{2} x_3 y_2 + \frac{1 - \rho_1}{2} x_3 y_3 + \frac{\rho_2}{2} x_3 y_4 \\
y_4' &= x_2 y_1 + \rho_1 x_2 y_3 + (1 - \rho_2) x_2 y_4 \\
&\quad + \frac{1}{2} x_3 y_1 + \frac{\rho_1}{2} x_3 y_3 + \frac{1 - \rho_2}{2} x_3 y_4
\end{aligned}
\tag{2.3.8}
$$

After some simplifications

$$x_1' = p_X p_Y^X, \ x_2' = q_X q_Y^X, \ x_3' = p_X q_Y^X + q_X p_Y^X,$$

$$y_1' = p_X p_Y^Y, \ y_2' = q_X q_Y^Y, \ y_3' = p_X q_Y^Y, \ y_4' = q_X p_Y^Y \qquad (2.3.9)$$

where

$$p_X = x_1 + \frac{1}{2}x_3, \ q_X = x_2 + \frac{1}{2}x_3$$

$$p_Y^X = y_1 + (1 - \rho_1)y_3 + \rho_2 y_4, \ q_Y^X = y_2 + \rho_1 y_3 + (1 - \rho_2)y_4 \qquad (2.3.10)$$

$$p_Y^Y = y_1 + \rho_1 y_3 + (1 - \rho_2)y_4, \ q_Y^Y = y_2 + (1 - \rho_1)y_3 + \rho_2 y_4$$

Let us relate the introduced notations to their genetic meaning. The
letter p with one or another index corresponds, as before, to the gene A and
the letter q corresponds to the gene a. A lower index X or Y corresponds
to the male or female part of the gamete pool. The male part in its turn,
is split into two sections: male or female sperm cells. An attribution to the
female (male) section of the male part is marked by the upper index X (Y).
In such a way every state of the gamete pool is a triple of probability distributions (p_X, q_X), (p_Y^X, q_Y^X), (p_Y^Y, q_Y^Y). But not every triple of distributions
are representable in form (2.3.10), i.e. the space of states of the gamete
pool is imbedded into the cube $\Delta^1 \times \Delta^1 \times \Delta^1$ but does not coincide with it.
Indeed, the quantities

$$p_X = x_1 + \frac{1}{2}x_3$$

$$p_Y^X = y_1 + (1 - \rho_1)y_3 + \rho_2 y_4 \qquad (2.3.11)$$

$$p_Y^Y = y_1 + \rho_1 y_3 + (1 - \rho_2)y_4$$

can be used as independent parameters of a state of the gamete pool. The coordinate p_X runs over the segment $[0,1]$ independent of the pair (p_Y^X, p_Y^Y). The latter runs in the plane over the convex hull Γ of images of the extremal points of the polyhedron $y_1 + y_3 + y_4 \leq 1$, $y_i \geq 0$ $(i = 1, 3, 4)$. These points and their images are shown in Table 2.6.

y_1	y_3	y_4	p_Y^X	p_Y^Y
0	0	0	0	0
1	0	0	1	1
0	1	0	$1 - \rho_1$	ρ_1
0	0	1	ρ_2	$1 - \rho_2$

Table 2.6

Thus, the set Γ is either a triangle or a quadrangle (Fig. 2.2) depending on whether the points $(1 - \rho_1, \rho_1)$, $(\rho_2, 1 - \rho_2)$ lie on the same side (i.e. $(2\rho_1 - 1) \cdot (2\rho_2 - 1) \leq 0$)

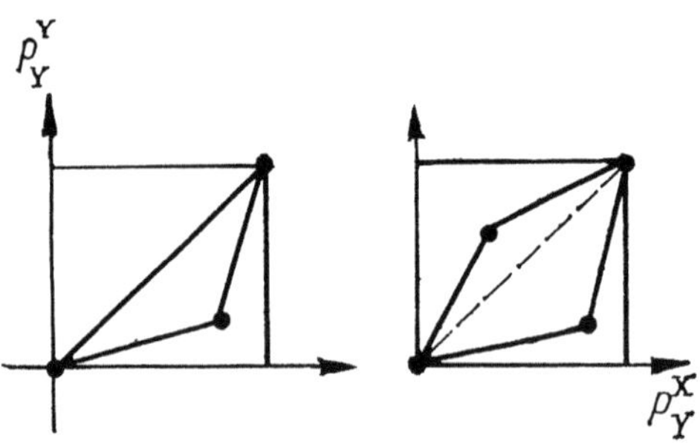

Fig. 2.2

or on the different sides (i.e. $(2\rho_1 - 1)(2\rho_2 - 1) > 0$) of the diagonal. In the degenerate case ($\rho_1 = \rho_2 = \frac{1}{2}$) one obtains the segment (the diagonal of the square).

Let us note that in the case of equal probabilities of the direct and inverse recombination, i.e. for $\rho_1 = \rho_2$, the set Γ is a triangle (or a segment). In all cases the space of the gamete pool is $\Delta^1 \times \Gamma$.

Formulas (2.3.9) defining the operator of fertilization correspond, as before, to independent random combination of gametes. For instance, the female genotype Aa arises after a union of the egg A and female sperm cell a, or th egg a and female sperm cell A. Correspondingly, $x_3' = p_X q_Y^X + q_X p_Y^X$.

Formulas (2.3.10) defining the operator of meiosis, as before, correspond to the averaging of meiosis schema over the population. For example, the female sperm cell A is produced by the male genotypes AA, Aa, aA with the conditional probabilities $1, 1 - \rho_1, \rho_2$ respectively. So the full probability of this gamete for the probabilities y_1, y_2, y_3 of aforesaid genotypes in the population will be $p_Y^X = y_1 + (1 - \rho_1)y_3 + \rho_2 y_4$.

Writing down the *evolutionary operator of the gamete pool* we will retain the coordinates p_X, p_Y^X, p_Y^Y only. Similarly to the X-linked locus we have

$$p_X' = \frac{p_X + p_Y^X}{2} \qquad (2.3.12)$$

and this is interpreted as before. Further on,

$$(p_Y^X)' = y_1' + (1 - \rho_1)y_3' + \rho_2 y_4' = p_X p_Y^Y + (1 - \rho_1)p_X q_Y^Y + \rho_2 q_X p_Y^Y.$$

Substituting $q_Y^Y = 1 - p_Y^Y$ and $q_X = 1 - p_X$, we obtain after some simple transformations

$$(p_Y^X)' = (1 - \rho_1)p_X + \rho_2 p_Y^Y + (\rho_1 - \rho_2)p_X P_Y^Y. \tag{2.3.13}$$

In the same way

$$(p_Y^Y)' = \rho_1 p_X + (1 - \rho_2)p_Y^Y + (\rho_2 - \rho_1)p_X p_Y^Y. \tag{2.3.14}$$

The equilibrium conditions are, by (2.3.12)-(2.3.14), as follows:

$$p_Y^X = p_X, \; p_Y^Y = \frac{\rho_1 p_X}{\rho_2 + (\rho_1 - \rho_2)p_X} = \frac{\rho_1 p_X}{\rho_1 p_X + \rho_2 q_X}. \tag{2.3.15}$$

The former condition means that the gene A (and consequently a) has an equal probability to occur in the X-gametes of both sexes. For $\rho_1 = \rho_2$ these conditions take on the intuitively clear form

$$p_X = p_Y^X = p_Y^Y \tag{2.3.16}$$

i.e. in this case for an equilibrium it is necessary and sufficient that the gene A (and hence a) would have the same probability for all gametes independent of their attribution to any section of the gamete pool.

The set of equilibrium states on the zygote level is determined by the equations

$$x_1 = p^2, \; x_2 = q^2, \; x_3 = 2pq$$

$$y_1 = \frac{\rho_1 p^2}{\rho_1 p + \rho_2 q}, \; y_2 = \frac{\rho_2 q^2}{\rho_1 p + \rho_2 q} \tag{2.3.17}$$

$$y_3 = \frac{\rho_2 pq}{\rho_1 p + \rho_2 q}, \; y_4 = \frac{\rho_1 pq}{\rho_1 p + \rho_2 q}.$$

As it was for the X-linkage, *the Hardy-Weinberg Law is valid for the entire female sex in an equilibrium state of the population.* For the male sex the distribution turns out to be more complicated but for $\rho_1 = \rho_2$ it takes on a simple form

$$y_1 = p^2, \; y_2 = q^2, \; y_3 = y_4 = pq \tag{2.3.18}$$

This is the *separate-sex analogue of the Hardy-Weinberg Law.*

For $\rho_1 = \rho_2(\equiv \rho)$ the convergence to an equilibrium may be studied in the same way as for the X-linkage case, since the equations (2.3.12)-(2.3.14) are reduced to the linear form:

$$p'_X = \frac{p_X + p_Y^X}{2}$$

$$(p_Y^X)' = (1 - \rho)p_X + \rho p_Y^Y \qquad (2.3.19)$$

$$(p_Y^Y)' = \rho p_X + (1 - \rho)p_Y^Y.$$

Here the transition matrix is

$$T = \begin{pmatrix} \frac{1}{2} & \frac{1}{2} & 0 \\ 1 - \rho & 0 & \rho \\ \rho & 0 & 1 - \rho \end{pmatrix}$$

and (2.3.19) is written as $P' = TP$, P being the column of the element p_X, p_Y^X, p_Y^Y. In the t-th generation $P^{(t)} = T^t P^{(0)}$.

The characteristic equation of the matrix T is

$$\lambda^3 + (\rho - \frac{3}{2})\lambda^2 + (\frac{1}{2} - \rho) = 0.$$

Its roots are

$$\lambda_1 = 1, \ \lambda_{2,3} = \frac{1}{2}(\frac{1}{2} - \rho \pm \sqrt{(\frac{1}{2} - \rho)(\frac{9}{2} - \rho)}).$$

If $0 < \rho < \frac{1}{2}$ then λ_2 and λ_3 are real, different and $|\lambda_3| < \lambda_2$, $0 < \lambda_2 < 1$, $-(\frac{1}{2}) < \lambda_3 < 0$. For $\rho = \frac{1}{2}$ the roots are equal: $\lambda_2 = \lambda_3 = 0$. For $\frac{1}{2} < \rho < 1$ the roots are complex (non-real), $|\lambda_3| = |\lambda_2| = \sqrt{\rho - (\frac{1}{2})} < 1/\sqrt{2}$. In all cases $\lim \lambda_2^t = \lim \lambda_3^t = 0$ ($t \to \infty$). Therefore, there exists $T^\infty = \lim T^t (t \to \infty)$, i.e. *all trajectories converge.* The rate of convergence is of the order $O(|\lambda_2|^t)$, the stabilization occurring monotonically for $\rho < \frac{1}{2}$ and in an oscillatory way for $\rho > \frac{1}{2}$. If $\rho = \frac{1}{2}$, then $T^2 = T$ i.e. on the gamete level an equilibrium is reached in the first generation of offspring and on the zygote level—in the second one, since the population evolutionary operator V is such that $V^3 = V^2$. In this particular case it is of the form (2.3.9) where

$$p_X = x_1 + \frac{1}{2}x_3, \ q_X = x_2 + \frac{1}{2}x_3$$

$$p_Y^X = p_Y^Y = y_1 + \frac{1}{2}y_3 + \frac{1}{2}y_4 \qquad (2.3.20)$$

$$q_Y^X = q_Y^Y = y_2 + \frac{1}{2}y_3 + \frac{1}{2}y_4.$$

As stated before, the space of states of the gamete pool reduces to $\Delta^1 \times \Delta^1$. Putting, for the sake of brevity, $p_Y^Y \equiv p_Y$, $q_Y^Y \equiv q_Y$ and making use of the identities $p_Y^X = p_Y^Y$, $q_Y^X = q_Y^Y$ we obtain from (2.3.9)

$$x_1' = p_X p_Y, \quad x_2' = q_X q_Y, \quad x_3' = p_X q_Y + q_X p_Y$$

$$y_1' = p_X p_Y, \quad y_2' = q_X q_Y, \quad y_3' = p_X q_Y, \quad y_4' = q_X p_Y \tag{2.3.21}$$

That these equations are the same for the female sex as for the X-linkage case is natural, since the difference between the probabilities of male and female sperm cells disappears.

Let us note that for an arbitrary ρ it would be possible to calculate T^t writing down every element of the matrix in a form $\alpha + \beta\lambda_2^t + \gamma\lambda_3^t$ and making use of the initial data $t = 0, 1$. Since the explicit formula for the state in the t-th generation derived in such a way is cumbersome, we omit it. But it is desirable to find the limit state $p^{(\infty)}$ which originates from the given initial state $p^{(0)}$. To this end, let us turn to the *Gene Conservation Law*

$$\frac{1}{2}p_X + \frac{1}{4}p_Y^X + \frac{1}{4}p_Y^Y = \text{const.} \tag{2.3.22}$$

Its validity follows from (2.3.19) and, also, is clear from its genetic interpretation. If the initial state (p_X, p_Y^X, p_Y^Y) passes to the state $(p^{(\infty)}, p^{(\infty)}, p^{(\infty)})$, in the limit (cf. (2.3.16)), then by (2.3.22)

$$p^{(\infty)} = \frac{1}{2}p_X + \frac{1}{4}p_Y^X + \frac{1}{4}p_Y^Y \tag{2.3.23}$$

For $\rho_1 \neq \rho_2$ the evolutionary equations (2.3.14) and (2.3.13) are non-linear, but the conservation law (2.3.22) is linear, i.e. the motion takes place in a plane. It is possible to use $\xi = p_X$ and $\eta = p_Y^Y$ as coordinates. Then $p_Y^X = -2\xi - \eta + 2c$ ($c = \text{const.}$) and by (2.3.12) and (2.3.14)

$$\xi' = -\frac{1}{2}\xi - \frac{1}{2}\eta + c$$

$$\eta' = \rho_1\xi + (1 - \rho_2)\eta + (\rho_2 - \rho_1)\xi\eta. \tag{2.3.24}$$

Here $0 \leq \xi \leq \eta \leq 1$. Let us show that map (2.3.24) is strongly contracting. To this end it is sufficient to prove that the differential

$$d\xi' = -\frac{1}{2}d\xi - \frac{1}{2}d\eta$$

$$d\eta' = (\rho_1 + (\rho_2 - \rho_1)\eta)d\xi + (1 - \rho_2 + (\rho_2 - \rho_1)\xi)d\eta$$

is strongly contracting. The matrix of this linear operator is of the form

$$\Phi = \begin{pmatrix} -\frac{1}{2} & -\frac{1}{2} \\ \alpha & \beta \end{pmatrix}$$

where $\rho_1 \leq \alpha \leq \rho_2$, $1 - \rho_2 \leq \beta \leq 1 - \rho_1$ (for the sake of the definiteness we assume $\rho_1 < \rho_2$). Let us introduce into the two dimensional space with the coordinates $d\xi, d\eta$ the norm $|d\xi| + |d\eta| + |d\xi + d\eta|$. Then for any $z = (d\xi, d\eta)$ one has

$$\|\Phi z\| = \frac{1}{2}|d\xi + d\eta| + |\alpha d\xi + \beta d\eta| + |(\alpha - \frac{1}{2})d\xi + (\beta - \frac{1}{2})d\eta|.$$

The unit "circle" $\|z\| = 1$ is the hexagon (Fig 2.3) with the vertices at the points $z_1 = (\frac{1}{2}, 0)$, $z_2 = (0, \frac{1}{2})$, $z_3 = (-\frac{1}{2}, \frac{1}{2})$ and $(-z_1)$, $(-z_2)$, $(-z_3)$. The convex even function $\|\Phi z\|$ (for $\|z\| \leq 1$) attains its maximum at one of the points z_1, z_2, z_3.

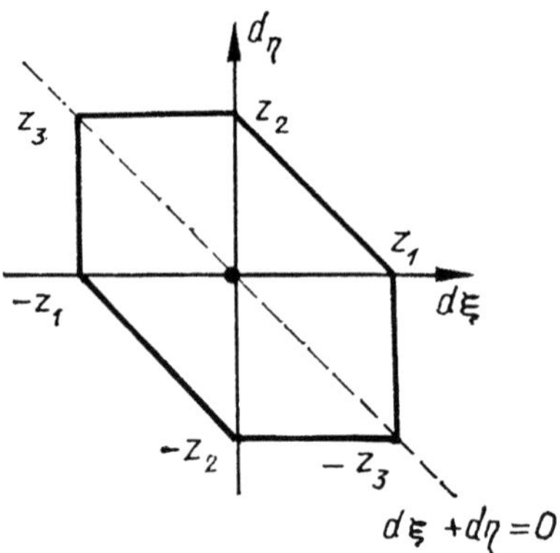

Fig. 2.3

But

$$\|\Phi z_1\| = \frac{1}{2}(\alpha + \frac{1}{2} + |\alpha - \frac{1}{2}|) = \max(\alpha, \frac{1}{2})$$

$$\|\Phi z_2\| = \frac{1}{2}(\beta + \frac{1}{2} + |\beta - \frac{1}{2}|) = \max(\beta, \frac{1}{2})$$

$$\|\Phi z_3\| = |\alpha - \beta|.$$

Hence, $\|\Phi\| \le \max(\alpha, \beta, \frac{1}{2}) \le \max(\rho_2, 1 - \rho_1) < 1$ as is required. It means that *for the XY-linkage with, the probabilities of recombinations being* $0 < \rho_1 < 1$, $0 < \rho_2 < 1$, $\rho_1 \ne \rho_2$, *the convergence to an equilibrium takes place with the rate* $O(\overline{\kappa}^t)$ *where* $\overline{\kappa} = \max(\rho_2, 1 - \rho_1)$.

2.4   Selection at a Two Allele, Autosomal Locus

Let us denote by $\lambda_1, \lambda_2, \lambda_3$ the survival coefficients of the zygotes AA, aa, Aa. Then at the reproductive stage

$$x'_j = \frac{\lambda_j \overline{x}_j}{\lambda_1 \overline{x}_1 + \lambda_2 \overline{x}_2 + \lambda_3 \overline{x}_3} \quad (j = 1, 2, 3), \tag{2.4.1}$$

(cf. (1.2.5)) where $\overline{x}_j$ are the probabilities of zygotes at the moment of conception

$$\overline{x}_1 = p^2, \ \overline{x}_2 = q^2, \ \overline{x}_3 = 2pq \tag{2.4.2}$$

and p, q are the probabilities of the genes (gametes) A and a in the gene pool in the preceding generation:

$$p = x_1 + \frac{1}{2}x_3, \ q = x_2 + \frac{1}{2}x_3. \tag{2.4.3}$$

On the gamete level

$$p' = \frac{p(\lambda_1 p + \lambda_3 q)}{W}, \ q' = \frac{q(\lambda_3 p + \lambda_2 q)}{W} \tag{2.4.4}$$

Here

$$W = \lambda_1 p^2 + \lambda_2 q^2 + 2\lambda_3 pq \tag{2.4.5}$$

is the mean population fitness. Let us assume $\lambda_1 > 0$, $\lambda_2 > 0$. Then $W > 0$. Also, let us exclude the selectively neutral case $\lambda_1 = \lambda_2 = \lambda_3$ studied in Section 2.1. Then the map (2.4.4) is not the identity, i.e. the

Conservation Law fails. It is quite clear from a biological point of view that the pressure of selection is different for the different genes.

Let us find the equilibrium states from the equations

$$p = \frac{p(\lambda_1 p + \lambda_3 q)}{W}, \; q = \frac{q}{W}(\lambda_3 p + \lambda_2 q). \tag{2.4.6}$$

There exist two *homozygote* equilibrium states: (1,0) and (0,1). In the state (1,0) the gene pool of the population contains A allele only, and on the zygote level the population consists of the homozygotes AA. A similar pattern is observed at the state (0,1). Now we need only find those equilibrium states for which $p > 0$, $q > 0$. They are called *polymorphic*, since, according to (2.4.2), more than one genotype is present on the zygote level. As distinguished from a selectively neutral population, where the equilibrium polymorphisms form a continuum, *two-allele selection leads to a unique polymorphic equilibrium* if one exists at all. *The set of equilibrium states is finite* (and consists of two or three points). To prove it we notice that the system (2.4.6) for $p > 0$, $q > 0$ is equivalent to the system

$$\lambda_1 p + \lambda_3 q = W, \; \lambda_3 p + \lambda_2 q = W \tag{2.4.7}$$

which, in its turn, is equivalent to the system

$$\lambda_1 p + \lambda_3 q = \lambda_3 p + \lambda_2 q,; p + q = 1$$

i.e.

$$(\lambda_1 - \lambda_3)p + (\lambda_3 - \lambda_2)q = 0, \; p + q = 1. \tag{2.4.8}$$

If the determinant

$$\begin{vmatrix} \lambda_1 - \lambda_3 & \lambda_3 - \lambda_2 \\ 1 & 1 \end{vmatrix} = \lambda_1 + \lambda_2 - 2\lambda_3$$

is different from zero, i.e.

$$\lambda_3 \neq \frac{\lambda_1 + \lambda_2}{2} \tag{2.4.9}$$

then there exists the unique solution

$$p^* = \frac{\lambda_2 - \lambda_3}{\lambda_1 + \lambda_2 - 2\lambda_3}, \; q^* = \frac{\lambda_1 - \lambda_3}{\lambda_1 + \lambda_2 - 2\lambda_3}. \tag{2.4.10}$$

If $\lambda_3 = \frac{1}{2}(\lambda_1 + \lambda_2)$, then system (2.4.8) is inconsistent (since not all $\lambda_1, \lambda_2, \lambda_3$ are equal).

The solution of (2.4.10) satisfies the conditions for a polymorphism, i.e. $p^* > 0$, $q^* > 0$ in two cases:

1) *superdominance*, i.e. $\lambda_3 > \max(\lambda_1, \lambda_2)$, when a heterozygote is fitter than both homozygotes;

2) *superrecessivity*, i.e. $\lambda_3 < \min(\lambda_1, \lambda_2)$, when a heterozygote is less fit than either homozygote.

In the *intermediate* case, i.e. $\lambda_2 \leq \lambda_3 \leq \lambda_1$ ($\lambda_2 < \lambda_1$ for the sake of definiteness) the population has no polymorphic equilibrium.

We will prove the convergence of all trajectories for a more general dynamic system, describing the process of selection in a diallele stationary gene structure, where instead of (2.4.2) the Extended Hardy-Weinberg Law is valid (see Section 2.1). In this situation

$$p' = \frac{pW_1}{W}, \; q' = \frac{qW_2}{W} \tag{2.4.11}$$

where

$$W_1 = \lambda_1 p + 2(\lambda_1 a_{12} + \lambda_3 c_1 b)q, \; W_2 = 2(\lambda_2 a_{21} + \lambda_3 c_2 b)p + \lambda_2 q$$

and the mean fitness of the population is

$$W = pW_1 + qW_2 = \lambda_1 p^2 + \lambda_2 q^2 + 2(\lambda_1 a_{12} + \lambda_2 a_{21} + \lambda_3 b)pq.$$

Formulas (2.4.10) are replaced with the following more general ones:

$$p^* = \frac{(\lambda_2 - \lambda_3) - 2a_{12}(\lambda_1 - \lambda_3)}{2b(\lambda_1 c_1 + \lambda_2 c_2 - \lambda_3)}, \; q^* = \frac{(\lambda_1 - \lambda_3) - 2a_{21}(\lambda_2 - \lambda_3)}{2b(\lambda_1 c_1 + \lambda_2 c_2 - \lambda_3)} \tag{2.4.12}$$

Now $p^* > 0$, $q^* > 0$ if and only if λ_3 lies outside the segment $[\lambda_2, \lambda_1]$ and the inequalities are fulfilled

$$2a_{21} < \frac{\lambda_1 - \lambda_3}{\lambda_2 - \lambda_3} < \frac{1}{2a_{12}}$$

(note that $4a_{12}a_{21} < 1$).

To study the dynamic system (2.4.11) let us introduce the projective coordinate $\theta = p/q$, $0 \leq \theta \leq \infty$. Then

$$\theta' = \theta g(\theta) = \theta \frac{\alpha_{11}\theta + \alpha_{12}}{\alpha_{21}\theta + \alpha_{22}}. \tag{2.4.13}$$

The coefficients

$$\alpha_{11} = \lambda_1, \; \alpha_{12} = 2(\lambda_1 a_{12} + \lambda_3 c_1 b)$$

$$\alpha_{21} = 2(\lambda_2 a_{21} + \lambda_3 c_2 b), \ \alpha_{22} = \lambda_2$$

are positive (excluding the case $a_{12} = a_{21} = 0$, $\lambda_3 = 0$, i.e. $\theta' = (\alpha_{11}/\alpha_{22})\theta^2$ and, obviously, the iterations $\theta^{(t)}$ tend to zero or infinity if $\theta^{(0)} \neq \alpha_{22}/\alpha_{11}$).

Note that for all $\theta > 0$

$$\frac{d\theta'}{d\theta} = \frac{\alpha_{11}\alpha_{21}\theta^2 + 2\alpha_{11}\alpha_{22}\theta + \alpha_{12}\alpha_{22}}{(\alpha_{21}\theta + \alpha_{22})^2} > 0. \qquad (2.4.14)$$

The existence of a polymorphic equilibrium is equivalent to the existence of a fixed point θ^*, $0 < \theta^* < \infty$. Obviously $g(\theta^*) = 1$ and

$$\theta^* = \frac{\alpha_{12} - \alpha_{22}}{\alpha_{21} - \alpha_{11}} = \frac{a_{12}(\lambda_2 - \lambda_1) + c_1 b(\lambda_2 - \lambda_3)}{a_{21}(\lambda_1 - \lambda_2) + c_2 b(\lambda_1 - \lambda_3)} \qquad (2.4.15)$$

In the superdominance case the inequalities

$$\alpha_{12} > \alpha_{22}, \ \alpha_{11} < \alpha_{21} \qquad (2.4.16)$$

are satisfied, by which $g(\theta) > 1$ in $(0, \theta^*)$ and $g(\theta) < 1$ in (θ^*, ∞), i.e. $\theta' > \theta$ in $(0, \theta^*)$ and $\theta' < \theta$ in (θ^*, ∞). From (2.4.14) $\theta' < \theta^*$ in $(0, \theta^*)$ and $\theta' > \theta^*$ in (θ^*, ∞), i.e. these intervals are invariant with respect to the map (2.4.13). It is clear that the iterations of the map in this case converge to θ^*.

In the superrecessive case the inequalities

$$\alpha_{12} < \alpha_{22}, \ \alpha_{11} > \alpha_{21} \qquad (2.4.17)$$

are valid, by which $g(\theta) < 1$ in $(0, \theta^*)$ and $g(\theta) > 1$ in (θ^*, ∞), i.e. $\theta' < \theta$ in $(0, \theta^*)$ and $\theta' > \theta$ in (θ^*, ∞). Again these intervals turn out to be invariant, but now the iterations in the former interval converge to zero, and in the latter to infinity.

In the intermediate case the map has no fixed points in $(0, \infty)$. So either $\theta' < \theta$ for all $\theta > 0$ or $\theta' > \theta$ for all $\theta > 0$. Thus, either the iterations converge to zero or to infinity.

It follows from the above reasoning that the coordinate θ (together with the coordinate $p = \theta/1 + \theta$) changes monotonically during the process, but the direction of its change depends, generally speaking, on the initial datum.

Let us estimate the rate of convergence to an equilibrium. By (2.4.14) and the relation $\alpha_{21}\theta^* + \alpha_{22} = \alpha_{11}\theta^* + \alpha_{12}$ we have $(d\theta'/d\theta)_{\theta^*} = (\alpha_{11}\theta^* + \alpha_{22})/(\alpha_{21}\theta^* + \alpha_{22})$. It follows, in view of (2.4.16), $(d\theta'/d\theta)_{\theta^*} < 1$ in the superdominance case. In the superrecessive case, due to (2.4.17) $(d\theta'/d\theta)_{\theta^*} > 1$. In the same case $(d\theta'/d\theta)_{\theta=0} = \alpha_{12}/\alpha_{22} < 1$, $(d\theta'/d\theta)_{\infty} = \alpha_{11}/\alpha_{21} > 1$. In both cases the convergence rate is exponential. Passing to the intermediate case let us suppose, for the sake of definiteness, that the iterations converge to zero. Then it follows that $\alpha_{12} \leq \alpha_{22}$. If $\alpha_{12} < \alpha_{22}$, then the rate of convergence is exponential. If $\alpha_{12} = \alpha_{22}$ i.e. $\lambda_1 = \lambda_2$, then in a neighborhood of zero

$$\theta' = \theta - \frac{\alpha_{21} - \alpha_{11}}{\alpha_{22}}\theta^2 + \cdots$$

It is necessary that $\alpha_{21} \geq \alpha_{11}$. If $\alpha_{21} > \alpha_{11}$, then the convergence rate is of the first degree (see Appendix). If $\alpha_{21} = \alpha_{11}$, then the map degenerates to the identity which was ruled out by our assumptions.

Let us consider the important issue of the mean fitness dynamics. The unique critical point of the function $W(p)$ in the interval $0 < p < 1$ is $\overline{p}$ for which

$$\overline{\theta} = \overline{p}/\overline{q} = \frac{a_{12}(\lambda_2 - \lambda_1) + b(\lambda_2 - \lambda_3)}{a_{21}(\lambda_1 - \lambda_2) + b(\lambda_1 - \lambda_3)}$$

and the condition $0 < \overline{\theta} < \infty$ should be met. If the locus is Mendelian, i.e. $a_{12} = a_{21} = 0$, $b = 1$, $c_1 = c_2 = \frac{1}{2}$ then $\overline{\theta} = \theta^*$. In this case

$$\frac{1}{2}\frac{d^2 W}{dp^2} = \lambda_1 + \lambda_2 - 2\lambda_3$$

i.e. in the superdominance case W attains its absolute maximum at the point p^*, the polymorphic equilibrium. In the remaining cases the absolute maximum is attained at the homozygote state corresponding to the fitter homozygote. It follows now that *in the presence of selection in a diallele autosomal Mendel locus the mean fitness increases in the process of evolution* (see Fig. 2.4).

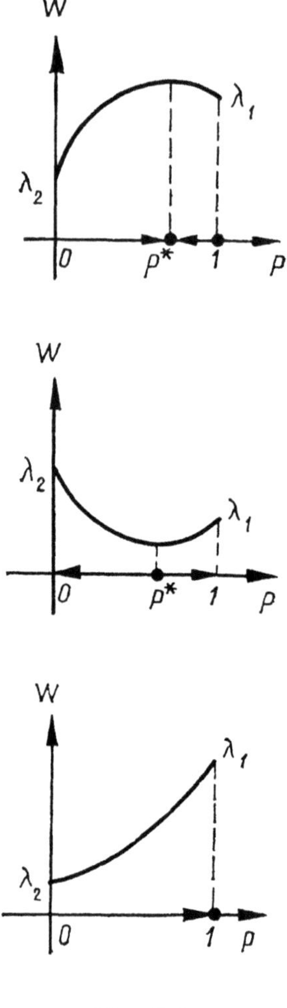

Fig. 2.4

This fact is called The Fundamental Theorem. It may be considered as a mathematical description of Darwin's conception of "survival of the fittest". But The Fundamental Theorem is valid in the presence of selection in a diallele stationary gene structure if and only if $0 < \bar{\theta} < \infty \Rightarrow \bar{\theta} = \theta^*$.

Remark It is possible to prove the convergence of all trajectories using the Fundamental Theorem in the case of its validity. We use this approach in chapter 9.

2.5 Mutation

Let $A_1, \ldots, A_m$ be the alleles of an autosomal locus. We assume that mutations occur at the gamete stage changing A_i into A_j with probability τ_{ji} (τ_{ii} is the probability of no change for A_i). So $\sum_j \tau_{ji} = 1$. The column stochastic matrix $T = (\tau_{ji})$ is called *the matrix of mutation rates.* Usually its nondiagonal elements are small ($\sim 10^{-6}$ or less). In any case it is reasonable to assume that $\tau_{ii} > 0$ for $i = 1, \ldots, m$. The evolutionary operator at the gamete level is

$$p'_j = \sum_{i=1}^{m} \tau_{ji} p_i \; (1 \le j \le m) \tag{2.5.1}$$

or in a vector form $P' = TP$ which yields $P^{(t)} = T^t P^{(0)}$.

If $|\lambda| = 1$ but $\lambda \neq 1$, then the inequality

$$|\tau_{ii} - \lambda| > 1 - \tau_{ii} = \sum_{j \neq i} \tau_{ji}$$

holds, i.e. the point 1 is closer to $\tau_{ii} > 0$ than all other points of the unit circle. Hence, the diagonal is dominant in the matrix $T - \lambda I$, and by the Hadamard Theorem there are no eigenvalues of T on the unit circle other than $\lambda = 1$. It follows then that $T^\infty = \lim T^t (t \to \infty)$ exists, i.e. *every trajectory converges to an equilibrium* which is generally speaking not unique: $P^{(\infty)} = T^\infty P^{(0)}$. The convergence rate is $O(t^{\nu-1} \rho^t)$ where ρ is the maximum of moduli of the eigenvalues not equal to 1, ν is the size of the largest Jordan blocks corresponding to an eigenvalue on the circle $|\lambda| = \rho$.

Now we will study the simplest case of interaction of mutation with selection for two alleles A and a. Suppose that A mutates into a with a rate τ and there are no inverse mutations. Then

$$T = \begin{pmatrix} 1 - \tau & 0 \\ \tau & 1 \end{pmatrix}.$$

Further, let A dominate over a and be fitter than a, but let nevertheless, a not be lethal: $\lambda_1 = \lambda_3 > \lambda_2 > 0$. The evolutionary operator is the composition of the selection operator and the mutation operator. Passing to $\theta = p/q$, as in section 2.4, we have

$$\theta' = \frac{\tilde{\theta}(\tilde{\theta} + 1)}{\tilde{\theta} + \omega}, \; \tilde{\theta} = \frac{(1 - \tau)\theta}{\tau \theta + 1} \; (\theta \ge 0). \tag{2.5.2}$$

Here $\omega = \lambda_2\lambda_1^{-1} < 1$. Since

$$\theta = \frac{\tilde{\theta}}{(1-\tau)-\tau\tilde{\theta}} \tag{2.5.3}$$

all equilibrious values $\tilde{\theta}$ are found from the equation

$$\frac{\tilde{\theta}(\tilde{\theta}+1)}{\tilde{\theta}+\omega} = \frac{\tilde{\theta}}{(1-\tau)-\tau\tilde{\theta}} \;\; (0 \le \tilde{\theta} \le \frac{1-\tau}{\tau}).$$

For $\tilde{\theta} \neq 0$ it follows

$$\tau\tilde{\theta}^2 + 2\tau\tilde{\theta} + (\omega + \tau - 1) = 0. \tag{2.5.4}$$

Equation (2.5.4) has the positive root

$$\tilde{\theta}^* = \sqrt{\frac{1-\omega}{\tau}} - 1 \tag{2.5.5}$$

for $\tau < 1-\omega$ and in this case only. Let us assume this condition of smallness of the mutation rate is fulfilled. The inequality $\theta^* < \frac{1-\tau}{\tau}$ is always satisfied. Thus, we have the polymorphic equilibrium

$$p^* = \frac{\theta^*}{1+\theta^*}, \; q^* = \frac{1}{1+\theta^*}$$

where $\theta^* = \tilde{\theta}^*/((1-\tau)-\tau\tilde{\theta}^*)$ according to (2.5.3). When $\tau << 1$ then $\tilde{\theta}^* >> 1$, but $\tau\tilde{\theta}^* < \sqrt{\tau(1-\omega)} << 1$. Then, approximately, $\theta^* \approx \tilde{\theta}^* \approx \sqrt{(1-\omega)\tau^{-1}}$ and

$$q^* \approx (\tilde{\theta}^*)^{-1} \approx \sqrt{(1-\omega)^{-1}\tau} \tag{2.5.6}$$

(*Haldane's Formula*). The error of this approximation is of the order of τ.

Let us show that the polymorphic equilibrium is global attractor. Indeed, for $0 < \theta < \theta^*$ one has $\theta < \theta' < \theta^*$, as, firstly, $\theta' \neq \theta$ in the interval $(0, \theta^*)$ and near zero $\theta' \sim \frac{1-\tau}{\omega}\theta > \theta$ and, secondly,

$$\frac{d\theta'}{d\theta} = \frac{d\theta'}{d\tilde{\theta}} \cdot \frac{d\tilde{\theta}}{d\theta} = \frac{(1-\tau)(\tilde{\theta}^2 + 2\omega\tilde{\theta} + \omega)}{(\tilde{\theta}+\omega)^2(\tau\theta+1)^2} > 0$$

i.e. θ' is the increasing function of θ. For $\theta > \theta^*$ we have $\theta^* < \theta' < \theta$, since again $\theta' \neq \theta$ on (θ^*, ∞) and θ' tends to a limit at infinity, i.e. surely $\theta' < \theta$.

Therefore, for $\theta^{(0)} \neq 0$ the sequence $\{\theta^{(t)}\}$ monotonically tends to θ^*. The rate of convergence is exponential as

$$\kappa = \frac{d\theta'}{d\theta}\Big|_{\theta^*} = \frac{1 - 2\sqrt{\tau(1-\omega)} + \tau}{1 - \tau} < 1$$

by virtue of the condition $\tau < 1 - \omega$. In the presence of the mutation $A \to a$ the selection accelerates, as the mutation process weakens the disguising action of dominance, because it leads to the formation of recessive homozygotes. But the value $\kappa \approx 1 - 2\sqrt{\tau(1-\omega)}$ is close to 1, i.e. the exponential κ^t decreases rather slowly.

2.6 Appendix: Asymptotic Estimates in One-Dimensional Dynamics

Let f be a continuous function on the interval $[0, a)$ for some positive a and suppose that

$$0 < f(x) < x \qquad x \in (0, a). \tag{2.6.1}$$

Any trajectory $\{x^{(t)}\}$, i.e. $x^{(t+1)} = f(x^{(t)})$, is therefore a monotone decreasing sequence for any initial value $x^{(0)}$ in $(0, a)$. As usual the limiting value $x^{(\infty)}$ is a fixed point for f and so $x^{(\infty)}$ must equal 0. It is often important to estimate the rate of approach to 0.

Theorem 2.6.1 *Let $f(x) = \lambda x + O(\phi(x))$ be a positive function on $(0, a)$ with $0 < \lambda < 1$, $x^{-1}\phi(x)$ nondecreasing in x and*

$$\int_0 x^{-2}\phi(x) < \infty \tag{2.6.2}$$

Then with $x^{(t+1)} = f(x^{(t)})$ for $x^{(0)}$ in $(0, a)$

$$x^{(t)} \sim C\lambda^t \tag{2.6.3}$$

with the positive constant C depending on $x^{(0)}$. In particular, the conditions of the theorem are satisfied with $\phi(x) = x^\alpha$ for $\alpha > 1$.

Proof Let $w(x) = f(x) - \lambda x$ so that $w(x)$ is $O(\phi(x))$ meaning for some constant K, $|w(x)| \leq K\phi(x)$. Evidently:

$$x^{(t)} = \lambda^t x^{(0)} \prod_{k=0}^{t-1} \left(1 + \frac{1}{\lambda}\frac{w(x^{(k)})}{x^{(k)}}\right).$$

It is thus sufficient to show that the infinite product

$$\prod_{k=0}^{\infty}\left(1+\frac{1}{\lambda}\frac{w(x^{(k)})}{x^{(k)}}\right)$$

converges. In turn it suffices to show that the series

$$\sum_{k=0}^{\infty}\frac{\phi(x^{(k)})}{x^{(k)}} < \infty.$$

From the assumptions on ϕ we have $x^{(k)-1}\phi(x^{(k)}) \to 0$ because $x^{(k)} \to 0$. So from the definition of f we have

$$\lim_{k\to\infty}\frac{x^{(k)}}{x^{(k-1)}} = \lambda.$$

Furthermore

$$\int_{x^{(k)}}^{x^{(k-1)}} x^{-2}\phi(x)dx \geq x^{(k)-1}\phi(x^{(k)})log\frac{x^{(k+1)}}{x^{(k)}} \geq Cx^{(k)-1}\phi(x^{(k)})$$

where C is a positive constant. So

$$\sum_{k=0}^{\infty}x^{(k)-1}\phi(x^{(k)}) \leq C^{-1}\int_{0}^{x^{(0)}} x^{-2}\phi(x)dx.$$

$\square$

To formulate the next theorem we denote by D_1 the interior of the unit disc in the complex plane together with the number 1. Note that $\lambda \in D_1$ if and only if the geometric progression $\{\lambda^t\}$ converges.

Theorem 2.6.2 *Let $\{w_t\}$ be a convergent sequence of complex numbers and λ be a fixed complex number. Define the sequence $\{x_t\}$ by the linear recurrence equation:*

$$x_{t+1} = \lambda x_t + w_t \qquad\qquad (t = 0,1,2,\ldots) \qquad\qquad (2.6.4)$$

In order that all solutions $\{x_t\}$ of (2.6.4) converge it is necessary and sufficient that λ lie in D_1 and, in addition, if $\lambda = 1$ that the series $\sum_{t=0}^{\infty}w_t$ converge.

Proof Necessity: Let $x_0^{(0)} = 0$ and $x_0^{(1)} = 1$. The difference $z_t = x_t^{(1)} - x_t^{(0)}$ satisfies the homogeneous equation $z_{t+1} = \lambda z_t$ with $z_0 = 1$. So $z_t = \lambda^t$ and convergence implies $\lambda \in D_1$. Furthermore, if $\lambda = 1$ then the solution of (2.6.4) is given by

$$x_{t+1} = x_0 + \sum_{k=0}^{t} w_k \qquad (2.6.5)$$

and so convergence of $\{x_t\}$ requires convergence of the series $\sum w_k$.

Sufficiency: When $\lambda = 1$ sufficiency is evident from (2.6.5). Assume $|\lambda| < 1$. Then by induction

$$x_{t+1} = \sum_{k=0}^{t} \lambda^{t-k} w_k + \lambda^{t+1} x_0 \qquad (t = 0, 1, \ldots).$$

Assume that w_t converges to w. We show that x_t converges to $w^* \equiv w/1-\lambda$. Observe that w^* is the limit of

$$y_{t+1} = \sum_{k=0}^{t} \lambda^{t-k} w$$

So it suffices to show that the difference $z_t = x_t - y_t$ converges to zero. Define $W_K = \max\{|w_k - w| : k \geq K\}$ so that W_K converges to zero and write the difference

$$z_{t+1} = \sum_{k=0}^{N-1} \lambda^{t-k}(w_k - w) + \sum_{k=N}^{t} \lambda^{t-k}(w_k - w) + \lambda^{t+1} x_0.$$

It follows that for $t \geq N$:

$$|z_{t+1}| \leq \frac{|\lambda|^{t-N}}{1 - |\lambda|} W_0 + \frac{1}{1 - |\lambda|} W_N + |\lambda|^{t+1} |x_0|.$$

With $t \geq 2N$ we see that each of the three terms tends to zero as N increases. $\square$

The previous cases are somewhat crude though sufficient for many applications. The following is a more delicate result.

Theorem 2.6.3 *Assume that on $(0, a)$ the function $f(x) = x - \phi(x) + \theta(x)$ with $\phi(x)$ differentiable and satisfying*

$$\phi(x) > 0 \text{ on } (0, a), \ \lim_{x \to 0} \phi(x) = 0$$

$$\phi'(x) \geq 0 \text{ on } (0,a), \ \lim_{x \to 0} \phi'(x) = 0$$

and $\theta(x)$ is $o(\phi(x))$, i.e.

$$\lim_{x \to 0} \frac{\theta(x)}{\phi(x)} = 0.$$

Define $\psi(t)$ $(0 \leq t < \infty)$ to be the inverse function of

$$t(x) = \int_x^a \frac{d\xi}{\phi(\xi)}$$

and assume that $\psi(t)$ satisfies

$$\underline{\lim}_{l \to \infty} \frac{t\psi'(t)}{\psi(t)} > -\infty.$$

Then for a trajectory $x^{(t+1)} = f(x^{(t)})$

$$x^{(t)} \sim \psi(t) \text{ as } t \to \infty.$$

Proof By monotonicity of $\phi(x)$:

$$\int_{x^{(k)}}^{x^{(k-1)}} \frac{d\xi}{\phi(\xi)} \geq \frac{x^{(k-1)} - x^{(k)}}{\phi(x^{(k-1)})} = 1 - \frac{\theta(x^{(k-1)})}{\phi(x^{(k-1)})}$$

and so we get by summation

$$t(x^{(k)}) \geq k - \sum_{i=0}^{k-1} \frac{\theta(x^{(i)})}{\phi(x^{(i)})} + \int_{x^{(0)}}^a \frac{d\xi}{\phi(\xi)}.$$

Consequently,

$$\underline{\lim}_{k \to \infty} \frac{t(x^{(k)})}{k} \geq 1$$

because $\lim_{i \to \infty} \theta(x^{(i)})/\phi(x^{(i)}) = 0$.

Define

$$P_k = \frac{\phi(x^{(k-1)})}{\phi(x^{(k)})}, \ Q_k = \frac{\theta(x^{(k-1)})}{\phi(x^{(k)})}.$$

By the Mean Value Theorem

$$P_k = 1 + \frac{\phi'(\xi_k)(x^{(k-1)} - x^{(k)})}{\phi(x^{(k)})}$$

where $x^{(k)} < \xi_k < x^{(k-1)}$. It follows as above that:

$$P_k = 1 + \phi'(\xi_k)[1 - \frac{\theta(x^{(k-1)})}{\phi(x^{(k-1)})}]P_k$$

whence it follows that $\lim_{k\to\infty} P_k = 1$ as $\lim \phi'(\xi_k) = 0$. It then follows that

$$\lim_{k\to\infty} Q_k = \lim_{k\to\infty} P_k \frac{\theta(x^{(k-1)})}{\phi(x^{(k-1)})} = 0$$

But observe that

$$\int_{x^{(k)}}^{x^{(k-1)}} \frac{d\xi}{\phi(\xi)} \leq \frac{x^{(k-1)} - x^{(k)}}{\phi(x^{(k)})} = P_k - Q_k$$

and from this we have

$$t(x^{(k)}) \leq k + \sum_{i=0}^{k-1}(P_i - 1 - Q_i) + \int_{x^{(0)}}^{a} \frac{d\xi}{\phi(\xi)}$$

and consequently

$$\overline{\lim}_{k\to\infty} \frac{1}{k} t(x^{(k)}) \leq 1 \quad .$$

Thus, we have that $\{\frac{1}{k} t(x^{(k)})\}$ converges to 1, that is

$$\int_{x^{(k)}}^{a} \frac{d\xi}{\phi(\xi)} \sim k \qquad\qquad (k \to \infty)$$

It remains to reverse this asymptotic formula. Notice that the function

$$\lambda(t) = \frac{t\psi'(t)}{\psi(t)} \qquad\qquad (t > 0)$$

is bounded below by hypothesis, say $\lambda(t) > -c > -\infty$. Also $\lambda(t) < 0$ because $\psi'(t) = -\phi(\psi(t)) < 0$. Because $\psi(0) = a$ we have

$$\psi(t) = a \exp[\int_0^t \frac{\lambda(\tau)}{\tau} d\tau]$$

and so

$$\frac{\psi(t + h)}{\psi(t)} = \exp[\int_t^{t+h} \frac{\lambda(\tau)}{\tau} d\tau].$$

As $0 > \lambda(t) > -c$ we have

$$(\frac{t}{t+h})^c < \frac{\psi(t+h)}{\psi(t)} < 1 \qquad\qquad (h > 0)$$

$$1 < \frac{\psi(t+h)}{\psi(t)} < (\frac{t}{t+h})^c \qquad\qquad (h < 0).$$

Consequently, if h is $o(t)$ and $t \to \infty$

$$\frac{\psi(t+h)}{\psi(t)} \to 1.$$

In particular

$$\lim_{k\to\infty} \frac{\psi(t(x^{(k)}))}{\psi(k)} = 1$$

which says $x^{(k)} \sim \psi(k)$. $\square$

Example If $f(x) = x - ax^\alpha + O(x^\beta)$ with $\beta > \alpha > 1$ and $a > 0$ then the hypotheses of Theorem 2.6.3 are satisfied. Here

$$\psi(t) = [a(\alpha - 1)(t + c)]^{-\frac{1}{\alpha-1}}$$

where c is constant. Consequently,

$$x^{(t)} \sim [a(\alpha - 1)t]^{-\frac{1}{\alpha-1}} \qquad\qquad (\text{as } t \to \infty).$$

In particular, for the important case $\alpha = 2$

$$x^{(t)} \sim \frac{1}{at} \qquad\qquad (t \to \infty).$$

If condition (2.6.1) is not valid the asymptotic behavior of the iterates $x^{(t+1)} = f(x^{(t)})$ may be very complicated.There can appear limit cycles of an arbitrary order as well as fully chaotic behavior. However,as a rule the genetic dynamics is regular.The reader has seen this above and will find more confirmations in chapters 6,9 where we will consider the populations having very complicated inner structure but the regular asymptotic behavior.

Chapter 3

Algebra Foundations

3.1 Nonassociative Algebras. Idempotents and Nilpotents

By the term *algebra* in this book we will understand a finite dimensional vector space $\mathcal{A}$ over the field of real numbers $\mathbf{R}$ (or in some cases, when explicitly specified, the complex numbers $\mathbf{C}$) equipped with a bilinear multiplication mapping $\mathcal{A} \times \mathcal{A}$ to $\mathcal{A}$. Bilinearity implies the distributive law for the multiplication. Also, with the exception of the operator algebras and Lie algebras discussed below we will assume that our algebras are all commutative. On the other hand, we do not assume associativity.

As usual, a *subalgebra* S of $\mathcal{A}$ is a subspace which is closed under multiplication, i.e. the image of $S \times S$ under multiplication lies in S. A subspace S is an *ideal* if it is closed under multiplication by all elements in S, i.e. the image of $\mathcal{A} \times S$ lies in S (thus, every ideal is the subalgebra). For example, the *square* of the algebra:

$$\mathcal{A}^2 = [xy : x, y \in \mathcal{A}] = [x^2 : x \in \mathcal{A}]$$

is an ideal, where $[\ldots]$ denotes linear span, i.e. the subspace of linear combinations on the enclosed set. The equivalence of these two descriptions of $\mathcal{A}^2$ follows immediately from the identity (cf.Section 1.2):

$$4xy = (x + y)^2 - (x - y)^2. \tag{3.1.1}$$

We will repeatedly use this identity to deduce descriptions about all products from given information about the squares. This procedure is called *bilinearization.* Recall that the squares arise naturally in our genetics applications because there we focus upon the dynamical system associated with the evolutionary operator $V(x) = x^2$ which generates the evolutionary algebra $\mathcal{A}_V$.

Let I be an ideal of the algebra $\mathcal{A}$, x_1 and x_2 are elements of $\mathcal{A}$. We write $x_1 \equiv x_2 \pmod{I}$, if $x_1 - x_2 \in I$. This relation is a congruence which divides the algebra on the classes. The set of the classes is called the *quotient* (or *factor*) *algebra* and it is denoted by $\mathcal{A}/I$. For instance, $\mathcal{A}/\mathcal{A}^2$ is the algebra with zero multiplication.

If $\mathcal{A}_1$ and $\mathcal{A}_2$ are algebras then a linear map $h : \mathcal{A}_1 \to \mathcal{A}_2$ for which $h(xy) = h(x)h(y)$ is called an *algebra homomorphism.*Two simplest examples are: 1) *embedding* of a subalgebra S into the algebra $\mathcal{A}$, $ix = x$ for all $x \in S$; 2)*quotient map* of the algebra $\mathcal{A}$ onto a quotient algebra $\mathcal{A}/I$, $\pi x = \tilde{x}$ where $\tilde{x}$ is the congruence class of the element x.

If $h : \mathcal{A}_1 \to \mathcal{A}_2$ is an algebra homomorphism then its image $\mathrm{Im}h \subset \mathcal{A}_2$ is a subalgebra and its kernel $\ker h = \{x : x \in \mathcal{A}_1 \text{ and } h(x) = 0\}$ is an ideal.A rather simple but fundamental fact is that the induced map $\tilde{h} : \mathcal{A}_1/\ker h \to \mathrm{Im}h$ defined correctly by $\tilde{h}\tilde{x} = hx$ is bijective algebra homomorphism, i.e. *isomorphism* . If there exists an algebra homomorphism $h : \mathcal{A}_1 \to \mathcal{A}_2$ then the algebras are called *isomorphic* ($\mathcal{A}_1 \approx \mathcal{A}_2$).

In this case the inverse map $h^{-1} : \mathcal{A}_1 \to \mathcal{A}_2$ is an isomorphism as well.

For each $x \in \mathcal{A}$ we have a linear operator $M_x : \mathcal{A} \to \mathcal{A}$ defined by $M_x(y) = xy$. As with any vector space, the set of linear operators on $\mathcal{A}$ forms an associative—but noncommutative—algebra under composition. We will refer to this space, End $\mathcal{A}$, and its subalgebras as *operator algebras* to emphasize the absence of commutativity. The map $M : \mathcal{A} \to$ End $\mathcal{A}$ defined by $x \mapsto M_x$ is a linear homomorphism but it is not an algebra homomorphism. In fact, it is easy to check that the identity

$$M_x M_y = M_{xy}$$

is true for all x, y in $\mathcal{A}$ exactly when the multiplication in $\mathcal{A}$ is associative.

For each x the kernel of M_x is a subspace of $\mathcal{A}$ but not necessarily a subalgebra, much less an ideal. However, the *annihilator* of the algebra:

$$\mathrm{ann}\, \mathcal{A} \equiv \cap_{x \in \mathcal{A}} \ker M_x = \{y \in \mathcal{A} : xy = 0\ \forall x \text{ in } \mathcal{A}\}$$

is an ideal. Also, we can describe the square of the algebra as

$$\mathcal{A}^2 = [\cup_{x \in \mathcal{A}} \mathrm{Im}\, M_x].$$

The linear forms on $\mathcal{A}$ which annihilate $\mathcal{A}^2$ we will call *disappearing*. The set of disappearing forms is a subspace, $N \equiv (\mathcal{A}^2)^\perp$, of the dual space $\mathcal{A}^*$, consisting of all linear forms on $\mathcal{A}$.

The family of operators $\{M_x : x \in \mathcal{A}\}$, that is, the subspace, Im M, generates a subalgebra Op $\mathcal{A}$ of End $\mathcal{A}$. Thus, Op $\mathcal{A}$ is the smallest subalgebra of End $\mathcal{A}$ including all the maps M_x. We will call Op $\mathcal{A}$ the *operator algebra* of $\mathcal{A}$. Any associative algebra is a Lie algebra with respect to the Lie bracket (commutator) multiplication: $[M_1, M_2] \equiv M_1 M_2 - M_2 M_1$. Each *Lie algebra* is not commutative as $[M_1, M_2] = -[M_2, M_1]$. It is not associative either, satisfying, instead, the Jacobi identity:

$$[M_1, [M_2, M_3]] + [M_2, [M_3, M_1]] + [M_3, [M_1, M_2]] = 0.$$

We will denote by Lie $\mathcal{A}$ *the Lie algebra of* $\mathcal{A}$, the smallest Lie subalgebra of End $\mathcal{A}$ containing all the operators M_x. Clearly, Lie $\mathcal{A}$ is contained in Op $\mathcal{A}$.

Consider the characteristic polynomial

$$\det(\lambda I - M_x) = \lambda^n - \sigma_1(x)\lambda^{n-1} + \sigma_2(x)\lambda^{n-2} - \ldots + (-1)^n \sigma_n(x)$$

(in this context the letter I denotes the unit operator, i.e. it is the identity map). Because $\det(\lambda I - M_x)$ is homogeneous of degree n in the pair (λ, x) the coefficients are homogeneous polynomial functions of x with $k = $ degree $\sigma_k(x)$. Observe that the multiplication in $\mathcal{A}$ is used here only to define the map M_x. By a polynomial function of x we mean an ordinary polynomial in the coordinates of x with respect to any basis.

From the Hamilton-Cayley Theorem we have, for all x in $\mathcal{A}$ that:

$$M_x^n - \sigma_1(x)M_x^{n-1} + \ldots + (-1)^n \sigma_n(x)I = 0$$

that is,

$$M_x^n y - \sigma_1(x)M_x^{n-1}y + \ldots + (-1)^n \sigma_n(x)y = 0 \quad (x, y \in \mathcal{A}) \tag{3.1.2}$$

It is called *the characteristic equation of the algebra*. Clearly,

$$M_x^k y = \underbrace{x(x(\ldots(xy)\ldots))}_{k \text{ copies of } x} \qquad (k = 1, 2, 3 \ldots)$$

In particular,

$$M_x^k x = \underbrace{x(x(\ldots(xx)\ldots))}_{k+1 \text{ copies of } x}$$

This expression is called the $k+1$ *principal power* of x and is denoted x^{k+1}. From (3.1.2) the sequence of principal powers satisfies the recurrence relation:

$$x^{n+r} - \sigma_1(x)x^{n+r-1} + \ldots + (-1)^n\sigma_n(x)x^r = 0 \ (r = 1, 2, \ldots) \qquad (3.1.3)$$

because $M_x^k(x^r) = x^{k+r}$.

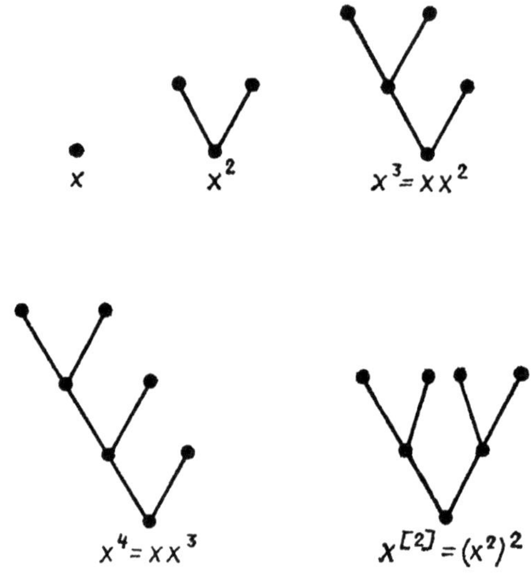

Fig. 3.1

In Fig. 3.1 the first four diagrams illustrate the first four principal powers of x. The last diagram shows the first alternative pattern of association. In general, we define the *plenary powers*, the successive squares, by induction: $x^{[k]} = (x^{[k-1]})^2$ $(k = 1, 2, \ldots; x^{[0]} = x)$.

We can describe all the alternative patters of association for each designated number of multiplications. We use binary tree diagrams as in Fig. 3.1.

T:

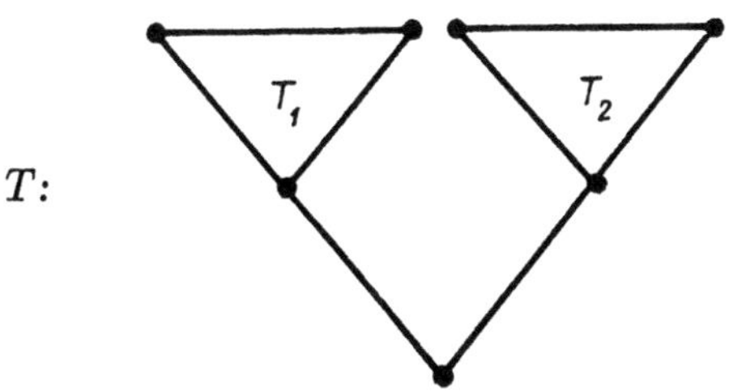

Fig. 3.2

Any nontrivial binary tree T can be split at its base to get two smaller trees T_1 and T_2 as in Fig. 3.2. We then define $x^T = x^{T_1} x^{T_2}$. Thus, we define x^T for any binary tree T by induction on the number $|T|$ of terminal nodes in the tree (= the number of multiplications). Observe that x^{T_1} and x^{T_2} are the two factors of the last multiplication in the formation of x^T. The tree T is called the *index* of the power x^T. For the principal power, x^k, we will identify the whole number k with the corresponding tree.

The idempotents of an algebra $\mathcal{A}$ are especially important for us because they are the fixed points of the evolution map $V(x) = x^2$ in the unit simplex of the evolutionary algebra $\mathcal{A}_V$. We denote by Id $\mathcal{A}$ the set of idempotents, $\{x \in \mathcal{A} : x^2 = x\}$. Clearly, 0 is in Id $\mathcal{A}$ and this set is an algebraic variety. If an element e of the algebra is *unity* , i.e. $ex = x$ for all $x \in \mathcal{A}$ then $e \in$ Id $\mathcal{A}$. If an unity exists then it is unique.

In contrast we call x (*principal*) *nilpotent* if the power $x^T = 0$ for some (principal) index T. The element x is called an *absolute nilpotent* if $x^2 = 0$. For an absolute nilpotent $x^T = 0$ for all nontrivial indices T. More generally $x^{T_1} = 0$ implies $x^T = 0$ if T_1 is one of the pieces of T as in Fig. 3.2. For each T the set $\mathrm{Nil}_T = \{x : x^T = 0\}$ of nilpotents with index T is an algebraic variety.

Theorem 3.1.1 *In any algebra there exists a nonzero absolute nilpotent or a nonzero idempotent.*

Proof Choose a Euclidean metric for the vector space A and consider the unit sphere $= \{x : \| x \| = 1\}$. If 0 is the only absolute nilpotent then we can define a map from the sphere to itself by

$$F(x) = \frac{x^2}{\| x^2 \|}$$

and observe that $F(-x) = F(x)$. We now apply a little topological argument. This last condition implies that the degree of F is even and so F is not homotopic to the identity. It follows that for some $0 < \tau < 1$ and some x on the sphere a point on the segment $\tau x + (1 - \tau)F(x)$ must equal 0. Then $x^2 = \lambda x$ with $\lambda = \tau \| x^2 \| /(\tau - 1)$ and so x/λ is a nonzero idempotent. $\square$

Observe from this argument that if $x^2 = \lambda x$ then either all multiples of x are absolute nilpotents ($\lambda = 0$ case) or there is a unique multiple of x which is an idempotent.

Let $h : A_1 \to A_2$ be an algebra homomorphism. Clearly, $h(x)$ is then idempotent if x is and so h maps Id A_1 into Id A_2.

Theorem 3.1.2 *Let $h : A_1 \to A_2$ be an algebra homomorphism. In each of the following cases h maps* Id A_1 *to* Id A_2 *bijectively, therefore, in these cases the two sets of idempotents have the same dimension as algebraic varieties.*

(1) h is surjective and $\ker h \subset \operatorname{ann} A_1$.

(2) h is injective and $\operatorname{Im} h \supset A_2^2$.

Proof (1) Let $y \in A_2$ with $y^2 = y$ and pull y back to $x \in A_1$, i.e. choose x such that $h(x) = y$. Observe that $z = x^2 - x$ maps to $y^2 - y = 0$ and so $z \in \ker h \subset \operatorname{ann} A_1$. Hence, xz and $z^2 = 0$, and so $(x + z)^2 = x^2$ which is $x + z$. That is, $x + z$ is idempotent and it maps to y because $z \in \ker h$. Hence $h :$ Id $A_1 \to$ Id A_2 is surjective.

Now suppose $x_1, x_2 \in$ Id A_1 with $h(x_1) = h(x_2)$. The latter implies $x_1 \equiv x_2 (\operatorname{mod} \ker h)$ and so by assumption $(\operatorname{mod} \operatorname{ann} A_1)$. This implies $x_1^2 = x_2^2$. Because x_1 and x_2 are idempotents $x_1 = x_2$. Hence $h :$ Id $A_1 \to$ Id A_2 is injective.

(2) If $y \in A_2$ with $y^2 = y$ then $y \in A_2^2$ and so is in the image of h. Pull y back to x and as before $z = x^2 - x$ lies in $\ker h$ which is 0 in this case. Thus, $x \in$ Id A_1. So $h :$ Id $A_1 \to$ Id A_2 is surjective. It is injective because h is injective on the whole space. $\square$

3.2 Nilalgebras

An algebra in which all the elements are (principal) nilpotents is called a (principal) *nilalgebra*.

Theorem 3.2.1 *In a (principal) nilalgebra $\mathcal{A}$ there exists a (principal) index T such that $x^T = 0$ for all x in $\mathcal{A}$.*

Proof If the algebraic variety Nil_T is a proper subset of $\mathcal{A}$ then its dimension is less than dimension of $\mathcal{A}$ and so $\mathcal{A}_T$ is a closed and nowhere dense set. By the Baire Category Theorem $\mathcal{A}$ is not the countable union of closed nowhere dense sets. But if $\mathcal{A}$ is a nilalgebra then it is the union of the countable family $\{\mathrm{Nil}_T\}$ and so $\mathcal{A} = \mathrm{Nil}_T$ for some T. For a principal nilalgebra $\mathcal{A}$ is the union of $\{\mathrm{Nil}_k\}$ and so $\mathcal{A}$ equals some Nil_k. $\square$

Recall that we construct the *complexification* $\mathcal{A}^{(c)}$ of a real vector space $\mathcal{A}$ by taking the tensor product (over $\mathbf{R}$) with the complex field $\mathbf{C}$. If $\mathcal{A}$ is an algebra the bilinear (over $\mathbf{R}$) multiplication map on $\mathcal{A}$ extends to a bilinear (over $\mathbf{C}$) multiplication on $\mathcal{A}^{(c)}$.

Corollary 3.2.2 *If $\mathcal{A}$ is a (principal) nilalgebra then its complexification $\mathcal{A}^{(c)}$ is also a (principal) nilalgebra.*

Proof By choosing a basis $\{e_1, \ldots, e_n\}$ for $\mathcal{A}$ we can describe a power like x^k by writing its coordinates as functions of the coordinates of x. Each is a homogeneous polynomial of degree k with real coefficients. So if for some $k > 0$ x^k is identically zero for all x in $\mathcal{A}$ these coefficients are 0. In the complexification the coefficients of the polynomials remain unchanged while the coordinate variables now extend over $\mathbf{C}$. Thus, for the same k, x^k is identically 0 on $\mathcal{A}^{(c)}$. For a general nilalgebra the argument is the same. $\square$

The algebra $\mathcal{A}$ is called *nilpotent* if for some $k \geq 1$ the product of operators:

$$M_{x_1} \ldots M_{x_k} = 0 \qquad (3.2.1)$$

for any choice of $x_1, \ldots x_k$ in $\mathcal{A}$ or equivalently

$$x_1(x_2(\ldots(x_k y)\ldots)) = 0 \quad (x_1, x_2, \ldots x_k, y \in \mathcal{A}). \qquad (3.2.2)$$

The following theorem will be important for the study of genetic algebras in section 3.5.

Theorem 3.2.3 *In order that $\mathcal{A}$ be nilpotent it is necessary and sufficient that $\mathcal{A}$ be a principal nilalgebra and that the Lie $\mathcal{A}$ be a solvable Lie algebra.*

Recall the Lie algebra $\mathcal{L}$ is called *solvable* if its square (or *derive algebra* $\mathcal{L}'$) is nilpotent.

Proof If $\mathcal{A}$ is nilpotent then from (3.2.2) $x^{k+1} = 0$ for all x in $\mathcal{A}$. Also Lie $\mathcal{A}$ is nilpotent and so is certainly solvable.

A well-known theorem of Lie says a Lie algebra of operators on a complex vector space is solvable if and only if there exists a basis with respect to which they are all simultaneously upper triangular.

Applying this theorem to the complexification $\mathcal{A}^{(c)}$ we choose a basis $\{e_1, \ldots e_n\}$ with respect to which M_z is given by the upper triangular matrix

$$\begin{pmatrix} \lambda_1(z) & & * \\ & \ddots & \\ 0 & & \lambda_n(z) \end{pmatrix} \tag{3.2.3}$$

Here the diagonal entries $\lambda_i(z)$ $(i = 1, \ldots n)$ are linear forms in z. We prove that they all vanish, for then (3.2.1) will hold with $k = n$. Assuming that $\mathcal{A}$, and hence $\mathcal{A}^{(c)}$, is a principal nilalgebra we prove the result by induction on the dimension n. The case $n = 1$ is an easy exercise and so we turn to the inductive step.

Let $z_1, \ldots, z_n$ be the coordinates of z in $\mathcal{A}^{(c)}$ with respect to the basis. From the triangular form (3.2.3) we see that the $n - 1$ dimensional subspace defined by $z_n = 0$ is invariant under all M_z's, i.e. it is an ideal. Regarding it as an algebra in itself with basis $\{e_1, \ldots e_{n-1}\}$ the inductive hypothesis applies to get that $\lambda_1, \ldots, \lambda_{n-1}$ all vanish on this subspace. Hence, as a function of z, each is just a constant times z_n.

Now the multiples of e_1, the subspace where $z_2 = \ldots = z_n = 0$, is an ideal as well by (3.2.3). Writing $\mathcal{A}_1$ for the $n - 1$ dimensional quotient space, we have that $\mathcal{A}_1$ is an algebra and the quotient map $\pi : \mathcal{A} \to \mathcal{A}_1$ is an algebra homomorphism. $\{\pi(e_2), \ldots, \pi(e_n)\}$ is a basis for $\mathcal{A}_1$ and each M_z factors to define a linear operator on $\mathcal{A}_1$ whose matrix with respect to $\{\pi(e_2), \ldots, \pi(e_n)\}$ is obtained by eliminating the first row and column from (3.2.3)and substituing $z_1 = 0$. This factored linear operator is precisely $M_{\pi(z)}$ and so by the induction hypothesis all the diagonal entries vanish.

Hence $\lambda_2(z) = \ldots = \lambda_{n-1}(z) = 0$ and $\lambda_n(z) = bz_1$ for all z in $\mathcal{A}$. If $z^r \equiv 0$ then $(bz_1)^{r-1}z_n \equiv 0$. Hence $b = 0$, i.e. $\lambda_n = 0$. It implies that the nth coordinate

$$(zw)_n = (M_z(w))_n = 0 \ \ (z, w \in \mathcal{A}). \tag{3.2.4}$$

It remains to prove that λ_1 vanishes. The previous subalgebra argument implies that for some complex number a, $\lambda_1(z) = az_n$, i.e.,

$$ze_1 = M_z(e_1) = az_ne_1 \ \ z \in \mathcal{A}. \tag{3.2.5}$$

In particular, $e_1^2 = 0$ and $e_1e_n = e_ne_1 = ae_1$.

We complete the proof by showing $a = 0$. Compute the powers of $e_1 + e_n$:

$$(e_1 + e_n)^2 = 2ae_1 + e_n^2.$$

Now by (3.2.4) the nth coordinate of e_n^k is 0 for $k \geq 2$ and so by (3.2.5) $e_n^k e_1 = 0$ for such k. Using induction we get

$$(e_1 + e_n)^k = 2a^{k-1}e_1 + e_n^k \qquad (k = 2, 3, \ldots)$$

We have $(e_1 + e_n)^r = 0$ and $e_n^r = 0$. Clearly, $2a^{r-1} = 0$ and so $a = 0$. $\square$

3.3 Baric Algebras and Spaces

A *character* for an algebra $\mathcal{A}$ is a nonzero multiplicative linear form on $\mathcal{A}$, that is, a nonzero algebra homomorphism from $\mathcal{A}$ to $\mathbf{R}$. Not every algebra admits a character. For example, an algebra space with the zero multiplication has no character. For any character χ ker χ is an ideal with codimension 1 as a vector subspace of $\mathcal{A}$. Also, it does not contain $\mathcal{A}^2$ because $\chi(x) \neq 0$ implies $\chi(x^2) = \chi(x)^2 \neq 0$.

Theorem 3.3.1 *The correspondence $\chi \mapsto$ ker χ is a bijection between the set of characters on an algebra $\mathcal{A}$ and the set of ideals I of codimension 1 such that $\mathcal{A}^2 \not\subseteq I$.*

Proof If I is a subspace of $\mathcal{A}$ with codimension 1 then I is the kernel of some nonzero linear form ϕ on $\mathcal{A}$ unique up to multiplication by a nonzero constant. If, in addition, $I = \ker \phi$ is an ideal then $\phi(x) = 0$ implies $\phi(xy) = 0$ for all y in $\mathcal{A}$. Hence, $\phi(x)$ divides the bilinear form $\phi(xy)$, that is, for each y there is a unique scalar $\psi(y)$ such that $\phi(xy) = \psi(y)\phi(x)$ as functions of x. By symmetry, ψ is just a multiple of ϕ itself. That is, there is a constant c such that

$$\phi(xy) = c\phi(x)\phi(y) \quad (x, y \in \mathcal{A}).$$

If $c = 0$ then ϕ vanishes on $\mathcal{A}^2$ (and conversely). Consequently, if the ideal $I = \ker \phi$ does not contain $\mathcal{A}^2$ c is nonzero and multiplying the above equation by c we see that $\chi = c\phi$ is the character with kernel I. If χ_1, χ_2 are the characters and $\ker \chi_1 = \ker \chi_2$ then $\chi_2 = a\chi_1$ ($a = $ const. $\neq 0$). But $a^2 = a$, hence $a = 1$ and $\chi_1 = \chi_2$. $\square$

Observe that if I is an ideal of codimension 1 which does contain $\mathcal{A}^2$ then the quotient algebra $\mathcal{A}/I$ is a one dimensional vector space but with the zero multiplication.

Corollary 3.3.2 *If we get the algebra $\tilde{\mathcal{A}}$ from an algebra $\mathcal{A}$ by adjoining an identity element then $\tilde{\mathcal{A}}$ admits a character.*

Proof $\mathcal{A}$ is an ideal in $\tilde{\mathcal{A}}$ with codimension 3one. Because $\tilde{\mathcal{A}}$ has an identity $\tilde{\mathcal{A}}^2 = \tilde{\mathcal{A}}$ and so is not contained in $\mathcal{A}$.

The direct proof is also easy. Write $x \in \tilde{\mathcal{A}}$ uniquely as $x = \chi e + y$ with e the adjoined identity and $y \in \mathcal{A}$. Clearly, the coefficient $\chi = \chi(x)$ is a character for $\tilde{\mathcal{A}}$. $\square$

An algebra can admit more than one character. For example, if $\mathcal{A} = \mathbf{R}^n$ with coordinatewise multiplication each coordinate projection map is a character.

Theorem 3.3.3 *The number of distinct characters for an algebra $\mathcal{A}$ is at most $n = \dim \mathcal{A}$.*

Proof Recall first that if $\chi_1, \ldots, \chi_k$ are distinct linear forms on a vector space $\mathcal{A}$ then there exists a vector x such that the values $\chi_1(x), \ldots, \chi_k(x)$ are all distinct. For if no such x existed $\mathcal{A}$ would be the finite union of the hyperplanes $\{\ker \chi_i - \chi_j : 1 \leq i < j \leq k\}$, contradicting the Baire Category Theorem.

Now suppose these forms are distinct characters on the algebra $\mathcal{A}$. For any character χ and vector $x \in \mathcal{A}$:

$$M_x^* \chi \equiv \chi \circ M_x = \chi(x)\chi.$$

This says that as an element of the dual space $\mathcal{A}^*$ χ is an eigenvector of the dual linear operator M_x^* with eigenvalue $\chi(x)$. With x chosen to distinguish the forms as in the previous paragraph we see that $\chi_1, \ldots, \chi_k$ are eigenvectors of M_x^* with distinct eigenvalues $\chi_1(x), \ldots, \chi_k(x)$. Consequently, the set $\{\chi_1, \ldots, \chi_k\}$ is linearly independent in $\mathcal{A}^*$ and so $k \leq \dim \mathcal{A}^* = n$. $\square$

A pair $(\mathcal{A}, \sigma)$ consisting of an algebra $\mathcal{A}$ and a character σ on $\mathcal{A}$ is called a *baric algebra*. The chosen character σ is called the *weight* and the hyperplane $H = \{x : \sigma(x) = 1\}$ is called the *unit hyperplane*. For $x \notin \ker \sigma$, i.e. $\sigma(x) \neq 0$, we can normalize to unit weight obtaining $x/\sigma(x)$ in H.

Example Because of its importance for genetics applications we first consider the evolutionary algebra $\mathcal{A}$ of a free population (N.B. for the remainder of chapter 3 we will consider only free populations and refer to them as populations), with multiplication given by:

$$(xy)_j = \sum_{i,k} p_{ik,j} x_i y_k$$

with

$$p_{ik,j} = p_{ki,j} \geq 0 \text{ and } \sum_j p_{ik,j} = 1.$$

The linear form $s(x) = \sum_i x_i$ is a character for this multiplication:

$$s(xy) = \sum_{ikj} p_{ik,j} x_i y_k = \sum_{ik} x_i y_k = s(x)s(y).$$

We fix this choice of weight for an evolutionary algebra and so regard it as a baric algebra. The weight s can be characterized as the only positive character (i.e. character taking positive values on the unit simplex Δ) on the algebra $\mathcal{A}$.

To prove this recall that each character χ is an eigenvector for the operator M_x^* with eigenvalue $\chi(x)$. But the matrix of this operator

$$\mu_{kj} = \sum_i p_{ik,j} x_i$$

is a stochastic matrix if x lies in Δ. So by the Perron-Frobenius Theorem for non-negative matrixes the eigenvalues of M_x^* for $x \in \Delta$ lie in the unit circle of the complex plane and so $|\chi(x)| \leq 1$ for $x \in \Delta$ and any character χ. Now suppose χ is a positive character and e a point of Δ at which χ takes it minimum value. Thus, for all $x \in \Delta$:

$$0 < \chi(e) \leq \chi(x) \leq 1.$$

We show that $\chi(e) = 1$ and so χ is constantly 1 on Δ. This means $\chi = s = 1$ on Δ and $\chi = s$ follows from linearity.

If, instead, $\chi(e)$ were less than 1 then

$$\chi(e^2) = \chi(e)^2 < \chi(e)$$

which contradicts the minimizing choice of e because e^2 lies in Δ when e does.

Example Let A be a linear operator in a vector space $\mathcal{A}$ and let σ be an A-invariant linear form on $\mathcal{A}$, i.e. $A^*\sigma = \sigma$. We make $\mathcal{A}$ into an algebra by defining the multiplication:

$$xy = \frac{\sigma(y)Ax + \sigma(x)Ay}{2} \qquad (3.3.1)$$

By A-invariance of σ, $\sigma(xy) = \sigma(x)\sigma(y)$ and so σ is a character. Thus, $(\mathcal{A}, \sigma)$ is a baric algebra. We will call it the algebra *induced by A and σ* (often not mentioning σ when it is understood). Instead of (3.3.1) we will often use the square equivalent:

$$x^2 = \sigma(x)Ax \qquad (3.3.2)$$

This, in turn, is equivalent to its restriction on the unit hyperplane:

$$x^2 = Ax \qquad\qquad (x \in H) \qquad (3.3.3)$$

Thus, the algebra induced by a linear operator is characterized by defining x^2 to be that operator on the hyperplane H. Observe that A-invariance of σ implies A-invariance of the hyperplane H (i.e. $A(H) \subset H$) and (3.3.1) is equivalent to

$$xy = A(\frac{x+y}{2}) \qquad\qquad (x, y \in H). \qquad (3.3.4)$$

Notice that here $\mathcal{A}^2 = \operatorname{Im} A$, $\operatorname{ann} \mathcal{A} = \ker A$.

If A is the identity operator, $A = I$, and σ is an arbitrary nonzero linear form then (3.3.1) is

$$xy = \frac{\sigma(y)x + \sigma(x)y}{2} \qquad (3.3.5)$$

which is equivalent to each of the identities:

$$x^2 = \sigma(x)x;$$

$$x^2 = x \qquad\qquad (x \in H)$$

$$xy = \frac{x+y}{2} \qquad\qquad (x, y \in H).$$

This baric algebra is called an *unit algebra*. Mendel n-allele gametic algebra is an unit evolutionary algebra. An evolutionary algebra $\mathcal{A}_V$ is unit if and only if $V = \mathrm{id}$. Observe that the associative law is false for the unit algebra.

Multiplication for an unit algebra is completely described by the form σ. If we forget about the multiplication we are left with a pair (E, σ) consisting of a vector space and a chosen nonzero linear form. Such a pair is called a *baric space* with weight σ.

At the opposite extreme from the unit algebras is the case where A is the projection onto multiples of a single vector, i.e. $A = \sigma \otimes e$ where e is a vector with $\sigma(e) = 1$, $Ax = \sigma(x)e$. Then (3.3.1) becomes

$$xy = \sigma(x)\sigma(y)e \qquad\qquad (3.3.6)$$

We call this baric algebra a *constant algebra*.

With $(\mathcal{A}, \sigma)$ a baric algebra, a subalgebra $\mathcal{A}_1$ of $\mathcal{A}$ is called a *baric subalgebra* if $\mathcal{A}_1 \not\subset \ker \sigma$, or, equivalently, $\sigma_1 \equiv \sigma|\mathcal{A}_1$ is not zero. Then $(\mathcal{A}_1, \sigma_1)$ is a baric algebra in its own right and we write $(\mathcal{A}_1, \sigma_1) \subset (\mathcal{A}, \sigma)$. For example, $\mathcal{A}^2$ is a baric subalgebra as is any subspace of $\mathcal{A}$ which contains $\mathcal{A}^2$. In the algebra induced by a linear operator A and σ the baric subalgebras are the A-invariant subspaces not contained in $\ker \sigma$ (c.f. (3.3.1)).

For a unit algebra every subspace is a (not necessarily baric) subalgebra. Motivated by this we call a subspace E_1 of a baric space (E, σ) a *baric subspace* if $E_1 \not\subset \ker \sigma$, i.e. $\sigma_1 \equiv \sigma|E_1 \neq 0$. So (E_1, σ_1) is a baric space and each such (and only such) subspace is a baric subalgebra with respect to the identity induced multiplication.

An ideal I is called a *baric ideal* if $I \subset \ker \sigma$, i.e. $\sigma|I = 0$ (and so a baric ideal cannot be a baric subalgebra). For any baric ideal the weight σ factors to the quotient algebra $\mathcal{A}/I$ to make the quotient a baric algebra in a natural way. (We write $(\mathcal{A}, \sigma)/I$). It is called the *baric quotient (factor) algebra*.

The largest baric ideal is the so-called *barideal* $\mathcal{B} \equiv \ker \sigma$. The weight on the baric quotient algebra gives an isomorphism of $(\mathcal{A}, \sigma)/\mathcal{B}$ with the ordinary algebra $\mathbf{R}$ of real numbers regarded as a baric algebra via the identity map on $\mathbf{R}$.

The annihilator ann $\mathcal{A}$ is a baric ideal as is any subspace of ann $\mathcal{A}$. Of course, the ideal $\mathcal{A}^2$ is not a baric ideal as it is a baric subalgebra.

For the algebra induced by a linear operator A the baric ideals are just the A-invariant subspaces contained in ker σ. In particular, for a unit algebra every subspace of ker σ (and only such one) is a baric ideal. A subspace E_0 of a baric space (E, σ) is called a *baric ideal* if $E_0 \subset$ ker σ. We can then define the *baric quotient space* $(E, \sigma)/E_0$. Observe that the baric subalgebras and baric quotient algebras of a unit algebra are unit algebras.

We now define the basic concept of *invariant linear form*. This is a linear form, f, on a baric algebra which satisfies the identity:

$$f(xy) = \frac{\sigma(y)f(x) + \sigma(x)f(y)}{2} \qquad (x, y \in \mathcal{A}) \qquad (3.3.7)$$

or equivalently:

$$f(x^2) = \sigma(x)f(x) \qquad (x \in \mathcal{A}) \qquad (3.3.8)$$

or again on the unit hyperplane:

$$f(x^2) = f(x) \qquad (x \in H). \qquad (3.3.9)$$

The latter equation explains the name "invariant form". In an evolutionary algebra $\mathcal{A}$ the invariant forms are those which satisfy $f(V(x)) = f(x)$ on the unit simplex and so such forms define conservation laws for the dynamical system. The gene conservation laws of section 2.1 are examples.

The set J of invariant linear forms is a subspace of the dual space $\mathcal{A}^*$. Since the weight σ itself is clearly invariant and $\sigma \neq 0$ we have dim $J \geq 1$. The members of J which are not multiples of σ are called *nontrivial invariant forms*.

For example, if $\mathcal{A}$ is the algebra induced by the operator A and form σ then (3.3.2) and (3.3.8) imply that invariance of a linear form f is equivalent to $f(Ax) = f(x)$ for all x, i.e. $A^*f = f$. Hence, J coincides with the eigenspace of the operator A^* associated with eigenvalue 1, i.e.

$$J = \ker (I - A^*).$$

From this it is clear that dim J can take on any integer value between 1 and dim $\mathcal{A}$. At one extreme, for a unit algebra all linear forms are invariant and so $J = \mathcal{A}^*$. At the other, for the constant algebra dim $J = 1$, that is, there are no nontrivial invariant forms.

The kernel ker f of an invariant form is a subalgebra and if f is nontrivial it is a baric subalgebra. More generally, if L is a linear subspace of J and

$\sigma \notin L$ then $L^{\perp} = \{x : f(x) = 0 \text{ for all } f \text{ in } L\} = \cap_{f \in L} \ker f$ is a baric subalgebra. (Proof: $L^{\perp\perp} = L$ and so $L^{\perp} \subset \ker \sigma$ would imply $\sigma \in L$). On the other hand if $\sigma \in L$ then $L^{\perp}$ is a baric ideal (c.f. (3.3.7)). In particular, $J^{\perp}$ is a baric ideal. Notice that

$$J^{\perp} \supset \text{ann } \mathcal{A} \qquad (3.3.10)$$

for if $x \in \text{ann } \mathcal{A}$ and $y \in H$ then $xy = 0$ and so for any $f \in J$ (3.3.7) implies

$$0 = \frac{f(x) + \sigma(x)f(y)}{2}. \qquad (3.3.11)$$

Applying this first with $f = \sigma$ we see that $\sigma(x) = 0$ and then for any f in J $f(x) = 0$.

The case where equality holds in (3.3.10) is important for genetic applications being closely related to the gene conservation laws, and so we will call a baric algebra *conservative* when

$$J^{\perp} = \text{ann } \mathcal{A}. \qquad (3.3.12)$$

Theorem 3.3.4 *In order that a baric algebra be conservative it is necessary and sufficient that the product xy depends only upon the values of the invariant forms on x and y. That is,*

$$x \equiv z \text{ and } y \equiv w \, (\text{mod} J^{\perp}) \text{ imply } xy = zw \qquad (3.3.13)$$

Proof It is easy to check that $x \equiv z$ and $y \equiv w$ (mod ann $\mathcal{A}$) imply $xy = zw$. Hence (3.3.13) is true for a conservative algebra. For the converse, apply (3.3.13) with $z = 0$ and $w = y$. We get $x \in J^{\perp}$ implies $xy = 0$. Hence, $J^{\perp} \subset \text{ann } \mathcal{A}$ which, together with (3.3.10), implies (3.3.12). $\square$

Corollary 3.3.5 *Let $\{f_1, \ldots, f_m\}$ be a basis for the subspace J of invariant forms for the algebra $(\mathcal{A}, \sigma)$. In order that $(\mathcal{A}, \sigma)$ be conservative it is necessary and sufficient that there exist vectors $u_{ik} \in \mathcal{A}$ $(1 \leq i, k \leq m)$ with $u_{ik} = u_{ki}$ so that the multiplication is given by the formula*

$$xy = \sum_{i,k=1}^{m} f_i(x)f_k(y)u_{ik} \, (x, y \in \mathcal{A}). \qquad (3.3.14)$$

Proof Clearly, if (3.3.14) holds the product xy depends only upon the values of the invariant forms $f_1, \ldots, f_m$ on x and y. So the algebra is conservative by the Theorem.

For the converse extend $\{f_1, \ldots, f_m\}$ to a basis for $\mathcal{A}^*$ and construct the dual basis for $\mathcal{A}$. The first m of these are vectors $u_1, \ldots, u_m$ in $\mathcal{A}$ such that $f_i(u_j) = \delta_{ij}$. Then

$$x \equiv \sum_i f_i(x) u_i \pmod{J^\perp},$$

with a similar equation for y. Because the product xy is determined by the $J^\perp$-congruence classes

$$xy = \sum_{i,k} f_i(x) f_k(y) u_i u_k$$

which is (3.3.14) with u_{ik} the product $u_i u_k$. $\square$

As usual formula (3.3.14) holds provided the diagonal version is true:

$$x^2 = \sum_{i,k=1}^{m} f_i(x) f_k(x) u_{ik} \tag{3.3.15}$$

and it is sufficient that this in turn be true on an open subset of the unit hyperplane H. Thus, in an evolutionary algebra it is sufficient to check the equation on the unit simplex. For example, the Mendel zygotic algebra is conservative by the Hardy-Weinberg Law.

It is useful to characterize conservative algebras by using an identity:

Theorem 3.3.6 *The baric algebra $(\mathcal{A}, \sigma)$ is conservative if and only if the following identity holds:*

$$x^2 y = \sigma(x) xy \qquad\qquad (x, y \in \mathcal{A}) \tag{3.3.16}$$

Proof By (3.3.8) $x^2 - \sigma(x)x$ lies in $J^\perp$ for any x in $\mathcal{A}$, i.e. $x^2 \equiv \sigma(x)x \pmod{J^\perp}$. Multiplying both sides by y we get (3.3.16) from (3.3.13) when the algebra is conservative. Conversely, (3.3.16) says that $x^2 - \sigma(x)x$ lies in ann $\mathcal{A}$. So if $f \in (\text{ann }\mathcal{A})^\perp$, (3.3.16) implies $f(x^2 - \sigma(x)x) = 0$ for all x in $\mathcal{A}$, i.e. $f \in J$. $(\text{ann }\mathcal{A})^\perp \subset J$ implies $J^\perp \subset \text{ann }\mathcal{A}$, the inclusion complementary to (3.3.10). Hence, $\mathcal{A}$ is conservative. $\square$

So each baric subalgebra and baric quotient algebra of a conservative algebra is conservative.

Corollary 3.3.7 *The algebra induced by a linear operator A is conservative if and only if $A^2 = A$, i.e. A is a projection. In particular, the unit and constant algebras are conservative.*

Proof We know that ann $\mathcal{A} = \ker A$ and $J = \ker (I - A^*)$ or equivalently $J^\perp = \mathrm{Im}(I - A)$. So the A-induced algebra is conservative exactly when $\mathrm{Im}(I - A) \subset \ker A$, and this is equivalent to $A(I - A) = 0$ or $A = A^2$. $\square$

Corollary 3.3.8 *In a conservative algebra $(\mathcal{A}, \sigma)$ the following identity holds:*

$$(x^2)^2 = \sigma^2(x)x^2 \qquad\qquad (x \in \mathcal{A}). \qquad (3.3.17)$$

Proof As in the proof of Theorem 3.3.6, $x^2 \equiv \sigma(x)x \bmod J^\perp$. Applying (3.3.13) we get (3.3.17). Alternatively, we can use (3.3.16) directly, first with $y = x^2$ and then with $y = x$ to get

$$(x^2)^2 = \sigma(x)x^3 = (\sigma(x))^2 x^2.$$

$\square$

The baric algebras in which (3.3.17) hold are called *stationary algebras* or *Bernstein algebras* because of the connection between this class and the problem of Bernstein (c.f. section 2.1 and chapters 4,5). For an evolutionary algebra condition (3.3.17) is equivalent to the Stationarity Principle

$$V^2 = V \qquad\qquad (3.3.18)$$

where $V(x) = x^2$ is the evolutionary operator on the unit simplex.

Thus, each conservative algebra is Bernstein but we will see that not every Bernstein algebra is conservative (see section 3.4). Furthermore, this difference has deep significance in biological applications (see chapters 4 and 5).

The baric subalgebras and baric quotient algebras of a Bernstein algebra are Bernstein.

The algebra induced by a linear operator A is Bernstein if and only if $A^2 = A$, i.e. A is a projection, and so in this case the algebra is necessarily conservative.

Recall that for any algebra $\mathcal{A}$ we denote by N, the subspace of disappearing forms, i.e. the linear forms which vanish on the subalgebra $\mathcal{A}^2$. Notice that

$$J \cap N = 0 \qquad\qquad (3.3.19)$$

for if $f(x^2) = \sigma(x)f(x)$ $(f \in J)$ and at the same time $f(x^2) = 0$ $(f \in N)$ for all x in $\mathcal{A}$ then f vanishes on the dense open subset of $\mathcal{A}$ where $\sigma \neq 0$. Consequently, $f = 0$.

It follows that

$$\dim J + \dim N \leq \dim \mathcal{A} \qquad\qquad (3.3.20)$$

If the algebra is induced by the linear operator A then $N = (\mathrm{Im}A)^\perp$.

Theorem 3.3.9 *A baric algebra* $(\mathcal{A}, \sigma)$ *is induced from some projection operator if and only if* $\dim J + \dim N = \dim \mathcal{A}$, *or equivalently,* $J \oplus N = \mathcal{A}^*$.

Proof If $\mathcal{A}$ is induced by a projection A then $J = \ker(I - A^*)$ which is $\mathrm{Im}\, A^*$. On the other hand, from (3.3.2) we see that $f(x^2)$ vanishes for all x if and only if $f(Ax)$ does. So $N = \ker A^*$. Because A^* is a projection its image and kernel split the space $\mathcal{A}^*$.

Conversely, if $\dim J + \dim N = \dim \mathcal{A}$ then $\mathcal{A}$ is the direct sum of $J^\perp$ and $N^\perp$. Let A be the projection on $\mathcal{A}$ with $\mathrm{Im}\, A = N^\perp$ and $\ker A = J^\perp$. For fixed $x \in \mathcal{A}$ apply a linear form f to $z = x^2 - \sigma(x)A(x)$. If f is in N then $f(x^2) = 0$ and $f(A(x)) = 0$. So $f(z) = 0$. If f is in J then

$$f(z) = f(x^2) - \sigma(x)f(A(x)) = \sigma(x)(f(x) - f(A(x)))$$

and this is 0 because $J = \mathrm{Im}\, A^* = \ker(I - A^*)$ by choice of A. Since $f(z) = 0$ for all f in J and N and $J + N = \mathcal{A}^*$, $z = 0$. For this operator A, (3.3.2) is true and so $(\mathcal{A}, \sigma)$ is induced from A. $\square$

Corollary 3.3.10 $(\mathcal{A}, \sigma)$ *is a unit algebra if and only if* $\mathcal{A}^* = J$.

Proof For a unit algebra we saw earlier that $\mathcal{A}^* = J$. Conversely, if $J = \mathcal{A}^*$ then $N = 0$. Thus $A = I$. $\square$

Given baric algebras $(\mathcal{A}_1, \sigma_1)$ and $(\mathcal{A}_2,, \sigma_2)$, a *baric algebra homomorphism* $h : (\mathcal{A}_1, \sigma_1) \to (\mathcal{A}_2, \sigma_2)$ is an algebra homomorphism $h : \mathcal{A}_1 \to \mathcal{A}_2$ such that $h^* \sigma_2 = \sigma_1$. For example, the embedding of a baric subalgebra and the quotient map to a baric quotient algebra are baric homomorphisms. Clearly, the composition of baric homomorphisms is baric. A bijective baric homomorphism is called a *baric isomorphism*. The inverse of a baric isomorphism is baric because $h^*(\sigma_2) = \sigma_1$ implies $\sigma_2 = (h^{-1})^*(\sigma_1)$. We write $(\mathcal{A}_1, \sigma_1) \approx (\mathcal{A}_2, \sigma_2)$ for isomorphism of baric algebras, that is, if there exists a baric isomorphism between them.

Proposition 3.3.11 *Let* $h : (\mathcal{A}_1, \sigma_1) \to (\mathcal{A}_2, \sigma_2)$ *be a baric homomorphism. $\mathrm{Im}\, h$ is a baric subalgebra of $\mathcal{A}_2$ and $\ker h$ is a baric ideal in $\mathcal{A}_1$. The usual bijection induced from h is a baric isomorphism* $(\mathcal{A}_1, \sigma_1)/\ker h \approx \mathrm{Im}\, h$.

Proof Because $h^*(\sigma_2) = \sigma_1$ we have $\sigma_2(h(x)) \neq 0$ if $\sigma_1(x) \neq 0$. Thus, $\mathrm{Im}\, h$ is a baric subalgebra. Also, $\sigma_1(x) = 0$ if $h(x) = 0$. Hence, $\ker h \subset \ker \sigma_1$ and the kernel is a baric ideal. It is easy to see that the induced bijection preserves the weight. $\square$

For all these baric algebra concepts there are obvious analogues for baric spaces. Thus, a *baric operator* $A : (E, \sigma) \to (E, \sigma)$ is a linear operator $A : E \to E$ such that $A^* \sigma = \sigma$.

Theorem 3.3.12 *Let $(\mathcal{A}_1, \sigma_1)$ and $(\mathcal{A}_2, \sigma_2)$ be baric algebras with spaces of invariant linear forms J_1 and J_2 respectively. Assume $h : (\mathcal{A}_1, \sigma_1) \to (\mathcal{A}_2, \sigma_2)$ is a baric homomorphism.*

(1) h^* transforms invariant forms to invariant forms, i.e. f invariant on $\mathcal{A}_2$ implies $h^* f$ is invariant on $\mathcal{A}_1$:

$$h^*(J_2) \subset J_1. \tag{3.3.21}$$

(2) If $\operatorname{Im} h \supset \mathcal{A}_2^2$ then h^* is injective on J_2.

(3) If h is surjective, then the subspace $(h^*)^{-1}(J_1)$ of $\mathcal{A}_2^*$—which contains J_2 by (3.3.21)—is equal to J_2. Thus, f is then invariant on $\mathcal{A}_2$ if and only if $h^* f$ is invariant on $\mathcal{A}_1$.

(4) If h is surjective and $\ker h \subset J_1^\perp$ then h^* maps J_2 bijectively onto J_1.

Proof We begin with $f_2 \in \mathcal{A}_2^*$ and $f_1 = h^*(f_2) \in \mathcal{A}_1^*$.

(1) Suppose $f_2 \in J_2$. Then $f_1(x^2) = f_2(h(x)^2) = \sigma_2(h(x))f_2(h(x)) = \sigma_1(x)f_1(x)$ and so $f_1 \in J_1$.

(2) Assume $\operatorname{Im} h \supset \mathcal{A}_2^2$. If $f_2 \in J_2$ and $f_1 = h^*(f_2) = 0$ then f_2 vanishes on $\operatorname{Im} h$ and so on $\mathcal{A}_2^2$. Thus $f_2 \in J_2 \cap N_2$ and $f_2 = 0$ by (3.3.19). So the kernel of $h^*|J_2$ is zero and h^* is injective on J_2.

(3) Assume h is surjective and $f_1 = h^*(f_2) \in J_1$. For $y \in \mathcal{A}_2$ choose x such that $h(x) = y$. Now as in (1): $f_2(h(x)^2) = f_1(x^2) = \sigma_1(x)f_1(x) = \sigma_2(h(x))f_2(h(x))$. Substituting $y = h(x)$ we get $f_2 \in J_2$.

(4) Assume h is surjective and $\ker h \subset J_1^\perp$. By duality it says $\operatorname{Im} h^* \supset J_1$. So for $f_1 \in J_1$ there exists $f_2 \in \mathcal{A}_2^*$ such that $h^*(f_2) = f_1$. By (3) $f_2 \in J_2$ and due to duality $\ker h^* = (\operatorname{Im} h)^\perp = 0$, i.e. h^* is injective. So h^* maps J_2 bijectively onto J_1. $\square$

The quotient map $h : A \to A/J^\perp$ is a baric homomorphism and $\ker h = J^\perp$; h^* is an isomorphism of $(\mathcal{A}/J^\perp)^*$ onto the subspace J of $\mathcal{A}^*$. But by (4) h^* maps the invariant forms on $\mathcal{A}/J^\perp$ bijectively onto J. Thus, every linear form on $\mathcal{A}/J^\perp$ is invariant and so by Corollary 3.3.10 the quotient algebra $\mathcal{A}/J^\perp$ is a unit algebra for every baric algebra $(\mathcal{A}, \sigma)$. This statement is very useful for some applications (see chapter 4).

If $(\mathcal{A}_1, \sigma_1)$ is a baric subalgebra of $(\mathcal{A}, \sigma)$ then every invariant form on $\mathcal{A}$ restricts to an invariant form on $\mathcal{A}_1$. This special case of (3.3.21) is also clear directly from the definition. In the conservative case we can sharpen the result.

Theorem 3.3.13 *For a baric algebra $(\mathcal{A}, \sigma)$, let L be a linear subspace of J which does not include σ. Then $\mathcal{A}_1 = L^\perp$ is a baric subalgebra with $\sigma_1 = \sigma|$*

A_1. *If (A, σ) is conservative then a linear form f on A is invariant if and only if its restriction f_1 to A_1 is invariant on A_1. Thus, if f_1 is an invariant linear form on A_1 then every extension to a linear form on A is invariant on A. Furthermore, if J_1 is the space of invariant linear forms on A, then*

$$\dim J_1 = \dim J - \dim L. \tag{3.3.22}$$

Proof If i is the embedding of $A_1 = L^\perp$ into A then the dual $i^* : A^* \to A_1^*$ associates to each linear form f on U its restriction $f_1 = f|A_1$. The kernel of $i*$ consists of the forms vanishing on A_1. Thus, i^* maps A^* onto A_1^* (every linear form on A_1 always extends to a form on A) with ker $i^* = L$. From (3.3.21) $i^*(J) \subset J_1$ and so it suffices to show $(i^*)^{-1}(J_1) \subset J$. For then i^* maps J onto J_1 with kernel L. This will give the extension result and (3.3.22).

$(i^*)^{-1}(J_1) \subset J$ says that $f_1 = f|A_1$ invariant, in J_1, for f in A^* implies f is invariant, $f \in J$. To prove this we observe first that the annihilator ann $A \subset J^\perp \subset L^\perp = A_1$ and if $x \in$ ann A then x annihilates A and a fortiori annihilates A_1. Hence, ann $A \subset$ ann $A_1 \subset J_1^\perp$. But A is conservative and so $J^\perp =$ ann A. Because f_1 is invariant its restriction to $J_1^\perp$ is 0. So the further restriction to $J^\perp =$ ann A is 0. But this means that the restriction of f to $J^\perp$ is 0 and so $f \in J$. $\square$

We now consider the idempotents in a baric algebra. The weight of an idempotent—in fact the value of any character on an idempotent—is either zero or one because $\sigma(x) = \sigma(x)^2$. Thus, Id $A \subset H \cup$ ker σ. Usually, although not always, the nonzero idempotents have unit weight, i.e. Id $A \subset H \cup 0$. For example,

Lemma 3.3.14 *In a Bernstein algebra (and a fortiori in a conservative algebra) the nonzero idempotents have unit weight.*

Proof When $x^2 = x$ identity (3.3.17) becomes $x = \sigma^2(x)x$. If $x \neq 0$, $\sigma^2(x) = 1$. So $\sigma(x) = 1$. $\square$

Corollary 3.3.15 *In a Bernstein algebra (A, σ)*

$$\text{Id } A = \{y : y = x^2 \text{ for } x \in H\} \cup \{0\} \tag{3.3.23}$$

and so the set of nonzero idempotents is nonempty and connected.

Proof $V(x) = x^2$ maps $H = \sigma^{-1}(1)$ continuously into itself. If x is a nonzero idempotent then $x \in H$ by the Lemma and clearly $V(x) = x$. On the other hand if $y = x^2$ and $\sigma(x) = 1$ then $y^2 = y$ by (3.3.17). Thus, V is a retraction of H onto the set of nonzero idempotents. $\square$

We will denote by $\mathrm{Id}_1 \mathcal{A}$ the idempotents of unit weight in a baric algebra $\mathcal{A}$, i.e. $\mathrm{Id}_1 \mathcal{A} = \mathrm{Id}\, \mathcal{A} \cap \sigma^{-1}(1)$.

It follows from (3.3.23) that in a Bernstein algebra the ideal $\mathcal{A}^2$ is generated as a subspace by the set of idempotents:

$$\mathcal{A}^2 = [\mathrm{Id}\, \mathcal{A}] = [\mathrm{Id}_1\, \mathcal{A}]$$

Proposition 3.3.16 *Let $(\mathcal{A}, \sigma)$ be a Bernstein algebra and $\mathcal{A}_1 = \mathcal{A}^2$ regarded as a baric subalgebra. Then $\mathcal{A}_1^2 = \mathcal{A}_1$ i.e 0 is the only disappearing form on $\mathcal{A}_1$.*

Proof Let ϕ be a disappearing form on $\mathcal{A}_1$. Then $\phi((x^2)^2) = 0$. Consequently, $\sigma^2(x)\phi(x^2) = 0$. Thus $\phi = 0$. $\square$

For an algebra induced by an operator A and linear form σ

$$\mathrm{Id}\, \mathcal{A} = \{x \in \mathcal{A} : Ax = x \text{ and } \sigma(x) = 1\} \cup \{0\} \qquad (3.3.24)$$

and so here again the set of nonzero idempotents is $\mathrm{Id}_1 \mathcal{A} = \mathrm{Id}\, \mathcal{A} \cap \sigma^{-1}(1)$. To prove this note that $x^2 = \sigma(x)A(x)$ and so if $x = x^2$ either $\sigma(x) = 0$ and hence $x = 0$ or $\sigma(x) = 1$ and so $x = Ax$. Thus, the nonzero idempotents consist of the intersection of the unit hyperplane H with the eigenspace of A corresponding to eigenvalue 1. Notice that 1 is always an eigenvalue because $A^*\sigma = \sigma$. However, nonzero idempotents need not exist, e.g. on $\mathbf{R}^2$ let $A(x_1, x_2) = (x_1, x_1 + x_2)$ with $\sigma(x_1, x_2) = x_1$. Note that $Ax = x$ exactly when $\sigma(x) = 0$.

In particular, for a unit algebra the nonzero idempotents consist of the entire unit hyperplane H.

We conclude this section with a little result which should be compared with the latter portion of the proof of Theorem 3.3.13.

Proposition 3.3.17 *If $(\mathcal{A}_1, \sigma_1)$ is a baric subalgebra of a conservative algebra $(\mathcal{A}, \sigma)$ then*

$$\mathrm{ann}\, \mathcal{A}_1 = \mathcal{A}_1 \cap \mathrm{ann}\, \mathcal{A}.$$

Proof It is obvious in any case that $\mathcal{A}_1 \cap \mathrm{ann}\, \mathcal{A}$ annihilates $\mathcal{A}_1$. If $x \in \mathrm{ann}\, \mathcal{A}_1$ and $f \in J$ then the restriction $f_1 = f|\mathcal{A}_1$ is invariant on $\mathcal{A}_1$ and so $f(x) = f_1(x) = 0$. Thus $x \in J^\perp$. Because $\mathcal{A}$ is conservative $x \in \mathrm{ann}\, \mathcal{A}$. $\square$

3.4　Bernstein Algebras

We now study the structure and properties of Bernstein algebras by using the idempotent elements as a tool. Let $(\mathcal{A}, \sigma)$ be a Bernstein algebra and $\mathcal{B} = \ker \sigma$ be its barideal. This is the subspace parallel to the unit hyperplane H. By Corollary 3.3.15 nonzero idempotents exist and we choose one and label it e. Any vector x in $\mathcal{A}$ can be represented uniquely as $x = \sigma e + y$ with $y \in \mathcal{B}$ and $\sigma = \sigma(x)$ because $\sigma(e) = 1$ by Lemma 3.3.14. We substitute this decomposition into the defining identity for a Bernstein algebra, (3.3.17):

$$(x^2)^2 = \sigma^2(x)x^2.$$

With $\sigma = \sigma(x)$ we multiply out and equate coefficients of the powers σ^k $(0 \leq k \leq 3)$ to get the list of identities which are (together with $e^2 = e$) equivalent to (3.3.17):

$$2e(2ey) = 2ey \tag{3.4.1}$$

$$2ey^2 + (2ey)^2 = y^2 \tag{3.4.2}$$

$$(y \in \mathcal{B})$$

$$(2ey)y^2 = 0 \tag{3.4.3}$$

$$(y^2)^2 = 0 \tag{3.4.4}$$

Because $\mathcal{B}$ is an ideal we see that the formula $L_e(y) = 2ey$ defines a linear operator $L_e : \mathcal{B} \to \mathcal{B}$. Our analysis of the structure of a Bernstein algebra will be built upon this operator.

Lemma 3.4.1 *For each nonzero idempotent e the operator $L_e : \mathcal{B} \to \mathcal{B}$ is a projection and the rank of L_e is independent of the choice of e.*

Proof (3.4.1) says precisely that $L_e^2 = L_e$, i.e. that L_e is a projection. The actual operator depends upon the choice of e, varying continuously with e. So the trace of L_e also varies continuously with e. But because L_e is a projection the trace of L_e is equal to the rank of L_e which is an integer number. Consequently, the rank d remains constant as e varies over the connected set of nonzero idempotents (see Corollary 3.3.15). $\square$

Corollary 3.4.2 *For each nonzero idempotent e the operator $\Lambda_e : \mathcal{A} \to \mathcal{A}$ defined by $\Lambda_e(x) = 2ex - \sigma(x)e$ is a projection and the rank of Λ_e is independent of e.*

Proof On $\mathcal{B}$, $\Lambda_e = L_e$ and $\Lambda_e(e) = e$. So $\Lambda_e(x) = \sigma e + L_e(y)$ when $x = \sigma e + y$ $(y \in \mathcal{B})$. Clearly, Λ_e is a projection and

$$\text{rank } \Lambda_e = \text{rank } L_e + 1.$$

$\square$

Further we will usually fix our choice e and drop it as a subscript, writing L and Λ. The number $m = \text{rank } \Lambda = \text{rank } L + 1$ is called the *rank* of the Bernstein algebra. The number $\delta = \text{def} L = \text{dimker} L$ $(= \text{dim ker} \Lambda)$ we call the *deficience* of the algebra. The pair (m, δ) is called the *type* of the algebra. Clearly, $1 \leq m \leq n$ and $0 \leq \delta \leq n - 1$ with $m + \delta = n$ $(n = \text{dim } \mathcal{A})$. Any integer pair satisfying these conditions is the type of some Bernstein algebra.

For example, if $(\mathcal{A}, \sigma)$ is induced by a projection operator A then, as we have seen, a nonzero idempotent e is an invariant vector $(A(e) = e)$ normalized to weight 1 $(\sigma(e) = e)$. Check from the definition that $\Lambda = A$ and so $m = \text{rank } A$, $\delta = \text{def} A$. Among these examples are all algebras of the extreme types:

Theorem 3.4.3 *For a Bernstein algebra $(\mathcal{A}, \sigma)$ the type is $(n, 0)$ if and only if $\mathcal{A}$ is a unit algebra and the type is $(1, n - 1)$ if and only if $\mathcal{A}$ is a constant algebra.*

Proof $m = n$ exactly when L is the identity operator on $\mathcal{B}$. So $2ey = L(y) = y$ for all y in $\mathcal{B}$ implies by (3.4.2) that $y^2 = 0$ for y in $\mathcal{B}$. Hence, $x^2 = (\sigma e + y)^2 = \sigma(\sigma e + y) = \sigma x$ which says that $\mathcal{A}$ is a unit algebra.

$m = 1$ exactly when L is the zero operator on $\mathcal{B}$, that is, $2ey = 0$ for all y in $\mathcal{B}$. Again by (3.4.2) $y^2 = 0$ and this time $x^2 = (\sigma e + y)^2 = \sigma^2 e$ which characterizes the constant algebra. $\square$

Corollary 3.4.4 *A Bernstein algebra of dimension $n \leq 2$ is either a unit or a constant algebra (both, in the trivial case $n = 1$).*

Corollary 3.4.5 *A Bernstein algebra of dimension $n > 1$ does not contain a unit element.*

Proof If we choose e to be the unit then the rank m equals n and so the algebra is a unit algebra. But then $x = ex = \frac{1}{2}(e + x)$ (when $\sigma(x) = 1$) and hence $x = e$. So the unit hyperplane is 0 dimensional. $\square$

The reason for the common result $y^2 = 0$ in the two parts of the proof of the Theorem is revealed by:

Theorem 3.4.6 *A Bernstein algebra $(\mathcal{A}, \sigma)$ is induced by a projection operator if and only if $y^2 = 0$ for all y in $\mathcal{B}$ i.e. $\mathcal{B}^2 = 0$. In that case, Λ_e is that projection operator independent of the choice of nonzero idempotent e.*

Proof Observe that with $x = \sigma e + y$ $(y \in \mathcal{B})$

$$\sigma(x)\Lambda_e(x) = \sigma(\sigma e + 2e(x - \sigma e)) = \sigma(\sigma e + 2ey)$$

and this last expression equals x^2 if and only if $y^2 = 0$. So when $y^2 = 0$ for all y in $\mathcal{B}$ we have (3.3.2) with $\Lambda_e = A$. For the converse $\Lambda_e = A$ follows from (3.3.1) when $(\mathcal{A}, \sigma)$ is induced from A $(A(e) = e)$. So from (3.3.2) we do have $x^2 = \sigma(x)\Lambda_e(x)$ in the case, and $y^2 = 0$. $\square$

Now we can split the barideal $\mathcal{B}$ by using the projection L. $\mathcal{B}$ is the direct sum $U \oplus W$ where $U = \mathrm{Im}L$ and $W = \ker L$. By definition $\dim U = m - 1$ and $\dim W = \delta$.

Lemma 3.4.7 $U^2 \subset W$, $W^2 \subset U$ and $UW \subset U$.

Proof We can rewrite (3.4.2) as:

$$L(y^2) + (L(y))^2 = y^2 \qquad\qquad (y \in \mathcal{B}). \qquad\qquad (3.4.5)$$

In particular, if $L(y) = 0$, i.e. $y \in W$, then $L(y^2) = y^2$ meaning $y^2 \in U$. On the other hand, if $L(y) = y$, i.e. $y \in U$ then from (3.4.5) $L(y^2) = 0$ and so $y^2 \in W$. Using bilinearization we see that $y_1 y_2 \in U$ (or W) if y_1 and y_2 are both in W (respectively, in U). Similarly,

$$L(y_1 y_2) + L(y_1)L(y_2) = y_1 y_2 \qquad (y_1, y_2 \in \mathcal{B}). \qquad (3.4.6)$$

In particular if $L(y_2) = 0$, $L(y_1 y_2) = y_1 y_2$. Thus, $y_2 \in W$ implies $y_1 y_2 \in U$ and we have $\mathcal{B}W \subset U$. In particular, $UW \subset U$. $\square$

By applying this splitting to the decomposition $x = \sigma e + y$ we can establish the following main structure theorem.

Theorem 3.4.8 *If $x = \sigma e + u + w$ with $u \in U$ and $w \in W$ then*

$$x^2 = \sigma^2 e + (\sigma u + 2uw + w^2) + u^2 \qquad\qquad (3.4.7)$$

where the expression in the parenthesis belongs to $\mathcal{A}$ and the last summand belongs to W. Furthermore the following identities hold:

$$u^3 = 0, \; u(uw) = 0, \; uw^2 = 0 \qquad\qquad (3.4.8)$$

$$(u^2)^2 = 0, \; (uw)^2 = 0 \qquad\qquad (3.4.9)$$

$$(uw)u^2 = 0, \; u^2 w^2 = 0, \; (uw)w^2 = 0, \; (w^2)^2 = 0 \qquad (3.4.10)$$

Proof Square x and use $e^2 = e$, $2eu = L(u) = u$ and $2ew = L(w) = 0$ to get (3.4.7). By Lemma 3.4.7 σu, $2uw$, $w^2 \in U$ and $u^2 \in W$.

Now let $y = su + tw$ so that $2ey = su$ and (3.4.3) is equivalent to $u(su + tw)^2 = 0$. This identity is equivalent to the identities obtained by equating the vector coefficients of s^2, st and t^2 to zero separately. Thus $u^3 = 0$, $u(uw) = 0$ and $uw^2 = 0$, i.e. the list of (3.4.8).

Similarly, (3.4.4) $(y^2)^2 = 0$ is equivalent to the list of five identities obtained by equating the coefficients of s^4, s^3t, etc. to zero separately. This yields four of the six identities of (3.4.9) and (3.4.10). The coefficient of s^2t^2 is $2u^2w^2 + 4(uw)^2$ and so this sum is zero. But by Lemma 3.4.7 $u^2w^2 \in U$ and $(uw)^2 \in W$ and so each term is zero separately. $\square$

This theorem is reversible. Suppose e is a nonzero idempotent of $\mathcal{A}$ and $\mathcal{A}$ splits as a direct sum of subspaces $\mathcal{A} = [e] \oplus U \oplus W$ so that multiplication satisfies (3.4.7) (with the indicated decomposition $\sigma u + 2uw + w^2 \in U$ and $u^2 \in W$) and so that the identities of (3.4.8), (3.4.9) and (3.4.10) hold. Then $(\mathcal{A}, \sigma)$ is a Bernstein algebra of type (m, δ) with $m = \dim U + 1$ and $\delta = \dim W$. To see this notice that from (3.4.7) we derive that $L(u + w) = 2e(u + w) = u$ is a projection with image U and kernel W (and hence (3.4.1)). By reversing the argument of Lemma 3.4.7 we obtain (3.4.4) from $UW + W^2 \subset U$ and $U^2 \subset W$. Finally, from the above proof we see that (3.4.3) follows from the list (3.4.8) and (3.4.4) follows from (3.4.9) and (3.4.10). Furthermore, as we will now show, the identities of (3.4.10) are redundant. They follow from the decomposition and the identities of (3.4.8) and (3.4.9).

Two of these are easy. $u_1 = uw \in U$ and so $u_1(w^2) = 0$ (cf. (3.4.8)) says $(uw)w^2 = 0$. Similarly $u_2 = w^2 \in U$ and so $u_2(w^2) = 0$ says $(w^2)^2 = 0$. For the others we make the full linearization of the first and the second identities (3.4.8):

$$u_1(u_2u_3) + u_2(u_3u_1) + u_3(u_1u_2) = 0 \quad (u_1, u_2, u_3 \in U) \qquad (3.4.11)$$

$$u_1(u_2w) + u_2(u_1w) = 0 \quad (u_1, u_2 \in U) \qquad (3.4.12)$$

For (3.4.11) let $u = su_1 + tu_2 + ru_3$, expand $u^3 = 0$ and use the coefficient of str. For (3.4.12) let $u = sw_1 + tu_2$, expand $u(uw) = 0$ and use the coefficient of st.

Now apply (3.4.11) with $u_1 = u_2 = u$ and $u_3 = uw$. We get $(uw)u^2 + 2u(u(uw)) = 0$ and since $u(uw) = 0$, $(uw)u^2 = 0$. With $u_1 = u_2 = u$ and $u_3 = w^2$ we get $u^2w^2 + 2u(uw^2) = 0$ and since $uw^2 = 0$, $u^2w^2 = 0$.

Remark (3.4.11) is Jacobi identity.It is a fundamental property of Lie algebras but Bernstein algebras are commutative while Lie's are anticommutative.

Corollary 3.4.9 *The set of nonzero idempotents can be parametrized by the subspace U and so, in particular, $\dim \mathrm{Id}_1\mathcal{A} = m - 1$. To be precise associate to $u \in U$ the idempotent*

$$e(u) \equiv e + u + u^2 \qquad\qquad (u \in U). \qquad\qquad (3.4.13)$$

This bijection of U to $\mathrm{Id}_1\mathcal{A}$ has inverse map

$$u = 2ex - 2e \qquad\qquad (x \in \mathrm{Id}_1\mathcal{A}). \qquad\qquad (3.4.14)$$

Proof Observe that $2eu = u$, $2eu^2 = 0$, $u^3 = 0$ and $(u^2)^2 = 0$ for u in U. Hence any x of the form (3.4.13) is an idempotent. Conversely, if $x = \sigma e + u + w \, (\neq 0)$ by the decomposition (3.4.7) $x^2 = x$ implies $\sigma = 1$ and $w = u^2$. Then $u = L(x - e) = 2e(x - e) = 2ex - 2e$. $\square$

The splitting of $\mathcal{A}$ as $[e] \oplus U \oplus W$ and the associated decomposition of a typical element x as $\sigma e + u + w$ we will refer to as the *standard decomposition* of a Bernstein algebra. Of course the decomposition depends on the choice of idempotent e though the dimensions of the components do not. In particular, using the translation by e we can identify the unit hyperplane H with the barideal $\mathcal{B}$ and regard the latter as the Cartesian product $U \times W$. In that sense, the idempotent $e(u) = e + u + u^2$ is identified with the pair (u, u^2) in $U \times W$. Thus, the set of nonzero idempotents, $\mathrm{Id}_1\mathcal{A}$, is identified with the graph of the map from U to W associating to u in U u^2 in W.

We now use Corollary 3.4.9 to interpret the projection operator L_e dynamically. Recall that on the unit hyperplane H the extension of the evolution operator, $V(x) = x^2$, satisfies $V^2 = V$. Because H is just a translate of the barideal $\mathcal{B}$ we can identify $\mathcal{B}$ with the tangent space of H at every point x of H and so regard the tangent map, the derivative, as a linear operator $d_eV : \mathcal{B} \to \mathcal{B}$. If e is an idempotent, a fixed point of V, then by the chain rule applied to $V^2 = V$ we get that $(d_eV)^2 = d_eV$, that is, the linear operator d_eV is a projection on $\mathcal{B}$. If $y \in \mathcal{B}$ we compute $d_eV(y)$ by applying V to the path $e + ty$ through e in H and taking the derivative at $t = 0$:

$$V(e + ty) = (e + ty)^2 = e + 2tey + t^2y^2$$

and the derivative at $t = 0$ is $2ey$. Thus, d_eV is the projection L_e. In particular, $U = \mathrm{Im}L_e$ is the tangent space at e of the submanifold $\mathrm{Id}_1\mathcal{A}$ of

H. In general, for $x \in H$, $V(x) = x^2$ is some idempotent e and $(d_eV)(d_xV) = d_xV$ by the chain rule again. In other words, the image of d_xV lies in U. So we have that the rank of the map $V : H \to H$ is at most $m-1$ at every point, with equality at the points of $\mathrm{Id}_1\mathcal{A}$. If we define $\tilde{V} : \mathcal{A} \to \mathcal{A}$ by $\tilde{V}(x) = x^2$ then $\tilde{V}$ is homogeneous of degree 2 and so the rank of $\tilde{V}$ at x is one plus the rank of V at $x/\sigma(x)$ when $\sigma(x) \neq 0$. So we have the rank of $\tilde{V}$ is $\leq m$ on a dense open subset of $\mathcal{A}$ and so on all of $\mathcal{A}$.

On the other hand, the rank of $\tilde{V}$ is m at each point $x = \sigma(e + u + u^2)$ with $u \in U$, $\sigma \neq 0$. The condition {rank of $\tilde{V}$ is $< m$} defines a proper algebraic subset of the space $\mathbf{R}^n$. This proves the next statement.

Corollary 3.4.10 *For Bernstein algebra $(\mathcal{A}, \sigma)$ of type (m, δ) the rank of map $x \mapsto x^2$ is m at every point $x \in G$ for G a dense open set.*

Remark The projection operator L_e on the barideal $\mathcal{B}$ is independent of the choice of the idempotent e if and only if the standard decomposition (with respect to fixed choice of e) satisfies $u^2 = 0$ and $uw = 0$ for all $u \in U$, $w \in W$. Indeed, $L_e(y) = 2ey$ and so varying e along the path $e(t) = e + tu + t^2u^2$ (c.f. (3.4.13)) we see that $L_{e(t)}(y)$ constant in t implies, by differentiating at $t = 0$, $uy = 0$ for $u \in U$, $y \in \mathcal{B}$. Conversely, assuming this identity, write $\tilde{e} = e + u + u^2 = e + u$, $y = u_1 + w_1$, and compute

$$L_{\tilde{e}}(y) = 2(e + u)(u_1 + w_1) = u_1 + 2uu_1 + 2uw_1 = u_1 = L_e(y).$$

Thus, $L_{\tilde{e}}(y)$ is independent of u. Observe that, as usual, $u^2 = 0$ for all u in U is equivalent to $uu_1 = 0$ for u, u_1 in U.

Notice, too, that from (3.4.13) $u^2 = 0$ ($u \in U$) is equivalent to the e-independence of $\mathrm{Im}\, L_e = U$. The kernel may still vary unless $uw = 0$ as well.

Corollary 3.4.11 *If $\{u_1, \ldots u_{m-1}\}$ is a basis for U then the ideal $\mathcal{A}^2$ is spanned by $\{e, u_k, u_k^2, u_ku_j \; (1 \leq k < j \leq m - 1)\}$. Hence, $\dim \mathcal{A}^2 \leq \frac{1}{2}m(m + 1)$.*

Proof The elements $e = e^2$ and $u_k = 2eu_k$ belong to $\mathcal{A}^2$ and

$$\mathcal{A}^2 = [e + u + u^2]_{u \in U} \; . \qquad \qquad \square$$

We now use the standard decomposition to analyze invariant and disappearing forms. For the first, notice that the equation $f(x^2) = \sigma(x)f(x)$, characterizing invariance of f, is equivalent to:

$$\sigma^2 f(e) + \sigma f(u) + f((u+w)^2) = \sigma^2 f(e) + \sigma f(u+w). \qquad (3.4.15)$$

Analogously, the equation $g(x^2) = 0$ defining the disappearance of g can be written:

$$\sigma^2 g(e) + \sigma g(u) + g((u+w)^2) = 0. \qquad (3.4.16)$$

As usual we use these identities by equating coefficients of the real variable σ. The coefficient of σ yields that $f(u+w) = f(u)$, i.e. $f(w) = 0$, when f is invariant and $g(u) = 0$ when g is disappearing. In the latter case, $g(e) = 0$ as well. Thus, we have, first:

Corollary 3.4.12 *With respect to the standard decomposition $x = \sigma e + u + w$ an invariant form f depends only on σ and u while a disappearing form g depends only on w.*

From the last term of (3.4.15) we see that $f((tu+w)^2) = 0$ and so each coefficient of t is zero. Thus, $f(uw) = 0$ and $f(w^2) = 0$. We already know $f(u^2) = 0$ because $u^2 \in W$. The reasoning is reversible and shows that a linear form which vanishes on W and the subspace $UW + W^2$ of U is invariant. Thus,

$$J^\perp = (UW + W^2) \oplus W. \qquad (3.4.17)$$

Similarly, if g is disappearing, w^2, $uw \in U$ imply $g(w^2) = g(uw) = 0$ because g vanishes on U. But $g((u+w)^2) = 0$ and so $g(u^2) = 0$ as well. Conversely, if $g(e) = 0$ and g vanishes on U and $U^2 \subset W$ then g is disappearing. Thus,

$$\mathcal{A}^2 = N^\perp = [e] \oplus U \oplus U^2 \qquad (3.4.18)$$

Notice that this description of the ideal $\mathcal{A}^2$ is in effect a repetition of Corollary 3.4.11.

$\mathcal{A}^2$ is a baric subalgebra of $\mathcal{A}$ and so is a Bernstein algebra in its own right. (3.4.18) is the standard decomposition for $\mathcal{A}^2$ and so we see that $\mathcal{A}^2$ has type (m, δ') where $\delta' = \delta - \dim N$. From (3.4.17) and (3.4.18) we see that:

$$\mathcal{A}^2 \cap J^\perp = (UW + W^2) \oplus U^2 = \mathcal{B}^2 \qquad (3.4.19)$$

where $\mathcal{B}$ is a barideal $(= U + W)$. So we have proved:

Corollary 3.4.13 *In a Bernstein algebra* $(\mathcal{A}, \sigma)$ *the square of the barideal* $\mathcal{B}$ *is an ideal and*

$$\mathcal{B}^2 = \mathcal{A}^2 \cap J^\perp = N^\perp \cap J^\perp = (N \oplus J)^\perp.$$

In particular, the codimension of $\mathcal{B}^2$ *in* $\mathcal{A}$ *is* $\dim N + \dim J$.

Remark The sum $N + J$ is direct by (3.3.19).

This result together with Theorem 3.3.9 yields another proof of Theorem 3.4.6 which we can restate as: A Bernstein algebra is induced from a projection if and only if $\mathcal{B}^2 = 0$.

Corollary 3.4.14 *In a Bernstein algebra of type* (m, δ) *the spaces of invariant and disappearing forms satisfy:*

$$\dim J \leq m \text{ and } \dim N \leq \delta. \tag{3.4.20}$$

Proof From (3.4.17) $J^\perp$ contains W whose complement $[e] \oplus U$ has dimension m by definition. So we have:

$$\dim J = \operatorname{codim} J^\perp = m - \dim(UW + W^2). \tag{3.4.21}$$

Similarly, from (3.4.18)

$$\dim N = \operatorname{codim} N^\perp = \delta - \dim(U^2). \tag{3.4.22}$$

□

Special interest attaches to the cases where equality holds in one or the other of the (3.4.20) relationships.

Theorem 3.4.15 *Let* $(\mathcal{A}, \sigma)$ *be a Bernstein algebra with a chosen standard decomposition.*

1. The following conditions are equivalent:

   (a) $\dim J = m$
   (b) $UW + W^2 = 0$, *i.e.*

$$uw = 0 \text{ and } w^2 = 0 \quad (u \in U \text{ and } w \in W) \tag{3.4.23}$$

   (c) For $x = \sigma e + u + w$:

$$x^2 = \sigma^2 e + \sigma u + u^2. \tag{3.4.24}$$

Conversely, given a symmetric bilinear map from $U \times U$ to W we can use (3.4.24) to define an algebra structure on $\mathbf{R} \times U \times W$ with weight $(\sigma, u, w) \mapsto \sigma$. This baric algebra is Bernstein and satisfies the above conditions.

2. *The following conditions are equivalent and when they hold we call the algebra* exceptional.

   (a) *$\dim N = \delta$*

   (b) *$U^2 = 0$, i.e.*
$$u^2 = 0 \qquad (u \in U). \qquad (3.4.25)$$

   (c) *For $x = \sigma e + u + w$*
$$x^2 = \sigma^2 e + (\sigma u + 2uw + w^2). \qquad (3.4.26)$$

Conversely given a bilinear map $U \times W \to U$ and a symmetric bilinear map from W to U we can use (3.4.26) to define an algebra structure on $\mathbf{R} \times U \times W$ with weight $(\sigma, u, w) \mapsto \sigma$. This baric algebra is an exceptional Bernstein algebra.

Proof The equivalence of (a) and (b) in parts (1) and (2) follows from (3.4.21) and (3.4.22) respectively. The equivalence of (b) and (c) in each case is obvious. In constructing the algebras using (3.4.24) and (3.4.26) it is the Bernstein conditions which require checking. Just observe that (3.4.8) and (3.4.9) both follow from either (3.4.23) or (3.4.25). Then use the reverse version of Theorem 3.4.8 described after its proof. $\square$

We have not bothered to name the complex of properties associated with part (1) because we are about to prove they are equivalent to conservatism of the algebra. First, we need a lemma:

Lemma 3.4.16 *For any standard decomposition of a Bernstein algebra $(\mathcal{A}, \sigma)$ we have the inclusions:*
$$\operatorname{ann} \mathcal{A} \subset W \subset J^{\perp}. \qquad (3.4.27)$$

Proof $W \subset J^{\perp}$ follows from (3.4.17). On the other hand if $x \in \operatorname{ann} \mathcal{A}$ $x^2 = 0$ implies $\sigma(x) = 0$ and so $x \in \mathcal{B}$. Then $L(x) = 2ex = 0$ implies $x \in W$. $\square$

Theorem 3.4.17 *A baric algebra $(\mathcal{A}, \sigma)$ is conservative if and only if it is a Bernstein algebra of type (m, δ) with $\dim J = m$.*

Proof Conservative algebras are always Bernstein. From the lemma ann $\mathcal{A} = J^{\perp}$ implies $W = J^{\perp}$ and so $\dim J = \operatorname{codim} J^{\perp} = m$.

If $\dim J = m$ then by (3.4.17) and Theorem 3.4.15 $J^{\perp} = W$. Also, $w \in W$ implies $\mathcal{A}W = [e]W + UW + W^2 = 0$ by Theorem 3.4.15 as well. Thus, $W \subset \operatorname{ann} \mathcal{A}$ and we have $\operatorname{ann} \mathcal{A} = J^{\perp}$ by the lemma. So $\mathcal{A}$ is conservative. $\square$

Corollary 3.4.18 *A Bernstein algebra is both conservative and exceptional if and only if it is induced by a projection.*

Proof By Theorems 3.4.15 and 3.4.17 an algebra is conservative and exceptional exactly when $\mathcal{B}^2 = 0$. By Theorem 3.4.6 this is equivalent to $(\mathcal{A}, \sigma)$ being a projection algebra. $\square$

Notice that a Bernstein algebra is conservative exactly when $\operatorname{ann}\mathcal{A} = W$ and so equality for either inclusion of (3.4.27) implies equality for both. To prove the sufficiency of this equation, observe that

$$\operatorname{ann} \mathcal{A} = \{v : v \in W \text{ and } uv = 0(u \in U) \text{ and } wv = 0 \ (w \in W)\}. \quad (3.4.28)$$

Hence $\operatorname{ann} \mathcal{A} = W$ implies $UW + W^2 = 0$. Result by Theorem 3.4.15 (1).

We now make use of the dual version of multiplication. Instead of a bilinear map from $\mathcal{A} \times \mathcal{A}$ to $\mathcal{A}$, multiplication can be regarded as a real-valued trilinear form on $\mathcal{A}^* \times \mathcal{A} \times \mathcal{A}$ associating to $(f, u_1, u_2) \mapsto f(u_1 u_2)$. It is symmetric in the latter two variables because multiplication is commutative on $\mathcal{A}$. Thus, each fixed f in $\mathcal{A}^*$ yields a symmetric bilinear form on $\mathcal{A}$, S_f with $S_f(u_1, u_2) = f(u_1 u_2)$. Alternatively, we have a "multiplication" $\mathcal{A}^* \times \mathcal{A} \to \mathcal{A}^*$ associating to (f, u_1) the linear form $u \mapsto f(u_1 u)$. In this interpretation, S_f becomes a linear homomorphism from $\mathcal{A}$ to $\mathcal{A}^*$. Commutativity of multiplication means that $S_f : \mathcal{A} \to \mathcal{A}^*$ is self-adjoint:

$$(S_f)^* = S_f : \mathcal{A} = \mathcal{A}^{**} \longrightarrow \mathcal{A}^* \qquad (f \in \mathcal{A}^*) \qquad (3.4.29)$$

To see this, write $<,>: \mathcal{A}^* \times \mathcal{A} \to \mathbf{R}$ as the standard pairing, $< f, u > \mapsto f(u)$. To say that $S_f : \mathcal{A} \to \mathcal{A}$ is self-adjoint is to say that

$$< S_f(u_1), u_2 > = f(u_1 u_2) = f(u_2 u_1) = < S_f(u_2), u_1 > \ .$$

Now if $(\mathcal{A}, \sigma)$ is a Bernstein algebra with a chosen standard decomposition we can use $\mathcal{A}^2 \subset W$ to regard $f \mapsto S_f$ as a homomorphism

$S : W^* \to \sum_U$ where $\sum_U$ is the space of self-adjoint homomorphisms $U \to U^*$. We will use it to define the subspace $\overline{U}$ of U^*:

$$\overline{U}^\perp \equiv \sum \{\mathrm{Im}\, S_f : f \in W^*\} \subset U^*$$

so that $g \in U^*$ lies in $\overline{U}$ if and only if it is a linear combination of linear forms $u \mapsto f(u_1 u)$ for $(f, u_1) \in W^* \times U$.

Theorem 3.4.19 *The orthogonal complement in U of $\overline{U}$ is given by:*

$$\begin{aligned}
\overline{U}^\perp &= \cap \{\ker S_f : f \in W^*\} \\
&= U \cap \mathrm{ann}\, U = \{u \in U : u_1 u = 0\ u_1 \in U\} \\
&= U \cap \mathrm{ann}(U + U^2).
\end{aligned}$$

$\overline{U}^\perp$ *is a baric ideal in $\mathcal{A}$ and contains W^2.*

Proof By standard duality results $(\sum \mathrm{Im}\, S_f)^\perp = \cap(\mathrm{Im}\, S_f)^\perp = \cap \ker S_f^*$ and so the first characterization of $\overline{U}^\perp$ follows because each S_f is self-adjoint. Clearly, $g(u) = 0$ for all g in $\overline{U}$ precisely when $f(u_1 u) = 0$ for all f in W^* and u_1 in U, i.e. when $u_1 u = 0$ in W for all u_1 in U. The inclusion $W^2 \subset \overline{U}^\perp$ then follows from the identity $u_1 w^2 = 0$ (c.f. (3.4.8)).

Now $2eu = L(u) = u$ for any u in U and so $\overline{U}^\perp$ is closed under multiplication by e. Also, $U \cdot \overline{U}^\perp = 0 \subset \overline{U}^\perp$. Finally, if $u \in \overline{U}^\perp$ and $w \in W$ we apply (3.4.12):

$$u_1(uw) = -u(u_1 w)$$

and the latter is zero because $u_1 w \in U$.

Finally, if u annihilates U then it automatically annihilates U^2. For by (3.4.11)

$$uu_1^2 + 2u_1(u_1 u) = 0.$$

So $u_1 u = 0$ implies $uu_1^2 = 0$. $\square$

In this proof we used (3.4.12) to show that $\overline{U}^\perp$ is invariant under the multiplication map $M_w : \mathcal{A} \to \mathcal{A}$, $x \mapsto xw$. Since $UW \subset U$ we know that M_w maps $U \to U$ and so we can regard the dual M_w^* as mapping $U^* \to U^*$. With $f \in W^*$ (3.4.12) says, for the map $S_f : U \to U^*$ that

$$M_w^* S_f + S_f M_w = 0 \tag{3.4.30}$$

because $< S_f M_w(u), u_1 > = f(u_1(uw))$ while $< M_w^* S_f(u), u_1 > = f((u_1 w)u)$.

Exactly as we derived (3.4.11) and (3.4.12) we can use $(uw)^2 = 0$ to get

$$(u_1 w_1)(u_2 w_2) + (u_1 w_2)(u_2 w_1) = 0 \ (u_1, u_2 \in U, \quad w_1, w_2 \in W). \qquad (3.4.31)$$

This in turn can be interpreted to say

$$M_{w_1}^* S_f M_{w_2} + M_{w_2}^* S_f M_{w_1} = 0 \quad (f \in W^*, \ w_1, w_2 \in W). \qquad (3.4.32)$$

Now apply (3.4.30) twice to yield:

$$(M_{w_1}^* M_{w_2}^* + M_{w_2}^* M_{w_1}^*) S_f = 0. \qquad (3.4.33)$$

The purpose of all this formal shuffling is to obtain:

Lemma 3.4.20 *The subspace $\overline{U}$ is invariant with respect to the family of operators $\{M_w^* : w \in W\}$ on U^*. Furthermore, the family is anticommutative on $\overline{U}$.*

Proof By (3.4.30) M_w^* maps the image of S_f into the image of S_f. By (3.4.33) the identity $M_{w_1}^* M_{w_2}^* + M_{w_2}^* M_{w_1}^* = 0$ holds on the image of S_f. The invariance and equality remain true when we sum over $f \in W^*$. $\square$

We use this result by applying:

Proposition 3.4.21 *Let $\mathcal{T}$ be an anticommutative family of linear operators on a nonzero vector space E. Then $\cap\{\ker T : T \in \mathcal{T}\}$ is not 0 and $\sum\{\mathrm{Im}\, T : T \in \mathcal{T}\}$ is not the entire space E.*

Proof Observe that the subspace generated by $\mathcal{T}$ is still an anticommutative family so that we can assume $\mathcal{T}$ is a subspace of the space of linear operators on E. For any linear operator T on E $\dim \ker T + \dim \mathrm{Im}\, T = \dim E$. Now anticommutativity implies $T^2 = 0$, i.e. $\mathrm{Im}\, T \subset \ker T$. So for such an operator

$$\dim\, \mathrm{Im}\, T \leq \frac{1}{2} \dim E \leq \dim\, \ker T.$$

For the kernel result, proceed by induction on the dimension of the subspace $\mathcal{T}$ of $L(E, E)$. The above argument is the initial step.

If T_1 is a nonzero element of $\mathcal{T}$ then $T T_1 + T_1 T = 0$ for all T in $\mathcal{T}$ shows that $E_1 = \ker T_1$ is invariant with respect to the operators $T \in \mathcal{T}$, and, of course, T_1 is 0 on E_1. So the anticommutative family $\{T|E_1 : T \in \mathcal{T}\}$ has lower dimension than $\mathcal{T}$ and by induction, $\cap\{\ker T : T \in \mathcal{T}\} = \cap\{\ker T|E_1 : T \in \mathcal{T}\}$ is a nonzero subspace of E_1.

For the image result use duality and the kernel result applied to the family of operators $\{T^* : T \in \mathcal{T}\}$ on E^*. $\square$

Recall from Corollary 3.4.14 that when $\mathcal{A}$ has type (m, δ)

$$1 \leq \dim J \leq m$$

$$0 \leq \dim N \leq \delta.$$

If $\dim J$ is as large as these inequalities allow, $\dim J = m$, then $\mathcal{A}$ is conservative (c.f. Theorem 3.4.17), while if $\dim N = \delta$ we call $\mathcal{A}$ exceptional. By Corollary 3.4.18 both the extremes can hold at once and this occurs precisely when $\mathcal{A}$ is induced from a projection. The opposite extremes are also possible but cannot occur together (unless $m = 1$ and $\delta = 0$). In fact, we will show that $\dim J = 1$ implies $\mathcal{A}$ is exceptional and that $\dim N = 0$ often implies $\mathcal{A}$ is conservative.

Theorem 3.4.22 *If a Bernstein algebra $(\mathcal{A}, \sigma)$ is not exceptional then it admits a nontrivial invariant form and so $\dim J \geq 2$.*

Proof Because $\mathcal{A}$ is not exceptional, $U^2 \neq 0$ or equivalently, $\overline{U}^{\perp}$ is a proper subspace of U and so $\overline{U}$ is a nonzero subspace of U^*. Now apply Proposition 3.4.21 to the family of operators $\{M_w^* : w \in W\}$ on $\overline{U}$. There exists f a nonzero element of $\overline{U}$ with $M_w^*(f) = 0$ for all $w \in W$.

The linear form f in U^* vanishes on W^2 because $W^2 \subset \overline{U}^{\perp}$. Furthermore $M_w^*(f) = 0$ for all $w \in W$ means f vanishes on the subspace UW of U as well. Extend f to a form $\tilde{f}$ on $\mathcal{A}$ by defining $\tilde{f}(e) = 0$ and $\tilde{f}(w) = 0$ for $w \in W$. $\tilde{f}$ vanishes on $(UW + W^2) \oplus W$ and so lies in J by (3.4.17). Because $\tilde{f}$ vanishes on e, it is not a multiple of σ. $\square$

Corollary 3.4.23 *A Bernstein algebra of dimension greater than one admits a nontrivial invariant form or a disappearing form.*

Proof If $\dim J = 1$ then by Theorem 3.4.22 $\dim N = \delta$. So $\dim N = 0$ then implies $\delta = 0$. An algebra of type $(m, 0)$ is a unit algebra for which $J = \mathcal{A}^*$. So $\dim J = 1$ then implies $\dim \mathcal{A} = 1$ and $\sigma : \mathcal{A} \to \mathbf{R}$ is an algebra isomorphism. $\square$

Using the construction of Theorem 3.4.15 it is easy to build examples of exceptional algebras of type (m, δ) such that $\dim J = 1$. For $x = (\sigma, u_1, \ldots, u_{m-1}, w_1, \ldots, w_\delta)$ in $\mathbf{R}^n$ ($n = m + \delta$) we define the multiplication by describing the coordinates of the square $x' = x^2$:

$$\sigma' = \sigma^2, \quad u_i' = \sigma u_i + u_i w_1, \quad w_j' = 0$$

for $1 \leq i \leq m-1$ and $1 \leq j \leq \delta$. $U^2 = 0$ and so this is an exceptional algebra (it is Bernstein by Theorem 3.4.15). Because $UW = U$, $\dim J = 1$.

Corollary 3.4.24 *Every Bernstein algebra of type (m, δ) with $m \leq 2$ or $\delta \leq 1$ (and this includes all cases with $n \leq 4$) is either conservative or exceptional.*

Proof By Theorem 3.4.3 types $(n, 0)$ and $(1, n-1)$ are both induced by projections and so are both conservative and exceptional. If the type is $(2, n-2)$ then either the algebra is exceptional or by Theorem 3.4.22 $\dim J \geq 2$. In the latter case $\dim J \leq m$ implies $\dim J = 2$ and the algebra is conservative by Theorem 3.4.17.

Finally, if the type is $(n-1, 1)$ and $\mathcal{A}$ is not exceptional, i.e. $U^2 \neq 0$, then there exists $u \in U$ with $u^2 \neq 0$. As $\dim W = 1$, W consists of multiples of u^2. Hence, $W^2 = 0$ because $(u^2)^2 = 0$ c.f. (3.4.9). On the other hand, if $u_1 \in U$ then by (3.4.11) $(u_2 = u_3 = u)$

$$u_1 u^2 + 2u(u_1 u) = 0.$$

As $u_1 u \in W$ it is a multiple of u^2 and $u(u_1 u)$ is that multiple of $u^3 = 0$. Thus, $u_1 u^2 = 0$ and this means $UW = 0$. So $\dim J = m$ by Theorem 3.4.15 and the algebra is conservative by Theorem 3.4.17. $\square$

Example For $x = (\sigma, u_1, \ldots, u_{m-1}, w_1, \ldots, w_\delta)$ in $\mathbf{R}^n$ with $m \geq 3$ and $\delta \geq 2$ we get a Bernstein algebra which is neither conservative nor exceptional when the coordinates of $x' = x^2$ are:

$$\sigma' = \sigma^2, \ u_i' = \sigma u_i \ (1 \leq i \leq m-2), \ u_{m-1}' = \sigma u_{m-1} + u_1 w_2,$$

$$w_1' = u_1^2, \ w_j' = 0 \ (2 \leq j \leq \delta).$$

Check that $(x^2)^2 = \sigma^2 x^2$ so that the algebra is Bernstein. Neither U^2 nor UW vanish and so the algebra is neither exceptional nor conservative.

For the next result we first obtain some dimension estimates:

Lemma 3.4.25 *Let E be a vector space of dimension k, and $B : E \to E$ a nonzero linear map. The dimension of $\{A : E \to E^* \text{ such that } A^* = A\}$ is $k(k+1)/2$. The dimension of the solution space $\{A : E \to E^* \text{ such that } A^* = A \text{ and } B^*A + AB = 0\}$ is at most $k(k-1)/2$.*

Proof Let $\{v_1, \ldots, v_k\}$ be a basis for E and let $\{v_1^*, \ldots v_k^*\}$ be the dual basis for E^* defined by

$$v_i^*(v_j) = \delta_{ij}. \tag{3.4.34}$$

With respect to these bases $A : E \to E^*$ is represented by a $k \times k$ matrix and $A^* = A$ if and only if the matrix is symmetric. The first result follows because the dimension of symmetric $k \times k$ matrices is $k(k+1)/2$.

If B is a nonzero multiple of the identity then $B^*A + AB = 0$ implies $A = 0$ and the solution space estimate follows. Observe that $B(v)$ is a multiple of v for all v in E if and only if B is a multiple of the identity. Excluding this case there exists v_1 in E so that $\{v_1, v_2 = B(v_1)\}$ is linearly independent. We suppose our basis chosen so that v_1 and $v_2 = B(v_1)$ are the first two elements.

Define $L(A) = (B^*A + AB)(v_1)$. We will show that L maps the space of self-adjoint linear maps onto E^*. So the $\dim \ker L = k(k+1)/2 - k = k(k-1)/2$. As the $\ker L$ contains the solution space, this will complete the proof. Define $A_i : E \to E^*$ by $v \mapsto v_2^*(v)v_i^* + v_i^*(v)v_2^*$. As its matrix is symmetric, A_i is a self- adjoint map. $L(A_i) =$

$$v_2^*(B(v_1))v_i^* + v_i^*(B(v_1))v_2^* +$$
$$v_2^*(v_1)(B^*v_i^*) + v_i^*(v_1)(B^*v_2^*).$$

Because $B(v_1) = v_2$, this equation together with (3.4.34) implies

$$L(A_i) = v_i^* \ (i > 2)$$

$$L(A_2) = 2v_2^* \text{ and } L(A_1) = v_1^* + B^*v_2^*.$$

Also $(v_1^* + B^*v_2^*)(v_1) = v_1^*(v_1) + v_2^*(v_2) = 2$. Thus $\{L(A_1), L(A_2), \ldots, L(A_k)\}$ is a basis for E^* and so L is surjective. $\square$

Theorem 3.4.22 says that when J is extremely small ($\dim J = 1$) then the algebra is exceptional. We now prove a result which is roughly dual. Notice first that since $\dim U = m - 1$, $\dim U^2 \le m(m-1)/2$, i.e. if $\{u_1, \ldots u_{m-1}\}$ is a basis for U then the set $\{u_i u_j : 1 \le i \le j \le m - 1\}$ spans U^2. So by (3.4.22) we have

$$\dim N \ge \delta - \frac{m(m-1)}{2}. \tag{3.4.35}$$

We can sharpen this result by observing that the product $u_1 u_2$ for $u_1, u_2 \in U$ depends only on the congruence classes of u_1 and u_2 mod $\overline{U}^\perp$ because $\overline{U}^\perp$ is the annihilator of U in U by Theorem 3.4.19. So we can get a spanning set for U^2 by beginning with a basis for the quotient space $U/\overline{U}^\perp$. Thus, we have proved

Lemma 3.4.26 *In a Bernstein algebra of type* (m, δ) *suppose that* k *is the dimension of* $S^{\perp} = U \cap \operatorname{ann} U$. *Then*

$$\dim N \geq \delta - \frac{(m-k)(m-k-1)}{2}. \tag{3.4.36}$$

Theorem 3.4.27 *If in a Bernstein algebra of type* (m, δ) *the inequality*

$$\dim N < \delta - \frac{(m-1)(m-2)}{2} \tag{3.4.37}$$

holds, then the algebra is conservative.

Proof First, combine (3.4.37) with (3.4.36) to get $k = 0$ and so $\overline{U}^{\perp} = 0$. By Theorem 3.4.19 W^2 is contained in $\overline{U}^{\perp}$ and so $W^2 = 0$. We prove $UW = 0$ as well, i.e. for each $w \in W$ the operator M_w on U is zero. The result then follows from (3.4.21) and Theorem 3.4.17.

Recall the homomorphism S which associates to each f in W^* the self-adjoint homomorphism $S_f : U \to U^*$ defined by $S_f(u)(u_1) = f(uu_1)$. Clearly, the kernel of S is the subspace of W^* annihilating U^2, i.e. the kernel is N. Hence,

$$\dim\{S_f : f \in W^*\} = \delta - \dim N > \frac{(m-1)(m-2)}{2}.$$

But by (3.4.30) each S_f is a self-adjoint solution of the equation $M_w^* A + A M_w = 0$. Since the dimension of U is $m - 1$, Lemma 3.4.25 implies that the solution space has dimension at most $\frac{1}{2}(m-1)(m-2)$ when $M_w \neq 0$. Consequently, M_w is, in fact, 0. $\square$

Corollary 3.4.28 *If* (A, σ) *is a Bernstein algebra of type* (m, δ) *with*

$$\delta > \frac{(m-1)(m-2)}{2} \tag{3.4.38}$$

then $N = 0$ *(or equivalently,* $A^2 = A$*) implies that the algebra is conservative.*

Proof (3.4.38) and $\dim N = 0$ imply (3.4.37). $\square$

We can construct conservative examples with N equal 0 for any pair (m, δ) with $\delta \leq \frac{1}{2}m(m-1)$ (cf. (3.4.35)). In our now familiar notation let $\sigma' = \sigma^2$, $u_i' = \sigma u_i$ and $w_{jk}' = u_j u_k$ where the biindex jk runs over any set of pairs with $1 \leq j \leq k \leq m - 1$ up to a total of δ pairs.

Corollary 3.4.29 *For every Bernstein algebra of type (m, δ) with $m \leq 3$ or $\delta \leq 1$ (and this includes all cases with $n \leq 5$) $N = 0$ implies the algebra is conservative.*

Proof We go back to Corollary 3.4.24 and its proof. Types $(n, 0)$ and $(1, n - 1)$ are conservative in any case.

Now assume $\delta > 0$ and $N = 0$ so that the algebra is not exceptional. Then $m \leq 2$ or $\delta = 1$ implies the algebra is conservative by Corollary 3.4.24.

For the remaining case $m = 3$ and $\delta \geq 2$. Now inequality (3.4.38) holds and so the result follows from Corollary 3.4.28. $\square$

To see that the above result is sharp we construct an example of type (4,2). For $x = (\sigma, u_1, u_2, u_3, w_1, w_2)$ let $\phi_1(u_2, u_3)$ and $\phi_2(u_2, u_3)$ be linearly independent linear forms and $\psi(u_2, u_3)$ another nonzero linear form. Define

$$\sigma' = \sigma^2, \ u_1' = \sigma u_1 + w_1 \phi_1 + w_2 \phi_2, \ u_2' = \sigma u_2, u_3' = \sigma u_3$$

$$w_1' = -\phi_2 \psi, \ w_2' = \phi_1 \psi.$$

Check that $(x^2)^2 = \sigma^2 x^2$ as ϕ_1, ϕ_2 are invariant and $w_1' \phi_1 + w_2' \phi_2 = 0$. By linear independence of $\{\phi_1, \phi_2\}$ and $\psi \neq 0$ the set $\{u^2 : u \in U\}$ spans W and so $N = 0$. On the other hand, $W^2 = 0$ and UW is one dimensional. Thus, $\dim J = 3$ while $m = 4$ and so the algebra is not conservative. Observe also that by linear independence of $\{\phi_1, \phi_2\}$ $w \in W$ annihilates $\mathcal{A}$ only when $w = 0$. So ann $\mathcal{A} = 0$.

Analogously, for any prescribed m, δ, ν satisfying $1 \leq m, 2 \leq \delta, 0 \leq \nu \leq \delta - 2$ and

$$\nu \geq \delta - \frac{(m-1)(m-2)}{2}$$

we can construct a nonconservative Bernstein algebra for which $\dim N = \nu$.

Corollary 3.4.30 *If a Bernstein algebra of type (m, δ) satisfies*

$$\dim \mathcal{A}^2 > \frac{m(m-1)}{2} + 1 \tag{3.4.39}$$

then $(\mathcal{A}, \sigma)$ is conservative.

Proof Because N is essentially $(U^2)^\perp$ in W^* we have $\delta - \dim N = \dim U^2 = \dim \mathcal{A}^2 - m$ (see (3.4.18)). So (3.4.39) is a rewriting of (3.4.37). $\square$

Recall the ideal $\mathcal{A}^2$ is not baric. It is contained in the barideal $\mathcal{B}$ and so is a baric subalgebra. If $\mathcal{A}_1$ is any baric subalgebra of $\mathcal{A}$ which contains $\mathcal{A}^2$ then $\mathcal{A}_1$ is an ideal (not a baric ideal). The converse is also true.

Lemma 3.4.31 *If a baric subalgebra $\mathcal{A}_1$ of a Bernstein algebra is also an ideal then $\mathcal{A}_1 \supset \mathcal{A}^2$.*

Proof Let $\mathcal{A}_2$ be a complement of $\mathcal{A}_1$, a subspace of $\mathcal{A}$ such that $\mathcal{A}$ is the direct sum $\mathcal{A}_1 \oplus \mathcal{A}_2$. For $x \in \mathcal{A}$ write $x = x_1 + x_2$ and $\sigma = \sigma(x) = \sigma_1 + \sigma_2$ with $x_i \in \mathcal{A}_i$ and $\sigma_i = \sigma(x_i)$ $(i = 1, 2)$. Now $x^2 = x_1' + x_2^2$ where $x_1' = x_1^2 + 2x_1 x_2$ lies in the ideal $\mathcal{A}_1$. Also, $\sigma^2 = \sigma_1^2 + 2\sigma_1\sigma_2 + \sigma_2^2$. Because the algebra is Bernstein $(x^2)^2 = \sigma^2 x^2$. For the $\mathcal{A}_2$ components this says $(x_2^2)^2 = (\sigma_1^2 + 2\sigma_1\sigma_2 + \sigma_2^2)x_2^2$. But as we vary x_1 over $\mathcal{A}_1$, keeping x_2 fixed, $\sigma_1 = \sigma(x_1)$ varies over $\mathbf{R}$ because $\mathcal{A}_1$ is a baric subalgebra $(\mathcal{B} \not\supset \mathcal{A}_1)$. So x_2^2 must equal 0, i.e. $x^2 \in \mathcal{A}_1$ for all $x \in \mathcal{A}$. $\square$

By (3.4.35) the possibility $N = 0$ can only occur when

$$\delta \leq \frac{m(m-1)}{2}. \tag{3.4.40}$$

We can characterize the extreme cases where $N = 0$ and equality holds in (3.4.40).

Theorem 3.4.32 *If $\delta = \frac{1}{2}m(m-1)$ then a Bernstein algebra of type (m, δ) with $N = 0$ is isomorphic as a baric algebra to the Mendel m-allele zygotic algebra.*

Proof If $(\mathcal{A}, \sigma)$ is such a algebra of type (m, δ) with $N = 0$, then it is conservative by Corollary 3.4.28 and so by Theorems 3.4.15, 3.4.17, $UW + W^2 = 0$ and the multiplication on $\mathcal{A}$ is determined by the bilinear map $U \times U \to W$. The dual version of this multiplication yields a homomorphism $S : W^* \to \sum_U$ where $\sum_U$ is the subspace of self-adjoint maps from U to U^*. The kernel of S is N and since $N = 0$ this map is injective. By Lemma 3.4.25 the dimension of $\sum_U$ is $\frac{1}{2}(m-1)m = \delta$. Thus, S is an isomorphism.

Now suppose $(\mathcal{A}_1, \sigma_1)$ is another Bernstein algebra of type (m, δ) with $N = 0$. We construct $F : \mathcal{A}_1 \to \mathcal{A}$ by relating the chosen idempotents $e \mapsto e_1$ and by choosing an arbitrary isomorphism from U_1 to U (c.f. $\dim U = m - 1 = \dim U_1$). This in turn induces an isomorphism from $\sum_U$ to $\sum_{U_1}$. Then the isomorphisms $S : W^* \to \sum_U$ and $S_1 : W_1^* \to \sum_{U_1}$ are used to get an isomorphism from W_1^* to W^*. Its dual is the required isomorphism from W_1 to W. Because the S describes the product from $U \times U$ to W and S_1 is its analogue, the linear isomorphism from $\mathcal{A}$ to $\mathcal{A}_1$ we have constructed preserves the multiplication $U \times U$ to W. As e is mapped to e_1 the multiplications and weights correspond and the map is a baric algebra isomorphism.

Recall that for the Mendel m-allele algebra, $\mathcal{A}$, we have basis vectors $\{e_{ij} : 1 \leq i, j \leq m\}$ with the identification $e_{ij} \equiv e_{ji}$. So $\dim \mathcal{A} = \frac{1}{2}m(m+1)$. The multiplication is given by

$$e_{ij}e_{kl} = \frac{1}{4}e_{ik} + \frac{1}{4}e_{jl} + \frac{1}{4}e_{il} + \frac{1}{4}e_{jk} \qquad (3.4.41)$$

In particular, writing e_i for the homozygote e_{ii} we have

$$e_i e_j = e_{ij}. \qquad (3.4.42)$$

In particular, $\mathcal{A}^2 = \mathcal{A}$ and so $N = 0$. The algebra is baric ($\sigma(e_{ij}) = 1$) and Bernstein by the Hardy-Weinberg Law.

Finally, we compute the type by choosing $e = e_{11} = e_2$ and computing the projection map on the vectors $v_{ij} \equiv e_{ij} - e_{11}$ which, form a basis for the barideal when $v_{11} = 0$ is excluded.

$$L(v_{1j}) = 2e_{11}(e_{1j} - e_{11}) = (e_{11} + e_{1j}) - 2e_{11} = v_{1j}$$

and for $i, j \neq 1$:

$$L(v_{ij}) = 2e_{11}(e_{ij} - e_{11}) = (e_{1i} + e_{1j}) - 2e_{11} = v_{1i} + v_{1j}.$$

Thus, the image of L is spanned by $\{v_{12}, \ldots, v_{1m}\}$ and so has dimension $m - 1$. The type is thus (m, δ) with

$$\delta = \frac{m(m+1)}{2} - m = \frac{m(m-1)}{2}.$$

$\square$

In particular for a Bernstein algebra of type $(2,1)$ either $\dim N = 1$ and the algebra is exceptional, determined as described in part 2 of Theorem 3.4.15, or $N = 0$ and the algebra is isomorphic to the Mendel 2- allele zygote algebra. More generally:

Theorem 3.4.33 *Let e_1 and e_2 be distinct nonzero idempotents in a Bernstein algebra and let $e_{12} = e_1e_2$. $[e_1, e_2, e_{12}]$ is a subalgebra and either   (a) e_{12} is a linear combination of $\{e_1, e_2\}$: In this case the subalgebra is the unit algebra of type (2,0).   (b) $\{e_1, e_2, e_{12}\}$ is linearly independent: In this case the subalgebra is the Mendel 2-allele algebra with canonical basis $\{e_1, e_1, e_{12}\}$.*

Proof Substitute $x = x_1 e_1 + x_2 e_2$ with $\sigma = x_1 + x_2$ into the defining identity $(x^2)^2 = \sigma^2 x^2$ for a Bernstein algebra. Compare coefficients for $x_1^3 x_2, x_1 x_2^3$ and $x_1^2 x_2^2$ to get

$$e_1 e_{12} = \frac{1}{2} e_1 + \frac{1}{2} e_{12}, e_2 e_{12} = \frac{1}{2} e_2 + \frac{1}{2} e_{12}$$

$$e_{12}^2 = \frac{1}{4} e_1 + \frac{1}{4} e_2 + \frac{1}{2} e_{12}.$$

These identities show that the linear span of e_1, e_2 and e_{12} is a subalgebra. When $\{e_1, e_2, e_{12}\}$ is linearly independent this list shows that (3.4.41) holds and so the subalgebra is the Mendel 2-allele.

When e_{12} is a linear combination then $m = n = 2$ for the subalgebra and so it has type (2,0) and is a unit algebra by Theorem 3.4.3. So $e_{12} = e_1 e_2 = \frac{1}{2}(e_1 + e_2)$. $\square$

Theorem 3.4.34 *If in a conservative algebra the elements $e_1, \ldots e_m$ are idempotents and the list of products $e_{ik} \equiv e_i e_k$ ($1 \le i \le k \le m$) constitutes a linearly independent set, then their linear span is a subalgebra and it is the Mendel m-allele algebra with canonical basis $\{e_{ik}\}$.*

Proof Choosing any idempotent e we can use Corollary 3.4.9 to write each e_i as $e + u_i + u_i^2$ with a unique choice $u_i \in U$ ($1 \le i \le m$). Because the algebra is conservative, $UW + W^2 = 0$ and so

$$e_{ij} = e_i e_j = e + \frac{u_i + u_j}{2} + u_i u_j$$

$$(3.4.43)$$

$$e_{ij} e_{kl} = e + \frac{u_i + u_j + u_k + u_l}{4} + \frac{(u_i + u_j)(u_k + u_l)}{4}.$$

From these it is trivial to check directly that (3.4.41) holds. $\square$

Notice that we did not require linear independence to derive (3.4.43) or to get (3.4.41) from them. Hence we obtain

Theorem 3.4.35 *A baric algebra $(\mathcal{A}, \sigma)$ is the homomorphic image of a Mendel multiallele algebra, i.e. there exists a baric isomorphism of $(\mathcal{A}, \sigma)$ with a quotient of the zygotic algebra, if and only if $(\mathcal{A}, \sigma)$ is conservative and $N = 0$.*

Proof Conservatism and $\mathcal{A}^2 = \mathcal{A}$ are properties of the Mendel algebra and they are preserved under factoring by a baric ideal and under baric isomorphisms. So these conditions are necessary.

On the other hand, suppose $(\mathcal{A}, \sigma)$ has type (m, δ) and satisfies $UW + W^2 = 0$ (conservative) and $U^2 = W$ ($N = 0$). Let $u_1, \ldots, u_{m-1}$ be a basis for U. Let $e_i = e + u_i + u_i^2$ ($i = 1, \ldots m - 1$) and $e_m = e$. Then (3.4.43) and (3.4.41) hold (with $u_m = 0$). Also $\{e_{ik} : 1 \leq i \leq k \leq m\}$ span $\mathcal{A}$ because $U^2 = W$. By (3.4.43) $\tilde{e}_{ik} \mapsto e_{ik}$ defines a homomorphism of the Mendel algebra with canonical basis $\{\tilde{e}_{ik}\}$ onto $\mathcal{A}$. $\square$

3.5   Genetic Algebras and Train Algebras

Many algebras which arise in genetics are first, baric and, secondly admit
a basis with respect to which the multiplication map M_x has a triangular
matrix for each x, with the diagonal entries equal 0 when x is in the barideal.
Because this structure has deep consequences which are useful in genetics
such algebras are called genetic algebras.

In detail, a baric algebra $(\mathcal{A}, \sigma)$ is called a *genetic algebra* if its complexification $(\mathcal{A}^{(c)}, \sigma^{(c)})$ admits a basis $v_0, v_1, \ldots v_{n-1}$ with $\sigma(v_0) = 1$, $\sigma(v_1) =$
$\ldots = \sigma(v_{n-1}) = 0$, satisfying:

$$v_0^2 = v_0 + \sum_{k=1}^{n-1} \lambda_{00,k} v_k$$

$$v_0 v_i = \sum_{k=i}^{n-1} \lambda_{0i,k} v_k \qquad (1 \le i \le n - 1) \qquad (3.5.1)$$

$$v_i v_j = \sum_{k>i,j}^{n-1} \lambda_{ij,k} v_k \qquad (1 \le i, j \le n - 1).$$

Alternatively, if we define the coefficients $\lambda_{ij,k}$ by

$$v_i v_j = \sum_{k=0}^{n-1} \lambda_{ij,k} v_k$$

then these equations are equivalent to

$$\lambda_{00,0} = 1$$

$$\lambda_{0i,k} = 0 \qquad (0 \le k < i \le n - 1) \qquad (3.5.2)$$

$$\lambda_{ij,k=0} \qquad (0 \le k \le \max(i,j) \text{ with } 1 \le i, j \le n - 1).$$

Notice that the first equation follows from $\sigma(v_0^2) = \sigma(v_0) = 1$ and $\sigma(v_k) =$
$0\ (1 \le k \le n - 1)$.

A basis satisfying (3.5.1) is called a *canonical basis* for the genetic algebra.

Example Every algebra induced by a linear operator A is genetic. Indeed,we can choose a vector v_0 such that $A v_0 = v_0$ and $\sigma(v_0) = 1$ due to
the condition $A^* \sigma = \sigma$.Because the subspace $(\ker \sigma)^{(c)}$ is $A^{(c)}$-invariant there
exists such a basis $v_1, \ldots, v_{n-1}$ that the matrix of $A^{(c)}$ is lower triangular.It

is easy to check that $v_0, v_1, \ldots, v_n$ is the required basis. Moreover, $v_i v_j = 0$ in this case if $i, j \geq 1$, i.e. $\mathcal{B}^2 = 0$.

Directly from the definition it is easy to check that any baric quotient algebra of a genetic algebra is genetic. For if $h : (\mathcal{A}, \sigma) \to (\tilde{\mathcal{A}}, \tilde{\sigma})$ is the quotient map and $\{v_0, \ldots v_{n-1}\}$ is a canonical basis for $\mathcal{A}^{(c)}$ then, because the complexification of the quotient is the quotient of the complexification, we can define the spanning set $\{h(v_0), \ldots, h(v_{n-1})\}$ for $\tilde{\mathcal{A}}^{(c)}$. Because h is a baric homomorphism $h(v_0) \neq 0$. Then select a basis by eliminating $h(v_i)$ if it is a linear combination of $\{h(v_0), \ldots, h(v_{i-1})\}$. The result is a canonical basis for $\tilde{\mathcal{A}}^{(c)}$.

This sort of direct argument using original definition of genetic algebra won't work for subalgebras. It is nonetheless true that a baric subalgebra of a genetic algebra is genetic. This follows from—and illustrates the usefulness of—a pair of alternative characterizations of a genetic algebra.

Theorem 3.5.1 *Each of the following conditions on a baric algebra $(\mathcal{A}, \sigma)$ holds if and only if $(\mathcal{A}, \sigma)$ is a genetic algebra.*

(Schafer's Condition): Let $F(\mu_1, \ldots \mu_n)$ be a noncommutative polynomial over **C**. *For $x_1, \ldots x_k \in \mathcal{A}$ the characteristic polynomial of the linear operator $F(M_{x_1}, \ldots M_{x_k})$ on $\mathcal{A}$ depends only on the weights $\sigma(x_1), \ldots \sigma(x_k)$.*

(Holgate's Condition): The barideal $\mathcal{B}$ is a principal nilalgebra and the Lie algebra Lie $\mathcal{A}$ (in the space End $\mathcal{A}$) is a solvable Lie algebra.

Proof (Genetic $\Rightarrow$ Schafer): For $x \in \mathcal{A}$ the v_0 component of x is $\sigma(x)v_0$ because $\sigma(v_0) = 1$ and $\sigma(v_i) = 0$ $(l \leq i \leq n - 1)$. So with respect to a canonical basis the matrix for M_x is lower triangular. In fact,

$$xv_i \equiv \sigma(x)\tau_i v_i \quad \text{mod } [v_{i+1}, \ldots v_{n-1}]$$

with $\tau_0 = 1$ and $\tau_i = \lambda_{0i,i}$ $(1 \leq i \leq n-1)$. Hence the matrix of $F(M_{x_1}, \ldots M_{x_k})$ is lower triangular with ii entry $F(\sigma(x_1)\tau_i, \ldots, \sigma(x_k)\tau_i)$. As the τ_i's are constants, Schafer's condition follows.

(Schafer $\Rightarrow$ Holgate): If $\sigma(x) = 0$ then by Schafer's condition M_x and $M_0 = 0$ have the same characteristic polynomial in λ, namely λ^n. Hence, M_x is nilpotent (c.f. Cayley-Hamilton Theorem) and we have for all $y \in \mathcal{A}$ Engel's identity:

$$\underbrace{x(x(\ldots(xy)\ldots))}_{n \text{ copies of } x} = 0 \tag{3.5.3}$$

In particular, for all $x \in \mathcal{B}$, $x^{n+1} = 0$ and so $\mathcal{B}$ is a principal nilalgebra. To prove that the Lie algebra Lie $\mathcal{A}$ is solvable it requires to show that the derived (Lie $\mathcal{A}$)', the Lie subalgebra generated by the commutators, is a nilpotent Lie algebra. We prove this by including (Lie $\mathcal{A}$)' in an operator subalgebra $\mathcal{M}$ of the linear operators on $\mathcal{A}$ and then showing that $\mathcal{M}$ is a nilpotent (associative but not commutative) algebra of operators.

Choose e with $\sigma(e) = 1$ and write $x = \sigma(x)e + y$ with $y \in \mathcal{B}$. Because the commutator $[M_e, M_e] = 0$ we have for $x_1, x_2 \in \mathcal{A}$:

$$[M_{x_1}, M_{x_2}] = \sigma(x_1)[M_e, M_{y_2}] - \sigma(x_2)[M_e, M_{y_1}] + [M_{y_1}, M_{y_2}]$$

So if we define $\mathcal{M}$ to be the algebra of operators generated by all linear operators of the form

$$M_e^a M_y M_e^b \qquad\qquad (y \in \mathcal{B}, a, b = 0, 1, \ldots) \qquad (3.5.4)$$

then $\mathcal{M}$ contains the derived Lie algebra (Lie $\mathcal{A}$)'. An element P of $\mathcal{M}$ is a noncommutative polynomial in expressions of the form (3.5.4) with no constant term. By the assumption of Schafer's condition the characteristic polynomial of P is the same as that obtained by replacing each y by 0. This means that the characteristic polynomial is that of 0: λ^n again. Thus each element of $\mathcal{M}$ is nilpotent. From Engel's Theorem it follows that the operator algebra $\mathcal{M}$ is nilpotent.

(Holgate $\Rightarrow$ Genetic): We have the algebra of operators $\{M_z : z \in \mathcal{A}^{(c)}\}$ acting on the barideal, $\mathcal{B}^{(c)}$ by complexification and restriction. Holgate's condition says that the Lie hull of this family of operators is solvable. So Theorem 3.2.3 implies that each M_z with $z \in \mathcal{B}^{(c)}$ is nilpotent.

As in the proof of Theorem 3.2.3 Lie's Theorem implies we can choose a basis $\{v_1, \ldots v_{n-1}\}$ for $\mathcal{B}^{(c)}$ with respect to which the operators $\{M_z : z \in \mathcal{A}^{(c)}\}$ are simultaneously lower triangular. Thus

$$v_i v_j = M_{v_i}(v_j) = \sum_{\substack{k \geq j}}^{n-1} \lambda_{ij,k} v_k$$

Furthermore, M_{v_i} is nilpotent so that the $\lambda_{ij,j}$'s are all 0. Thus, we can restrict the sum to $k > j$. Finally, by commutativity of multiplication we can restrict the index k to $k > \max(i, j)$.

Now choose $v_0 \in \mathcal{A}$ with $\sigma(v_0) = 1$. Because M_{v_0} is lower triangular on $\mathcal{B}^{(c)}$ with respect to $\{v_1, \ldots v_{n-1}\}$ and because $\sigma(v_0^2) = \sigma(v_0) = 1$, the remaining equations of (3.5.1) follow. $\square$

Remark From the latter proof we see that it is sufficient to consider the restrictions of the operators $\{M_x|\mathcal{B} : x \in \mathcal{A}\}$ in Holgate's Condition.

If all the $\tau_i = \lambda_{0i,i}$'s are real then by the **R**-version of Lie's Theorem we can choose the basis $v_0, v_1, \ldots v_{n-1}$ in $\mathcal{A}$ rather than going to the complexification $\mathcal{A}^{(c)}$.

The barideal $\mathcal{B}$ of a genetic algebra is nilpotent. This is clear directly from the lower triangular form of the nilpotent operators $M_{v_1}, \ldots M_{v_{n-1}}$. However, this conditions is not sufficient. Define an algebra structure on $\mathbf{R}^3$ by $e_1^2 = e_1, e_1 e_3 = e_2, e_2^2 = e_3$ and $e_i e_j = 0$ for the remaining choices. Here $\{e_1, e_2, e_3\}$ is the standard basis for $\mathbf{R}^3$. $\sigma(x) = x_1$ is a character and so defines $\mathbf{R}^3$ as a baric algebra with $\mathcal{B} = [e_2, e_3]$. The matrix of M_x for $x = (x_1, x_2, x_3)$ is given by

$$\begin{pmatrix} x_1 & 0 & 0 \\ x_3 & 0 & x_1 \\ 0 & x_2 & 0 \end{pmatrix}$$

When $x_1 = 0$ these matrices are lower triangular and so the barideal is nilpotent. But the characteristic polynomial is given by

$$\det(\lambda I - M_x) = \lambda^3 - x_1\lambda^2 - x_1 x_2 \lambda + x_1^2 x_2$$

which depends on x_2 as well as $x_1 = \sigma(x)$. So the algebra is not genetic.

We now generalize the concept of genetic algebra by weakening Schafer's Condition. A baric algebra $(\mathcal{A}, \sigma)$ is called a *train algebra* if for each $x \in \mathcal{A}$ the characteristic polynomial of M_x on $\mathcal{A}$ depends only on the weight $\sigma(x)$. Equivalently, there are real constants $a_1, \ldots a_n$ such that

$$\det(\lambda I - M_x) = \lambda^n - a_1\lambda^{n-1} + \cdots + (-1)^n a_n \qquad (\sigma(x) = 1). \quad (3.5.5)$$

Beyond the unit hyperplane we have

$$\det(\lambda I - M_x) = \lambda^n - a_1\sigma\lambda^{n-1} + \cdots + (-1)^n a_n \sigma^n \ (\sigma(x) = \sigma). \quad (3.5.6)$$

In particular, the eigenvalues of M_x, the roots of $\det(\lambda I - M_x) = 0$, counted with multiplicity, are the same for all x in the unit hyperplane when $(\mathcal{A}, \sigma)$ is a train algebra. Writing these roots as $\tau_0, \tau_1, \ldots \tau_{n-1}$ we call them the *train roots* of the algebra. Because $M_x^* \sigma = \sigma$, when $\sigma(x) = 1$, 1 is an eigenvalue of M_x and so 1 is always one of the train roots. We will choose the numbering so that $\tau_0 = 1$. Notice that this notation coincides with that developed in the proof of Theorem 3.5.1 for a genetic algebra. By using a canonical basis we see that we can label $\tau_0 = 1$, $\tau_i = \lambda_{0i,i}$ $(1 \le i \le n-1)$ for a genetic algebra. If the algebra is induced by a linear operator A then the train roots coincide with the eiegenvalues of $\frac{1}{2}A$ except τ_0.

Lemma 3.5.2 *A baric algebra $(\mathcal{A}, \sigma)$ is a train algebra if and only if there exists a polynomial $p(\lambda)$ such that for all $x \in \mathcal{A}$ with $\sigma(x) = 1$*

$$p(M_x) = 0 \tag{3.5.7}$$

Proof (3.5.5) implies (3.5.7) for the characteristic polynomial by the Cayley-Hamilton Theorem. Conversely, $p(T) = 0$ implies the spectrum of T is contained amongst the roots of $p(\lambda) = 0$. There are only finitely many polynomials of degree n whose roots lie amongst these. On the other hand the polynomial $\det(\lambda I - T)$ varies continuously with T. So as x varies in the connected set H, the unit hyperplane, the characteristic polynomial of M_x must remain constant within this finite collection of polynomials. $\square$

The polynomial $P(\lambda) = \det(\lambda I - M_x)$ $(x \in H)$ is called the *characteristic polynomial* of the algebra. By definition it is independent of the choice of x in the unit hyperplane in a train algebra and contained in the ideal of polynomials $Q(\lambda)$ such that

$$Q(M_x) = 0 \quad (x \in H).$$

The monic polynomial generating this ideal is called the *minimal polynomial* of the train algebra. It is a divisor of the characteristic polynomial and is in turn divisible by the minimal polynomials of the operators M_x $(x \in H)$. It is, in fact, the least common multiple of these. If $\mathcal{A}$ is the train algebra and $\sigma(x) = 0$, then M_x is nilpotent, i.e. the barideal $\mathcal{B}$ satisfies the Engel condition (3.5.3). In particular, $\mathcal{B}$ is principal nilalgebra. But $\mathcal{A}$ need not be genetic.

Lemma 3.5.3 *If an algebra $\tilde{A}$ satisfies Engel's condition then the algebra A obtained by adjoining an unity is a train algebra with barideal $\tilde{A}$. All the train roots are 1 in this case.*

Proof With e the attached unity in $\mathcal{A}$ we can write $x \in \mathcal{A}$ uniquely as $x = \sigma e + y$ with $y \in \tilde{A}$. Clearly, $x \mapsto \sigma$ is a character and with this weight $\tilde{A}$ is the barideal.

Now let $\sigma(x) = 1$, i.e. $x = e + y$. $\tilde{A}$ is invariant for M_x and $M_x(x) = x^2 \equiv x \pmod{\tilde{A}}$. This says that M_x is bloc triangular with respect to the splitting $[e] \oplus \tilde{A}$ of $\mathcal{A}$:

$$M_x = \begin{pmatrix} 1 & 0 \\ y & \tilde{M}_x \end{pmatrix}$$

where $\tilde{M}_x$ is the restriction of M_x to $\tilde{A}$, namely $I + \tilde{M}_y$. Hence,

$$\det(\lambda I - M_x) = (\lambda - 1)\det(\lambda \tilde{I} - \tilde{M}_x) = (\lambda - 1)\det((\lambda - 1)\tilde{I} - \tilde{M}_y).$$

But by Engel's condition $\tilde{M}_y$ is nilpotent and so its characteristic polynomial is λ^{n-1} ($n - 1 = \dim \tilde{A}$). Because of the translation by 1 we see that the characteristic polynomial of M_x is $(\lambda - 1)^n$. $\square$

Example Using the standard basis on $\mathbf{R}^5$ we define $e_1 e_3 = e_4$, $e_1 e_5 = -e_3$, $e_2 e_3 = e_5$, $e_2 e_4 = e_3$ and the other products $e_i e_k = 0$ ($i \leq k$). Extend to a commutative multiplication on $\mathbf{R}^5$. For $x \in \mathbf{R}^5$:

$$M_x = \begin{pmatrix} 0 & 0 \\ S_x & D_x \end{pmatrix} \quad \text{with}$$

$$S_x = \begin{pmatrix} -x_5 & x_4 \\ x_3 & 0 \\ 0 & x_3 \end{pmatrix} \quad \text{and} \quad D_x = \begin{pmatrix} 0 & x_2 & -x_1 \\ x_1 & 0 & 0 \\ x_2 & 0 & 0 \end{pmatrix}.$$

$D_x^3 = 0$ and so $M_x^4 = 0$. Thus, $\tilde{A} = \mathbf{R}^5$ satisfies Engel's condition and so by adjoining an identity we get a 6-dimensional train algebra A. But

$$M_{e_1}(M_{e_2}(e_3)) = e_1(e_2 e_3) = -e_3.$$

So the operator $M_{e_1} M_{e_2}$ is not nilpotent and the algebra $\tilde{A}$ is not nilpotent. Hence A is not genetic.

By sharpening the demand that the barideal be nilpotent we can get a sufficient condition for an algebra to be genetic. Consider the principal powers of the barideal B, defined inductively by

$$B^1 = B \text{ and } B^k = B(B^{k-1}) \qquad\qquad (k \geq 2).$$

Thus, B^k is the linear subspace of A spanned by all products $y_1(y_2(\ldots(y_{k-1}y_k)\ldots))$ with $y_1, \ldots y_k \in B$. Because B is an ideal we have, by induction, that $A \supset B \supset B^2 \supset B^3 \ldots$. Also each B^k is a subalgebra (though not a baric subalgebra) of A. In fact, $B(B^k) = B^{k+1}$ shows that each B^k is an ideal of the algebra B. The powers B^k need not be ideals of A however. We call a baric algebra (A, σ) a *special train algebra* if each B^k is an ideal of A and $B^k = 0$ for some k. Observe that $B^k = 0$ for some k says precisely that the barideal B is nilpotent. Obviously, if $B^2 = 0$ then A is a special train algebra. Thus, every algebra induced by a linear operator is special train. For $n < 4$ every genetic algebra is special train as well.

Theorem 3.5.4 *Every special train algebra is a genetic algebra.*

Proof Choose $v_0 \in \mathcal{A}$ with $\sigma(v_0) = 1$ and the construct a basis for $\mathcal{A} : v_0, v_1, \ldots, v_{n_1}, v_{n_1+1}, \ldots, v_{n_2}, \ldots$ so that $v_{n_{i-1}+1}, \ldots, v_{n_i}$ lie in $\mathcal{B}^i$ and map to a basis in $\mathcal{B}^i/\mathcal{B}^{i+1}$. Because $\mathcal{B}^k = 0$ for k large enough we can construct such a basis for $\mathcal{A}$. Note that $\{v_{n_{i-1}+1}, \ldots v_n\}$ is a basis for $\mathcal{B}^i$. Because $\mathcal{B}(\mathcal{B}^i) = \mathcal{B}^{i+1}$ we see that with respect to this basis the matrix for M_v with $v \in \mathcal{B}$ is lower triangular with a vanishing diagonal. Because each $\mathcal{B}^i$ is an ideal, $M_{v_0}(\mathcal{B}^i) \subset \mathcal{B}^i$ and so M_{v_0} has a lower triangular matrix after the complexification. Thus, $(\mathcal{A}, \sigma)$ is a genetic algebra with $\{v_0, v_1 \ldots\}$ a canonical basis. $\square$

Notice that $M_{v_0}(\mathcal{B}^i) \subset \mathcal{B}^i$ fixed $v_0 \in \mathcal{A} \backslash \mathcal{B}$ is equivalent to the condition that $\mathcal{B}^i$ be an ideal. It was this apparently weaker condition that we used in the proof.

Corollary 3.5.5 *Every special train algebra is a train algebra.*

Example Using the standard basis on $\mathbf{R}^4$, define $e_0^2 = e_0$, $e_0 e_1 = e_1$, $e_0 e_2 = e_3$, $e_1^2 = e_2$, with the remaining products zero. This is a genetic algebra. But $\mathcal{B}^2 = [e_2]$ is not an ideal because $e_0 e_2 = e_3$. So this is not a special train algebra.

Theorem 3.5.6 *In a Bernstein algebra the principal power $\mathcal{B}^k$ of the barideal are ideals.*

Proof Let e be a nonzero idempotent. It is sufficient to prove $e(\mathcal{B}^k) \subset \mathcal{B}^k$ for all $k \geq 1$. Proceed by induction and use (3.4.2) in the form:

$$y_1 y_2 = 2e(y_1 y_2) + 4(e y_1)(e y_2).$$

We must show that for a typical element $y_1 y_2$ of $\mathcal{B}^{k+1}$, obtained from $y_1 \in \mathcal{B}$ and $y_2 \in \mathcal{B}^k$, the product $e(y_1 y_2)$ lies in $\mathcal{B}^{k+1}$. But $e y_1 \in \mathcal{B}$ and by induction $e y_2 \in \mathcal{B}^k$ and so $y_1 y_2 - 4(e y_1)(e y_2)$ lies in $\mathcal{B}^{k+1}$. $\square$

Corollary 3.5.7 *For a Bernstein algebra $(\mathcal{A}, \sigma)$ the following conditions are equivalent:*

1. *$(\mathcal{A}, \sigma)$ is a special train algebra.*

2. *$(\mathcal{A}, \sigma)$ is a genetic algebra.*

3. *The barideal $\mathcal{B}$ is nilpotent.*

Proof $1 \Rightarrow 2 \Rightarrow 3$ always. For a Bernstein algebra $3 \Rightarrow 1$ by the theorem.
$\square$

Corollary 3.5.8 *A conservative Bernstein algebra is genetic.*

Proof Use the standard decomposition. Because the algebra is conservative $(U + W)W = 0$. Hence, $\mathcal{B}^2 = U^2 \subset W$ and $\mathcal{A}W = 0$. It follows that $\mathcal{B}^3 = 0$. $\square$

In a conservative algebra we get a canonical basis by listing e, a basis for U and then a basis for W.

Corollary 3.5.9 *In order that an exceptional Bernstein algebra be a genetic algebra it is necessary and sufficient that the operators $M_w : U \to U$ $(M_w(u) = wu;\ w \in W)$ generate a nilpotent subalgebra in the algebra of linear operators of U.*

Proof In an exceptional algebra $U^2 = 0$. So $WU \subset \mathcal{B}^2 \subset U$. So multiplying by $\mathcal{B} = W + U$ we have by induction $\mathcal{F}^{(k-1)} \subset \mathcal{B}^k \subset \mathcal{F}^{(k-2)}$ where

$$\mathcal{F}^{(k)} = W(W(\ldots(WU))\ldots)$$

with k factors of W. Hence, $\mathcal{B}^k = 0$ for some k if and only if $\mathcal{F}^{(k)} = 0$ for some k. The latter condition says that the family of product operators of W on U generates a nilpotent algebra. $\square$

Recall that by Theorem 3.4.15 we can build an exceptional Bernstein algebra using any bilinear map $W \times U \to U$. We see that both genetic and nongenetic algebras occur amongst exceptional Bernstein algebras. In fact, with some M_w a nonnilpotent operator we obtain exceptional Bernstein algebras which are not train algebras.

Example For $x = (\sigma, u_1, u_2, u_3, w_1, w_2)$, define $\sigma' = \sigma^2$, $u_1' = \sigma u_1 + w_1 u_2 + w_2 u_3$, $u_2' = \sigma u_2$, $u_3' = \sigma u_3$, $w_1' = -u_2 u_3$, $w_2' = u_2^2$. As we saw in the remarks after Corollary 3.4.29 this defines a Bernstein algebra of type (4,2) which is neither conservative nor exceptional. It is genetic because $\mathcal{B}^2 = \{x : \sigma = u_2 = u_3 = 0\}$, $\mathcal{B}^3 = \{x : \sigma = u_2 = u_3 = w_1 = w_2 = 0\}$, $\mathcal{B}^4 = 0$.

Theorem 3.5.10 *If a Bernstein algebra $(\mathcal{A}, \sigma)$ satisfies $\mathcal{A}^2 = \mathcal{A}$, or equivalently $N = 0$, then it is a genetic algebra. Therefore, $\mathcal{A}^2$ is genetic.*

In the case that $\delta > \frac{1}{2}(m-1)(m-2)$ the result follows because the algebra is conservative by Corollary 3.4.28. Similarly, the result follows when $m \leq 3$ or $\delta \leq 1$ (and so when $n \leq 5$) by Corollary 3.4.29. We omit here the difficult proof of theorem 3.5.10 in the general case.

We now consider polynomial equations satisfied by the elements of a train algebra. If we begin with the characteristic equation for M_x in the unit hyperplane and apply it to x we get for the principal powers (see (3.1.3)):

$$x^{n+1} - a_1 x^n + \ldots (-1)^n a_n x = 0. \tag{3.5.8}$$

For each polynomial with 0 constant term $Q(x) = b_0 \lambda^r + \cdots + b_{r-1} \lambda$ we define $Q(x) = b_0 x^r + \ldots + b_{r-1} x$ for $x \in H$, i.e. for $\sigma(x) = 1$.

Now if $P(\lambda)$ is an arbitrary polynomial then $\hat{P}(\lambda) \equiv \lambda P(\lambda)$ is a polynomial without constant term and so $\hat{P}(x)$ is defined. Observe that

$$\hat{P}(x) = P(M_x)(x). \tag{3.5.9}$$

A polynomial $P(\lambda)$ is called a *train polynomial* for a train algebra $(\mathcal{A}, \sigma)$ if $\hat{P}(x) = 0$ for x in the unit hyperplane H.

Lemma 3.5.11 *The set of train polynomials is an ideal in the polynomial ring* $\mathbf{R}[\lambda]$ *and included in it is the characteristic polynomial of the train algebra.*

Proof Let $P(\lambda)$ be a train polynomial, $T(\lambda)$ be an arbitrary polynomial and $R(\lambda) = T(\lambda)P(\lambda)$.

$$\hat{R}(x) = T(M_x)(P(M_x)(x)) = T(M_x)(0) = 0$$

for $x \in H$ and so $R(\lambda)$ is a train polynomial. Similarly, the sum of two train polynomials is a train polynomial. $\square$

If $P(\lambda) = c_0 \lambda^r + \cdots + c_r$ is a train polynomial then $\lambda^{k-1}P(\lambda)$ are train polynomials for $k = 1, 2, \ldots$ and we have the linear recurrence relation (the *train equation*):

$$c_0 x^{r+k} + \cdots + c_r x^k = 0 \qquad (k = 1, 2, \ldots). \tag{3.5.10}$$

Observe that $P(M_x) = 0$ for all $x \in H$ implies P is a train polynomial. The converse is not true, i.e. the ideal of polynomials which annihilate all M_x's $(x \in H)$ may be a proper subset of the ideal of train polynomials. Thus, the monic polynomial generating the ideal of all train polynomials– we call it the *minimal train polynomial*—may be a proper divisor of the minimal polynomial of the algebra, the generator of the former ideal. The degree of the minimal train polynomial is called the *train rank* of the algebra.

Proposition 3.5.12 *The train rank of the train algebra $(\mathcal{A}, \sigma)$ is the maximum ρ such that for some $x \in \mathcal{A}$ $\{x, x^2, \ldots, x^\rho\}$ is a linearly independent set.*

Proof With train rank r we can write the minimal train polynomial as:

$$P(\lambda) = \lambda^r - \sum_{k=0}^{r-1} c_k \lambda^k,$$

and so for every $x \in H$:

$$x^{r+1} = \sum_{k=1}^{r} c_{k-1} x^k \tag{3.5.11}$$

In general if $\sigma(x) \neq 0$ we can apply this equation to $x/\sigma(x)$ and multiply by $\sigma(x)^{r+1}$:

$$x^{r+1} = \sum_{k=1}^{r} c_{k-1} \sigma^{r-k+1} x^k \qquad (\sigma = \sigma(x)). \tag{3.5.12}$$

Because $\sigma(x) \neq 0$ on a dense set, this equation holds when $\sigma(x) = 0$ as well, i.e.

$$x^{r+1} = 0 \qquad (x \in \mathcal{B}).$$

Thus, $\{x, \ldots, x^{r+1}\}$ is linearly dependent for all x in $\mathcal{A}$. For the other direction define for each x in H the subspace Λ_x of $\mathbf{C}^r$ consisting of all $\alpha = (\alpha_1, \ldots, \alpha_r)$ such that

$$0 = \sum_{k=1}^{r} \alpha_k x^k.$$

Multiply this equation by x and substitute from (3.5.11). We get

$$0 = \sum_{k=1}^{r} \alpha'_k x^k$$

where

$$\alpha'_k = \alpha_{k-1} + c_{k-1} \alpha_r \qquad (1 \leq k \leq r, \alpha_0 = 0). \tag{3.5.13}$$

This means that the subspace Λ_x is invariant with respect to multiplication by the matrix

$$T = \begin{pmatrix} 0 & 0 & 0 & . & . & . & c_0 \\ 1 & 0 & 0 & . & . & . & c_1 \\ 0 & 1 & 0 & . & . & . & c_2 \\ . & & & & & & \\ . & & & & & & \\ . & & & & & & \\ 0 & 0 & . & . & . & 1 & c_{r-1} \end{pmatrix} .$$

In particular, if Λ_x is a nontrivial subspace of $\mathbf{C}^r$ it contains an eigenvector of T. But for each eigenvalue μ of T the corresponding eigenspace in $\mathbf{C}^r$ is one dimensional. To see this write $T(\alpha) = \mu\alpha$, using (3.5.13) as

$$\alpha_{k-1} = \mu\alpha_k - c_{k-1}\alpha_r,$$

and observe that a solution α (if any) is determined, inductively, by the choice of α_r.

Let F be a finite subset of $\mathbf{C}^r$ obtained by choosing one eigenvector for each eigenvalue. What we have proved so far is $\{x, \ldots, x^r\}$ linearly dependent implies

$$\alpha \in \Lambda_x \text{ for some } \alpha \in F.$$

Now for each $\alpha \in \mathbf{C}^r$ define the closed algebraic variety Λ^α in $\mathcal{A}^{(c)}$ by:

$$\Lambda^\alpha = \{x : 0 = \sum_{k=1}^r \alpha_k x^k\}$$

so that:

$$x \in \Lambda^\alpha \Leftrightarrow \alpha \in \Lambda_x.$$

It follows that the set of x in H such that $\{x, \ldots x^r\}$ is linearly dependent is precisely the union $\cup\{\Lambda^\alpha : \alpha \in F\}$. Suppose first that this union is all of H. Then at least one of the varieties has dimension $n - 1$ and so equals H. But if $\Lambda^\alpha = H$ then $\sum_{k=1}^r \alpha_k \lambda^{k-1}$ is a train polynomial and so $P(\lambda)$ is not minimal. This contradiction means that the open set $H - \cup\{\Lambda^\alpha : \alpha \in F\}$ is nonempty (it is, in fact, dense) and for x in this set $\{x, \ldots, x^r\}$ is linear independent. $\square$

Now for each $x \in \mathcal{A}$ define the subspace:

$$E_x = [x, x^2, \ldots].$$

By the now familiar argument leading to (3.5.10) we see that we can choose a basis for E_x $\{x, x^2, \ldots x^l\}$. Thus, the dimension of E_x is the maximum k such that $\{x, \ldots x^k\}$ is linearly independent. In particular:

Corollary 3.5.13 *The train rank is* $\max(\dim E_x)$ *for* $x \in H$. $\square$

Let us use these results to analyze Bernstein train algebras.

Theorem 3.5.14 *Let* $(\mathcal{A}, \sigma)$ *be a Bernstein algebra of type* (m, δ). *If* $(\mathcal{A}, \sigma)$ *is a train algebra then its train rank is at most* $m - \dim J + 2$.

Before we prove the theorem we require some preliminary results.

Proposition 3.5.15 *In a Bernstein algebra* $(\mathcal{A}, \sigma)$ *each* x *in the unit hyperplane* H *satisfies:*

$$x^i x^k = \frac{x^i + x^k}{2} \qquad (i, k \geq 2). \qquad (3.5.14)$$

Thus, the subspace $E'_x = [x^2, x^3, \ldots]$ *is a unit subalgebra of* $(\mathcal{A}, \sigma)$ *for each* x *with* $\sigma(x) = 1$.

Proof We start with a variation of the identities (3.4.1)–(3.4.4). Begin with the defining identity (3.3.17) applied to $\sigma x + y$ where $\sigma(x) = 1$ and $\sigma(y) = 0$. Expanding out and comparing σ^3 coefficients we get:

$$2x^2(xy) = xy \qquad (y \in \mathcal{B}, x \in H) \qquad (3.5.15)$$

and from the σ^2 terms:

$$(2xy)^2 + 2x^2y^2 = y^2 \qquad (y \in \mathcal{B}, x \in H). \qquad (3.5.16)$$

Now let $y = x - x^{k-1}$ $(k \geq 2)$ so that $xy = x^2 - x^k$. From (3.5.15) and $(x^2)^2 = x^2$ we get (3.5.14) when $i = 2$:

$$x^2 x^k = \frac{x^2 + x^k}{2} \qquad (3.5.17)$$

For the general result we first prove that all the principle powers x^k $(k \geq 2)$ are idempotents. We proceed by induction. With $k = 2$ it is the $\sigma = 1$ version of the defining identity (3.3.17). Now assuming the result for k, let $y = x - x^k$. By inductive hypothesis, $y^2 = x^2 + x^k - 2x^{k+1}$.

Then from (3.5.17) $2x^2y^2 = y^2$. Then from (3.5.16) $(xy)^2 = 0$ follows, i.e. $(x^2 - x^{k+1})^2 = 0$. From (3.5.17) again, we get $(x^{k+1})^2 = x^{k+1}$.

We prove the general case of (3.5.14) by induction on k and suppose that $2 \le i \le k$. For the inductive step to $k+1$ let $y = x^i - x^k$. Then from previous results $y^2 = 0$, and (3.5.16) again says that $(xy)^2 = 0$. So $(x^{i+1} - x^{k+1})^2 = 0$. From this we have (3.5.14) with (i, k) replaced by $(i+1, k+1)$. So for $k+1$ the result is proved for values $i+1 = 3, \ldots, k+1$. Finally, the case $i+1 = 2$ is (3.5.17) itself. $\square$

Corollary 3.5.16 *For a Bernstein algebra the baric subalgebra $E_x' = [x^2, x^3, \ldots]$ contains all powers x^T of x with $|T| \ge 2$, not just the principal powers. Consequently:*

$$x^{T_1} x^{T_2} = \frac{x^{T_1} + x^{T_2}}{2} \tag{3.5.18}$$

for every pair of index trees T_1 and T_2 with $|T_1|, |T_2| \ge 2$.

Proof Use induction on $|T|$. With $|T| = 2$, $x^T = x^2$. With $k \ge 2$ and $|T| = k + 1$ we can write $x^T = x^{T_1} x^{T_2}$ with $|T_1|, |T_2| \le k$. Because E_x' is a subalgebra, it follows that $x^T \in E_x'$. (3.5.18) then follows because E_x' is a unit algebra. $\square$

The proof shows that x^T is a linear combination of the principal powers and only dyadic rationals are required as coefficients for $x^T = \Sigma \theta_k x^k$, $\theta_k \ge 0$, $\Sigma \theta_k = 1$.

Corollary 3.5.17 *A Bernstein algebra $(\mathcal{A}, \sigma)$ is power associative if and only if*

$$x^3 = \sigma(x)x^2 \qquad\qquad (x \in \mathcal{A}). \tag{3.5.19}$$

In particular, every conservative algebra is power associative.

Proof Note first that a unit algebra is associative exactly when it is one dimensional: for $x_1, x_2 \in \mathcal{A}$ with $\sigma(x_1) = \sigma(x_2) = 1$, $x_1 x_2^2 = \frac{1}{2}(x_1 + x_2)$ while $(x_1 x_2)x_2 = \frac{1}{4}(x_1 + 3x_2)$. So associativity implies $x_1 = x_2$. The unit hyperplane consists of one point and so the algebra is one dimensional. When $(\mathcal{A}, \sigma)$ is power associative we apply this remark to E_x' and use Proposition 3.5.15. So $\dim E_x' = 1$, i.e. $x^3 = x^2$ for all $x \in H$, and this is equivalent to (3.5.19). Conversely, (3.5.19) implies $x^3 = x^2$ and so E_x' is one dimensional. Thus, corollary 3.5.16 implies that $x^T = x^2$ for all T with $|T| \ge 2$ when $\sigma(x) = 1$ and in general

$$x^T = \sigma(x)^{|T|-2} x^2 \qquad\qquad (|T| \ge 3) \tag{3.5.20}$$

and this implies power associativity. When $\mathcal{A}$ is conservative (3.3.16) with $y = x$ yields (3.5.19). $\square$

The identity (3.5.19) which says that $\lambda^2 - \lambda$ is a train polynomial, is equivalent to the conditions $w^2 = 0$, $(uw)w = 0$ in the standard decomposition. These conditions are satisfies not only for conservative algebras but for some exceptional algebras as well and so the Bernstein algebras which are power associative include more than just the conservative algebras.

Corollary 3.5.18 *In any Bernstein algebra E_x is a baric subalgebra for any x in the unit hyperplane H. Furthermore, for each such x the following conditions are equivalent:*

1. *$x \in E_x'$*

2. *$E_x = E_x'$*

3. *x is an idempotent.*

4. *E_x is a unit algebra.*

5. *The Bernstein algebra E_x has type $(l, 0)$ where $l = \text{dimension } E_x$.*

6. *dimension $E_x = 1$.*

Alternatively, when these conditions do not hold, i.e. $x \notin E_x'$, then the Bernstein algebra E_x has type $(l - 1, 1)$ and is exceptional.

Proof E_x' is an M_x-invariant baric subalgebra and $x^2 \in E_x'$. So $E_x = [x] + E_x'$ is a baric subalgebra. We now prove $(1) \Rightarrow (2) \Rightarrow (4) \Rightarrow (3) \Rightarrow (6) \Rightarrow (1)$ and $(4) \Leftrightarrow (5)$.

$(1) \Rightarrow (2)$: $E_x = [x] + E_x'$.

$(2) \Rightarrow (4)$: E_x' is a unit algebra.

$(4) \Rightarrow (3)$: $\sigma(x) = 1$ implies x is idempotent in a unit algebra.

$(3) \Rightarrow (6)$: $x^k = x$ for $k = 2, 3, \ldots$ when x is an idempotent.

$(6) \Rightarrow (1)$: x is the only element of E_x with weight one. So $x = x^2 \in E_x'$.

$(4) \Leftrightarrow (5)$: A Bernstein algebra of dimension l is a unit algebra precisely when it has type $(l, 0)$ by Theorem 3.4.3.

In general, $(E_x)^2 = E_x'$ and so the dimension of $N_x = ((E_x)^2)^\perp$ in E_x is the codimension of E_x' in E_x. So when (1)–(6) do not hold $\dim N_x = 1$. On the other hand, because E_x' is always a unit algebra, δ for E_x is at most 1. Hence

$$1 = \dim N_x \leq \delta \leq 1$$

and E_x has type $(l - 1, 1)$. Because $\dim N = \delta$ E_x is exceptional. $\square$

We can reverse this procedure by starting with a unit algebra $(\mathcal{A}, \sigma)$ and attaching a new element e. We extend the weight function by defining $\sigma(\xi e + y) = \xi + \sigma(y)$. Then define e^2 to be an arbitrary element of $H \subset \mathcal{A}$ and $M_e(y) = ey$ to be an arbitrary linear map $M_e : \mathcal{A} \to \mathcal{A}$ such that $M_e^* \sigma = \sigma$. We then get an exceptional Bernstein algebra of type $(l, 1)$ where $l = \dim \mathcal{A}$.

Proof of Theorem 3.5.14 Let r denote the train rank and apply Proposition 3.5.12 to get $x \in H$ such that $\{x, x^2, \ldots, x^r\}$ is linearly independent. Let $e = x^2$ and apply Proposition 3.5.15 to get $2e(x^k - x^2) = x^k - x^2$ for $k = 3, 4, \ldots, r$. Hence, $\{x^3 - x^2, \ldots, x^r - x^2\}$ is a linearly independent set in U.

Now let f be an invariant form. Notice that for $x \in H$

$$f(x^k) = f(x) \qquad\qquad (k \geq 1). \qquad\qquad (3.5.21)$$

This is trivial for $k = 1$. Inductively we have from (3.3.7)

$$f(x^{k+1}) = \frac{1}{2}(f(x) + f(x^k))$$

whence (3.5.21) for $k + 1$ by inductive hypothesis.

In particular, $f(x^k - x^2) = 0$ for $k = 3, \ldots, r$ and so $[x^3 - x^2, \ldots, x^r - x^2]$ is an $r - 2$ dimensional subspace of $U \cap J^\perp$. So $r - 2 \leq m - \dim J$. $\square$

The train rank $r = m - d + 2$ can be achieved for any prescribed d with $1 \leq d \leq m$ and $d = \dim J$. For example when $d = m$ the algebra is conservative. For a conservative algebra $\lambda^2 - \lambda$ is a train polynomial by (3.5.19). But $\lambda - 1$ is a train polynomial if and only if $x^2 = x$ on H, only for a unit algebra. Since 1 is always a train root $\lambda^2 - \lambda$ is the minimal train polynomial for a conservative, nonunit Bernstein algebra.

Now let $d < m$ and start with an m dimensional unit algebra with $\{e_1, \ldots e_m\}$ a basis with unit weights. Now we attach e_0 as suggested above to get an exceptional algebra of type $(m, 1)$. We define

$$e_0 e_k = \begin{cases} e_{k+1} & 0 \leq k \leq m - d \\ e_{m-d+1} & k = m - d + 1 \\ \frac{e_1 + e_k}{2} & m - d + 2 \leq k \leq m. \end{cases}$$

Then a linear form f is invariant if and only if $f(e_0) = f(e_1) = \cdots = f(e_{m-d+1})$. (First, derive $f(e_0) = f(e_{m-d+1})$ and then examine $f(e_k)$ by downward induction.) So $\dim J = (m + 1) - (m - d + 1) = d$. On the other hand, $e_0^k = e_{k-1}$ for $k = 1, \ldots, m - d + 2$. Hence, $r \geq m - d + 2$ and so from the theorem $r = m - d + 2$.

For a Bernstein algebra which is train we can completely describe the train roots.

Theorem 3.5.19 *If a Bernstein algebra of type (m, δ) is a train algebra then the train roots are $1, \frac{1}{2}, 0$ with multiplicities $1, m - 1, \delta$ respectively.*

Proof Let e be an idempotent. With respect to the standard decomposition $[e] \oplus U \oplus W$ the operator M_e maps $e \mapsto e$, $u \mapsto \frac{1}{2}u$ and $w \mapsto 0$. So the characteristic polynomial is $(\lambda - 1)(\lambda - \frac{1}{2})^{m-1}\lambda^\delta$. $\square$

The above argument shows that $(\lambda - 1)(\lambda - \frac{1}{2})^{m-1}\lambda^\delta$ is the characteristic polynomial of M_x where x is a nonzero idempotent in a Bernstein algebra. If the algebra is train then this polynomial is independent of the choice of x in H. We now reverse this argument by computing the characteristic polynomial of M_x for an arbitrary x with $\sigma(x) = 1$ in a Bernstein algebra and we use the result to get conditions that the algebra be train.

With $\sigma(x) = 1$ let e be the idempotent x^2, and write $x = e + \overline{u} + \overline{w}$. We analyze M_x using the standard decomposition. Let $B_{\overline{w}} : U \to U$ denote the restriction to U of $M_{\overline{w}}$. Writing $\mathcal{A}$ as the direct sum $[e] \oplus W \oplus U$, (note the reversal of order), we get the bloc triangular formula:

$$M_x = \begin{pmatrix} 1 & 0 & 0 \\ \square & 0 & 0 \\ \square & \square & \frac{1}{2}I_U + B_{\overline{w}} \end{pmatrix} \tag{3.5.22}$$

where the entries marked $\square$ are irrelevant to the computation of the characteristic polynomial.

Proof $M_x(e) = x^3$ and $\sigma(x^3) = 1$. Hence, the e-coefficient of $M_x(e)$ is 1. The barideal $\mathcal{B}$ is M_x invariant so the e-component of $M_x(u)$, $M_x(w)$ are 0 for $u \in U$, $w \in W$.

Now let $\tilde{M}_x$ denote the restriction of M_x to the barideal $\mathcal{B} = W \oplus U$. Equation (3.5.15) can be rewritten in operator form as:

$$L\tilde{M}_x = \tilde{M}_x \tag{3.5.23}$$

where L is the projection operator defined by e. From (3.5.23):

$$\text{Im } \tilde{M}_x \subset \text{Im } L = U. \tag{3.5.24}$$

Consequently the W-components of $\tilde{M}_x(u)$ and $\tilde{M}_x(w)$ are 0 for all u in U and w in W. In particular, using $x = e + \overline{u} + \overline{w}$ we have

$$\overline{u}u = 0 \text{ and so}$$

$$\tilde{M}_x(u) = \frac{1}{2}u + \overline{w}u = (\frac{1}{2}I_U + B_{\overline{w}})(u). \tag{3.5.25}$$

$\square$

From (3.5.22) we have immediately:

Lemma 3.5.20 *Let x be an element of unit weight in a Bernstein algebra of type (m, δ). Writing $e = x^2$ and $x = e + \overline{u} + \overline{w}$ with $\overline{u} \in U$ and $\overline{w} \in W$, let $Q(\lambda)$ be the characteristic polynomial of the operator $B_{\overline{w}}$ on U $(B_{\overline{w}}(u) = \overline{w}u)$. The characteristic polynomial of M_x is given by:*

$$\det(\lambda I - M_x) = (\lambda - 1)\lambda^\delta Q(\lambda - \frac{1}{2}). \tag{3.5.26}$$

$\square$

Observe that $x - x^2 = \overline{u} + \overline{w}$ and $\overline{u} = L(\overline{u} + \overline{w}) = 2x^2(x - x^2) = 2x^3 - 2x^2$. So $\overline{w} = x + x^2 - 2x^3$. Thus, to be a train algebra it is necessary and sufficient that all such operators $B_{\overline{w}}$ be nilpotent so that $Q(\lambda) = \lambda^{m-1}$. However, the condition must be verified for all standard decompositions with $e = x^2$, $\sigma(x) = 1$ and $\overline{w} = \overline{w}(x)$. It is not clear that nilpotence of each member of $\{B_w : U \to U$ for $w \in W\}$ is sufficient when we restrict to the decomposition relative to a single idempotent e. However, we do have:

Theorem 3.5.21 *In order that a Bernstein algebra be a train algebra it is necessary and sufficient that the restriction $M_y|\mathcal{B}$ be nilpotent for all y in $\mathcal{B}$, i.e. that Engel's identity hold on the barideal $\mathcal{B}$:*

$$\underbrace{y(y(\ldots(y \ x))\ldots) = 0}_{n-1 \text{ copies of } y} \qquad (y, x \in \mathcal{B}). \tag{3.5.27}$$

Proof In any train algebra M_y is nilpotent on $\mathcal{A}$ for all y in $\mathcal{B}$. A fortiori $M_y|\mathcal{B}$ is nilpotent and (3.5.27) holds. If the algebra is Bernstein then with respect to any standard decomposition B_w is the further restriction of $M_w|\mathcal{B}$ to U. So the nilpotence of $M_y|\mathcal{B}$ for all $y \in \mathcal{B}$ implies that the polynomial $Q(\lambda)$ in (equation 3.5.26) is λ^{m-1}. Thus, the characteristic polynomial is independent of the choice of x in H, i.e. the algebra is train. $\square$

Lemma 3.5.22 *If u_0 is an eigenvector with nonzero eigenvalue for B_w acting on the complexification of U then u_0 annihilates $\mathcal{B}' = \mathcal{B} \cap \mathcal{A}^2$, the barideal of the subalgebra $\mathcal{A}^2$.*

Proof Suppose $u_0 w = \mu u_0$ with $\mu \neq 0$. Then by (3.4.31)

$$\mu u_0(uw) = (u_0 w)(uw) = 0 \qquad\qquad (u \in U)$$

and so $u_0(uw) = 0$. Then by (3.4.12)

$$\mu u u_0 = u(u_0 w) = -u_0(uw) = 0$$

and so u_0 annihilates U. As shown in Theorem 3.4.19 u_0 then automatically annihilates U^2. Hence, u_0 annihilates $\mathcal{B}'$ (c.f. (3.4.18)). $\square$

Theorem 3.5.23 *A Bernstein algebra $(\mathcal{A}, \sigma)$ such that $\mathcal{A}^2 = \mathcal{A}$, or equivalently $\dim N = 0$, is a train algebra.*

Proof We prove that each $B_w : U \to U$ is nilpotent by showing, after complexification, that it has no nonzero eigenvalues. For if $u_0 w = B_w(u_0) = \mu u_0$ with $\mu \neq 0$ then by the lemma above u_0 annihilates $\mathcal{B} \cap \mathcal{A}^2$ which equals $\mathcal{B}$ in this case. So $u_0 w = 0$ contradicting $\mu \neq 0$. $\square$

Corollary 3.5.24 *For any Bernstein algebra $(\mathcal{A}, \sigma)$ the baric subalgebra $\mathcal{A}^2$ is a train algebra.*

Proof By Corollary 3.3.16, $\mathcal{A}_1^2 = \mathcal{A}_1$ for $\mathcal{A}_1 = \mathcal{A}^2$. $\square$

Of course, the two former assertions follow from Theorem 3.5.10.

Example Consider an exceptional Bernstein algebra of the type $(2, n - 2)$, hence $\dim \mathcal{A} = 1$. Then $M_w(u) = \phi(w)u$ where ϕ is a linear form of $w \in W$. The operator M_w is nilpotent if and only if $\phi = 0$ and the algebra is train in this case for $M_y^2 = 0$. If $\phi \neq 0$ then the algebra is not train.

3.6 Duplication of an Algebra

If E_1 and E_2 are vector spaces, recall that the *tensor product $E_1 \otimes E_2$* consists of all linear combinations on symbols $x \otimes y$ with $x \in E_1$ and $y \in E_2$ subject to the relations:

$$(x_1 + x_2) \otimes y = x_1 \otimes y + x_2 \otimes y$$

$$x \otimes (y_1 + y_2) = x \otimes y_1 + x \otimes y_2$$

$$t(x \otimes y) = (tx) \otimes y = x \otimes (ty)$$

for $x, x_1, x_2 \in E_1$; $y, y_1, y_2 \in E_2$ and $t \in \mathbf{R}$. In particular, $E \otimes E$ is the tensor square of the space E.

The *symmetric tensor square $E * E$* is obtained from $E \otimes E$ by imposing the additional relations

$$x * y = y * x \quad (x, y \in E).$$

So by definition we obtain the bilinear mapping:

$$\otimes : E_1 \times E_2 \longrightarrow E_1 \otimes E_2$$

$$(x, y) \longmapsto x \otimes y \quad (x \in E_1, y \in E_2)$$

and the symmetric bilinear mapping:

$$* : E \times E \longrightarrow E * E$$

$$(x, y) \longmapsto x * y \quad (x, y \in E).$$

These bilinear maps are "universal": if $M : E_1 \times E_2 \to E_3$ is a bilinear map then there is a unique homomorphism $M^{\otimes} : E_1 \otimes E_2 \to E_3$ such that

$$M(x, y) = M^{\otimes}(x \otimes y) \quad (x \in E_1, y \in E_2).$$

Similarly if $M : E \times E \to E_3$ is a symmetric bilinear map then there is a unique homomorphism $M^0 : E * E \to E_3$ such that

$$M(x, y) = M^0(x * y) \quad (x, y \in E). \tag{3.6.1}$$

In particular, an algebra structure on $\mathcal{A}$ consists of a choice of multiplication, that is a symmetric bilinear map $\mu : \mathcal{A} \times \mathcal{A} \to \mathcal{A}$ or, equivalently, a homomorphism $\mu^0 : \mathcal{A} * \mathcal{A} \to \mathcal{A}$. Using product notation (3.6.1) becomes:

$$xy = \mu^0(x * y). \tag{3.6.2}$$

The *duplication* of the algebra $\mathcal{A}$ is $\mathcal{A}^D \equiv \mathcal{A} * \mathcal{A}$ together with the multiplication is defined by

$$XY = \mu^\circ(X) * \mu^\circ(Y) \ (X, Y \in \mathcal{A}^D). \qquad (3.6.3)$$

In particular:

$$(x_1 * x_2)(y_1 * y_2) \equiv (x_1 x_2) * (y_1 y_2) \qquad (x_1, x_2, y_1, y_2 \in \mathcal{A}) \qquad (3.6.4)$$

Notice from (3.6.2) and (3.6.3)

$$\mu^\circ(XY) = \mu^\circ(X)\mu^\circ(Y) \qquad (X, Y \in \mathcal{A}^D) \qquad (3.6.5)$$

(where the product on the left side is in $\mathcal{A}^D$ and the right is in $\mathcal{A}$). Thus, $\mu^\circ : \mathcal{A}^D \to \mathcal{A}$ is an algebra homomorphism. Notice that

$$\operatorname{Im} \mu^\circ = \mathcal{A}^2; \ \ker \mu^\circ = \operatorname{ann} \mathcal{A}^D. \qquad (3.6.6)$$

The first is clear from (3.6.2), while (3.6.3) obviously implies $\ker \mu^\circ \subset \operatorname{ann} \mathcal{A}^D$ with obvious equality when $\mu^\circ = 0$. On the other hand, if $\mu^\circ(Y) \neq 0$ then $XY = 0$ implies $\mu^\circ(X) * \mu^\circ(Y) = 0$ and so $\mu^\circ(X) = 0$.

From (3.6.6), μ° factors to yield an algebra isomorphism

$$\mathcal{A}^2 \cong \mathcal{A}^D/\operatorname{ann} \mathcal{A}^D. \qquad (3.6.7)$$

We now relate the idempotents of $\mathcal{A}$ and those of $\mathcal{A}^D$.

Theorem 3.6.1 *The canonical homomorphism $\mu^\circ : \mathcal{A}^D \to \mathcal{A}$ maps the set of idempotents* $\operatorname{Id} \mathcal{A}^D$ *bijectively onto* $\operatorname{Id} \mathcal{A}$.

Proof Any algebra homomorphism maps idempotents to idempotents. In this case $\mu^\circ : \operatorname{Id} \mathcal{A}^D \to \operatorname{Id} \mathcal{A}$ is onto because $(e*e)^2 = e^2 * e^2$ and $\mu^\circ(e*e) = e^2$. So if $e \in \mathcal{A}$ is idempotent $e * e$ is an idempotent mapping onto e. On the other hand, if $X^2 = X$ and $Y^2 = Y$ in $\mathcal{A}^D$ then $\mu^\circ(X) = \mu^\circ(Y)$ implies $X - Y \in \ker \mu^\circ = \operatorname{ann} \mathcal{A}^D$ and so $X - Y = X^2 - Y^2 = (X - Y)(X + Y) = 0$. $\square$

It follows from the proof that the inverse of the map $\mu^\circ : \operatorname{Id} \mathcal{A}^D \to \operatorname{Id} \mathcal{A}$ is the restriction of the quadratic diagonal map $\phi^\circ : \mathcal{A} \to \mathcal{A}^D$ defined by $\phi^\circ(x) = x * x$.

An obvious corollary of the theorem is:

$$\dim \operatorname{Id} \mathcal{A}^D = \dim \operatorname{Id} \mathcal{A}. \qquad (3.6.8)$$

Now let us consider the duplication of various special kinds of algebras.

If $f : A \to \mathbf{R}$ is a linear form then we get a real symmetric bilinear form on A by $(x, y) \mapsto f(x)f(y)$.

The associated linear form on $A * A = A^D$ is denoted f^D. Thus,

$$f^D(x * y) = f(x)f(y). \tag{3.6.9}$$

Technically, f^D is the linear map defined functorially by $f * f : A * A \to \mathbf{R} * \mathbf{R}$ where $\mathbf{R} * \mathbf{R}$ is identified with $\mathbf{R}$ via the isomorphism $\mu_{\mathbf{R}}^{\circ}$ ($\mu_{\mathbf{R}}$ is ordinary multiplication on $\mathbf{R}$). We can notice here that the duplication considered as the correspondence $A \mapsto A^D$, $f \mapsto F^D$ is a functor on the category of algebras. Here $F : A_1 \to A$ is an arbitrary algebra homomorphism, $F^D : A_1^D \to A_2^D$ is the same by the definition $F^D(x * y) = F(x) * F(y)$. In particular, we have the following statement.

Lemma 3.6.2 *f is a character on A if and only if f^D is the composition in $f \circ \mu^{\circ}$ in which case f^D is a character on A^D.*

Proof $(f \circ \mu^{\circ})(x * y) = f(xy)$. By (3.6.9) this equals $f^D(x * y)$ if and only if $f(xy) = f(x)f(y)$, i.e. f is a character. In that case $(f \circ \mu^{\circ})(X) = f^D(X)$ for all X in A^D because the set $\{x * y : x, y \in A\}$ spans A^D. So if f is a character f^D is the composition of algebra homomorphisms and so is an algebra homomorphism as well, i.e. a character on A^D. $\square$

So if (A, σ) is a baric algebra we can use weight σ^D on A^D and so obtain the duplication (A^D, σ^D) naturally as a baric algebra. Notice that σ is not zero on A^2 and so σ^D is not identically zero either. Since $\sigma^D = \sigma \circ \mu^{\circ}$ by the Lemma, μ° is a baric algebra homomorphism.

Example. Let (A, σ) be a unit algebra with $\{e_1, \ldots e_m\}$ a basis normalized by weight. Thus, $e_i e_j = \frac{1}{2}(e_i + e_j)$. Now define $e_{ik} \equiv e_i * c_k$ (so that $e_{ik} = e_{ik}$). Then by (3.6.3)

$$e_{ik} e_{jl} \equiv e_i e_k * e_j e_l = \frac{1}{4}(e_{ij} + e_{il} + e_{kj} + e_{kl}),$$

and $\{e_{ik} : 1 \leq i \leq k \leq m\}$ is a basis for A^D. Thus the duplication of a unit algebra of dimension m is the Mendel m-allele zygotic algebra.

From (3.6.3) and (3.6.2) we prove by induction

$$M_X^{k+1}(Y) = \mu^{\circ}(X) * M_{\mu^{\circ}(X)}^k(\mu^{\circ}(Y)) \ (X, Y \in A^D, k = 0, 1, 2, \ldots). \tag{3.6.10}$$

Theorem 3.6.3 *The duplication (A^D, σ^D) of a train algebra (A, σ) is a train algebra. The nonzero train roots of A^D are included among the train roots of (A, σ).*

Proof Let $\sigma^D(X) = 1$ and so with $x = \mu^\circ(X)$, $\sigma(x) = 1$. Now let $P(\lambda) = \sum_{k=0}^{n}(-1)^{n-k}a_{n-k}\lambda^k$ be a polynomial annihilating $M_x : \mathcal{A} \to \mathcal{A}$ for all $x \in \mathcal{A}$ with $\sigma(x) = 1$. Then from (3.6.10), $\lambda P(\lambda) = \sum_{k=0}^{n}(-1)^{n-k}a_{n-k}\lambda^{k+1}$ annihilates every M_X. So $\mathcal{A}^D$ is a train algebra by Lemma 3.5.2. $\square$

By the same argument using (3.6.10) with $X = Y$ we prove:

Corollary 3.6.4 *If $P(\lambda)$ is a train polynomial for a train algebra $(\mathcal{A}, \sigma)$ then $\lambda P(\lambda)$ is a train polynomial for $(\mathcal{A}^D, \sigma^D)$.*

Theorem 3.6.5 *If $(\mathcal{A}, \sigma)$ is a genetic algebra its duplication $(\mathcal{A}^D, \sigma^D)$ is a genetic algebra.*

Proof If $\{v_0, v_1, \dots v_{n-1}\}$ is a canonical basis for $\mathcal{A}$ more precisely $\mathcal{A}^{(c)}$ then define $v_{ij} \equiv v_i * v_j$ to get a basis via pairs $\{ij : 0 \le i \le j \le n - 1\}$. (Observe that $(\mathcal{A} * \mathcal{A})^{(c)} = \mathcal{A}^{(c)} * \mathcal{A}^{(c)}$ where the tensor product in the latter case is tensor product over the field $\mathbf{C}$). By using reverse lexicographical ordering: $ij < \overline{ij}$ when $j < \overline{j}$ or $j = \overline{j}$ and $i < \overline{i}$, we verify directly that the basis for $\mathcal{A}^D$ is canonical. $\square$

If the train roots of the genetic algebra $(\mathcal{A}, \sigma)$ and $1, \tau_1, \dots \tau_{n-1}$ then those of $(\mathcal{A}^D, \sigma^D)$ are $1, \tau_1, \dots \tau_{n-1}, 0, \dots, 0$ with $\frac{1}{2}n(n-1)$ 0's adjoined. Recall that for a genetic algebra the train roots are the diagonal elements $\lambda_{0i,i}$.

For an arbitrary algebra $\mathcal{A}$ we can regard the pair $(\mathcal{A}_D, \mathcal{A})$ as an algebraic analogue of a two-level population with the "meiosis operator" $\mu^0 : \mathcal{U}^D \to \mathcal{A}$, $\mu^0(x * y) = xy$, and the "fertilization operator" $\phi^0 : \mathcal{A} \to \mathcal{A}^D$, $\phi^0(x) = x * x$. However, there exist such two-level populations that cannot be described by duplication. For example, the Extended Hardy-Weinberg operator (see Section 2.1) has a two-level structure with the identity map on the gamete level. But by the duplication of the unit algebra we get the classical (not Extended) Hardy-Weinberg operator. So we introduce a general definition of *two-level algebra* as a pair of algebras $(\mathcal{A}_\zeta, \mathcal{A}_\gamma)$ connected by the commutative diagram

$$
\begin{array}{ccc}
\mathcal{A}_\zeta & \to & \mathcal{A}_\zeta \\
\phi\uparrow & \searrow\mu & \uparrow\phi \\
\mathcal{A}_\gamma & \to & \mathcal{A}_\gamma
\end{array}
\qquad\qquad (3.6.11)
$$

where the horizontal arrows are squaring, μ is a linear map and ϕ is a quadratic map (here we do not requiere of μ to be surjective). This definition means that $x^2 = (\phi\mu)(x)$ for $x \in \mathcal{A}_\zeta$ and $p^2 = (\mu\phi)(p)$ for $p \in \mathcal{A}_\gamma$. Thefore

$(\mu(x))^2 = \mu(x^2)$, i.e. μ is an algebra homomorphism. Similarly $(\phi(p))^2 = \phi(p^2)$ but ϕ is not an algebra homomorphism because it is non-linear.

The trivial two-level algebra is $(\mathcal{A},\mathcal{A})$ with $\mu = $ id and $\phi(x) = x^2$. The duplication is a non-trivial example. A general construction is following. Let $\mathcal{A}$ be an arbitrary algebra and I be an ideal contained in ann$\mathcal{A}$. Then the pair $(\mathcal{A},\mathcal{A}/I)$ is a two-level algebra where μ is quotient and $\phi : \mathcal{A}/I \to \mathcal{A}$ is defined correctly by $\phi(\mu(p)) = p^2$. In particular we get the two-level algebra $(\mathcal{A},\mathcal{A}/\mathrm{ann}\mathcal{A})$ as a minimal quotient construction.

3.7 Stochastic Spaces

In an evolutionary algebra the unit simplex plays a special role. In order to introduce this kind of structure into our algebraic theory we review some elementary results concerning convex cones in a finite dimensional vector space E over $\mathbf{R}$.

A convex cone, or more simply a *cone*, K in a vector space E is a nonempty subset closed under addition and multiplication by nonnegative scalars. In particular 0 lies in K. For any subset A of E we denote by $< A >$ the smallest cone in E containing A. $< A >$ consists of all linear combinations on elements of A with nonnegative coefficients. For a finite list $\{x_1,\ldots x_n\}$ we write $< x_1,\ldots x_m >$ for the cone $< \{x_1,\ldots,x_m\} >$. In particular, if $x \in E$ then $< x >$ is the ray consisting of all nonnegative multiples of x. K is called a *closed cone* if it is a closed subset of E.

Proposition 3.7.1 *In the vector space E let K, K_1 etc. denote cones.*

(a) $K_1 \cap K_2$, $-K \equiv \{-x : x \in K\}$, $K_1 + K_2 \equiv \{x_1 + x_2 : x_1 \in K_1$ and $x_2 \in K_2\}$, $K_1 - K_2 \equiv K_1 + (-K_2)$ are cones in E.

(b) The dual $K^+ = \{f \in E^* : f(x) \geq 0$ for all $x \in K\}$ is a closed cone in the dual space E^*.

$$(K_1 + K_2)^+ = K_1^+ \cap K_2^+$$

$$(K_1 \cap K_2)^+ \supset K_1^+ + K_2^+ \tag{3.7.1}$$

(c) The closure $\overline{K}$ is a closed cone and $K^+ = (\overline{K})^+$. Using the canonical identification of E with E^{**}, we have

$$K^{++} = \overline{K}. \tag{3.7.2}$$

(d) A cone K is a linear subspace if and only if $K = -K$. In particular, the subspace $[K]$ spanned by K, is the cone $K - K$, while $K \cap (-K)$ is the largest subspace contained in K.

When K is a subspace, the dual cone K^+ is the annihilator of K in E^*. So

$$K^+ \cap (-K)^+ = [K]^\perp$$

$$[K^+] \subset (K \cap -K)^\perp, \tag{3.7.3}$$

with equality in the latter case when K is closed.

(e) K is called a polyhedral *cone if it is generated by a finite set, i.e. if $K = < x_1, \ldots x_m > = < x_1 > + \cdots + < x_n >$ for some list $\{x_1, \ldots, x_m\}$ in E. The minimal m is called the* order *of the cone K and it is denoted by* ord K.

Every polyhedral cone is closed. $K_1 + K_2$ and $K_1 \cap K_2$ are polyhedral if K_1 and K_2 are. A closed cone K is polyhedral if and only if its dual K^+ is polyhedral.

(f) A cone K is called solid *if it satisfies the following three equivalent conditions:*

$$[K] = E;$$

Int $K \neq \emptyset$ (where Int denotes interior with respect to E); $\tag{3.7.4}$

$$K^+ \cap (-K)^+ = 0.$$

A cone is called pointed *if it contains no nontrivial subspace, i.e. if $K \cap (-K) = 0$. K is solid if and only if $\overline{K}$ is solid if and only if K^+ is pointed. A closed cone is pointed if and only if its dual is solid. In general,*

$$\text{Int } K^+ = \{f : f(x) > 0 \ \forall x \in \overline{K} \backslash 0\} \tag{3.7.5}$$

where the interior is taken with respect to E^*.

A closed, pointed, solid cone is called proper. *A closed cone K is a proper cone if and only if K^+ is proper.*

Proof It is easy to check that the constructions of (a) yield cones. Since 0 lies in any cone $K_1 \cup K_2 \subset K_1 + K_2$ and the former set generates the latter, i.e.

$$< K_1 \cup K_2 > = K_1 + K_2. \tag{3.7.6}$$

The dual map operator is monotone, i.e.

$$K_1 \subset K_2 \ \Rightarrow \ K_2^+ \subset K_1^+. \tag{3.7.7}$$

(3.7.1) follows easily from monotonicity and the observation that for a linear form f, $f(x_1+x_2) = f(x_1)+f(x_2)$ is nonnegative for all x_1, and x_2 in K_1, K_2 exactly when $f(x_1)$ and $f(x_2)$ are each nonnegative for all x_1, x_2 in K_1, K_2.

That the closure of a cone is still a cone and that K^+ is closed are obvious. Also, $K^+ = (\overline{K})^+$ by continuity of linear forms.

For (3.7.2) the inclusion $K \subset K^{++}$ is clear and so $\overline{K} \subset K^{++}$ because the latter is closed. For the other direction, suppose the vector y does not lie in $\overline{K}$. By the Separating Hyperplane Theorem applied to $\{y\}$ and $\overline{K}$ compact and closed convex sets, respectively, there exists $f \in E^*$ and $t \in \mathbf{R}$ such that $f(x) \geq t$ for all x in $\overline{K}$ while $f(y) < t$. As the zero vector lies in $\overline{K}$, $0 \geq t$ and so $f \in K^+$ with $f(y) < 0$. Hence, $y \notin K^{++}$.

We notice that the following sharpening of (3.7.1) holds:

$$(\overline{K}_1 \cap \overline{K}_2)^+ = \overline{K_1^+ + K_2^+}. \tag{3.7.8}$$

To prove this, apply (3.7.2) and the first part of (3.7.1):

$$\overline{K_1^+ + K_2^+} = (K_1^+ + K_2^+)^{++} = (K_1^{++} \cap K_2^{++})^+ = (\overline{K}_1 \cap \overline{K}_2)^+.$$

If $K = -K$ then for $f \in K^+$ and $x \in K$ $f(x) \geq 0$ and $f(-x) \geq 0$ imply that $f(x) = 0$. So K^+ is the annihilator of the subspace K. The remainder of (d) easily follows from (b).

The proofs, which we omit, that a polyhedral cone is closed and has a polyhedral dual can be found, for example, in Gale (1960) section 2.5. It then follows conversely, that if K^+ is polyhedral then $K^{++} = \overline{K} = K$ is polyhedral. If K_1 and K_2 are polyhedral then $K_1 + K_2$ is obviously polyhedral and the intersection $K_1 \cap K_2$ is polyhedral by duality (3.7.8).

Any nonempty open subset spans E and so Int $K \neq \emptyset$ implies $[K] = E$. Conversely, if K spans E then it contains a basis $\{e_1, \dots e_n\}$ and the positive linear combinations of $e_1, \dots e_n$ form an open subset of K. The subspace $[K]$ is all of E if and only if its annihilator is 0 in E^*. So by (3.7.3) $[K] = E$ if and only if $K^+ \cap (-K)^+ = 0$. Thus, K is solid if and only if K^+ is pointed. As $K^+ = (\overline{K})^+$, this occurs if and only if $\overline{K}$ is solid. By applying this result to K^+ we see that a closed cone is pointed if and only if its dual is solid in E^*. Thus, a closed cone is proper if and only if its dual is proper.

Choose a norm for E and let S be the unit sphere $= \{x \in E : \| x \| = 1\}$. Because every element of $\overline{K} \backslash 0$ is a positive multiple of a vector in $S \cap \overline{K}$, K^+ is $\{f \in E^* : f(x) \geq 0 \text{ for all } x \in S \cap \overline{K}\}$, and f is positive on $\overline{K} \backslash 0$ if and only if it is positive on $S \cap \overline{K}$. Because $S \cap \overline{K}$ is closed and bounded it is compact and so $\{f \in E^* : f(x) > 0 \text{ for all } x \in S \cap \overline{K}\}$ is open. On the other hand if $f \in K^+$ and $f(x) = 0$ for some $x \in S \cap \overline{K}$ then choose $g \in <x>^+ \setminus 0$. For all $t > 0$, $f - tg$ is outside of $\overline{K}^+ = K^+$. So f does not lie in the interior of K^+. This proves (3.7.5). $\square$

Lemma 3.7.2 *Let K be a closed cone such that K^+ is solid (i.e. K is pointed). If $f \in \text{Int } K^+$ denote by H the unit hyperplane $f^{-1}(1)$, parallel to the kernel of f. The set $\Delta = K \cap H$ is a compact convex set which generates K.*

Conversely, let Δ be a compact convex set such that the affine subspace generated by Δ does not contain 0. The cone K generated by Δ is then a closed, pointed cone.

Proof There exists f in $\text{Int } K^+$ and by (3.7.5) f is positive on $K \cap S$ where S is the unit sphere with respect to some norm. As a closed bounded set $K \cap S$ is compact and so $f \geq \epsilon$ on $K \cap S$ for some $\epsilon > 0$. If $x \in \Delta = K \cap f^{-1}(1)$ then

$$1 = f(x) = \parallel x \parallel f(x/ \parallel x \parallel) \geq \parallel x \parallel \epsilon.$$

Δ is thus a subset of the ball of radius ϵ^{-1}. As a closed, bounded set it is compact. It is clearly convex. Obviously, Δ generates K.

For the converse, let $e \in \Delta$. We can restate the assumption by saying that Δ is a compact subset of the translate via e of a proper subspace W of E. Choose $f \in E^*$ such that f lies in $W^\perp$ and $f(e) = 1$. Then $\Delta \subset f^{-1}(1)$. Because Δ is convex the cone $K = < \Delta >$ can be described $\{tx : t \geq 0$ and $x \in \Delta\}$. Because $\Delta \subset f^{-1}(1)$ each nonzero $y \in K$ can be described uniquely as $y = f(y)x$ with $f(y) > 0$ and $x = y/f(y)$ in Δ. So if $\{y_n\}$ is a sequence in K converging to $y \neq 0$ then $f(y_n)$ converges to $f(y)$ and $f(y) > 0$ because $x_n = y_n/f(y_n)$ is bounded. Then $\{x_n\}$ converges to x in Δ. As $y = f(y)x$ it lies in K and so K is closed. Because f is positive in $K \setminus 0$, $f \in \text{Int } K^+$ and so K is pointed. $\square$

Corollary 3.7.3 *Let Δ be a compact convex subset of a vector space E. Assume that the affine subspace generated by Δ, H, does not contain the zero vector. The following four conditions are equivalent*

1. *dimension H = dimension $E - 1$.*

2. *Δ spans E, i.e. $[\Delta] = E$.*

3. *There exists a unique σ in E^* such that $\sigma = 1$ on Δ.*

4. The cone K generated by Δ is a proper cone, i.e. $K = <\Delta>$ is a closed, pointed, solid cone.

A pair (E, Δ) where Δ satisfies the above conditions is called a stochastic space *and Δ is then called the* basic body *of the space. The dual cone K^+ is then a proper cone in E^* and*

$$K^+ = \{f \in E^* : f(x) \geq 0 \; ; \forall x \in \Delta\}.$$

$$\text{Int } K^+ = \{f \in E^* : f(x) > 0 \;\; \forall x \in \Delta\}. \tag{3.7.9}$$

Let $\overset{\circ}{\Delta}$ denote the interior with respect to H of the basic body, Δ. Then we have:

$$\overset{\circ}{\Delta} = H \cap \text{Int } K = \{x \in \Delta : f(x) > 0 \;\; \forall f \in K^+ \backslash 0\}. \tag{3.7.10}$$

Proof H is the translate by e in Δ of H_0 a proper subspace of E. Clearly,

$$\dim H = \dim H_0 = \dim E^* - \dim H_0^\perp. \tag{3.7.11}$$

So H is a hyperplane if and only if $H_0^\perp$ is one dimensional in which case $\sigma \in H_0^\perp$ such that $\sigma(e) = 1$ is unique.

By the lemma the cone K is always closed and pointed. It is solid exactly when Δ spans E which says

$$\dim E = \dim(H_0 \oplus [e]) = \dim H + 1.$$

This proves the equivalence among the conditions (1)—(4). K^+ is then proper by part (f) of Propoposition 3.7.1 Furthermore, as $K \backslash 0$ consists of positive multiples of Δ, f in E^* is nonnegative on K if and only if it is nonnegative on Δ and it is positive on $K \backslash 0$ if and only if it is positive on Δ. So (3.7.9) follows from (3.7.5).

Because K is closed, $K^{++} = K$, (3.7.2), and so by (3.7.5) applied to the dual

$$\text{Int } K = \{x \in K : f(x) > 0 \;\; \forall f \in K^+ \backslash 0\}. \tag{3.7.12}$$

Finally, $H \cap \text{Int } K$ is open relative to H and is contained in $H \cap K = \Delta$. So $H \cap \text{Int } K \subset \overset{\circ}{\Delta}$. On the other hand, if $x \in \Delta \backslash \text{Int } K$ then by (3.7.12) $f(x) = 0$ for some $f \in K^+ \backslash 0$. As $f \neq 0$ we can choose $y \in E$ with $f(y) < 0$. Define the path

$$x(t) = \frac{1}{1 + t\sigma(y)}(x + ty)$$

for $t \geq 0$ and small enough that $1 + t\sigma(y) > 0$. It is a continuous path in $H = \sigma^{-1}(1)$, $x(0) = x$ and $f(x(t)) < 0$ for $t > 0$. Hence, $x \notin \overset{\circ}{\Delta}$. $\square$

A stochastic space (E, Δ) is naturally a baric space with the weight function $\sigma \in E^*$ uniquely characterized by $\Delta \subset \sigma^{-1}(1)$. In particular, σ is in the interior of the dual of the proper cone K generated by Δ. Conversely, a stochastic space structure on E can be described by the choice of a proper cone K and an element σ of Int K^+. For then by Lemma 3.7.2 and its corollary (E, Δ) is a stochastic space with $\Delta = K \cap \sigma^{-1}(1)$.

If (E, Δ) is the stochastic space then each vector $x \in \Delta$ is called *stochastic*. Each vector $x \in K$ is called *nonnegative* (*positive* if $x \in$ Int$K$).

Using the proper cone K we can define a partial order on E by $x \geq y$ when $x - y \in K$. The ordering is reflexive, antisymmetric and transitive because $0 \in K$, $K \cap (-K) = 0$ and $K + K = K$, respectively. By definition, $x \geq y$ implies $x + z \geq y + z$. We write $x > y$ for $x \geq y$ and $x \neq y$, i.e. $x - y \in K \backslash 0$ and write $x >> y$ for $x - y \in$ Int K. So $x \geq 0 \Leftrightarrow x$ is nonnegative; $x > 0 \Leftrightarrow x$ is nonnegative and $x \neq 0$; $x >> 0 \Leftrightarrow x$ is positive. Obviously, $x > 0 \Leftrightarrow \sigma(x) > 0$ and $x/\sigma(x) \in \Delta$; $x \gg 0 \Leftrightarrow \sigma(x) > 0$ and $x/\sigma(x) \in \overset{\circ}{\Delta}$. So if $x \geq 0$ and $\sigma(x) = 0$ then $x = 0$.

A homomorphism $h : E_1 \to E_2$ between stochastic spaces (E_1, Δ_1) and (E_2, Δ_2) is called *nonnegative* if $h(K_1) \subset K_2$, or equivalently if $x \geq y$ implies $h(x) \geq h(y)$. It is called *positive* if $h(K_1 \backslash 0) \subset$ Int K_2, or equivalently if $x > y$ implies $h(x) >> h(y)$. Thus, for a positive homomorphism $K_1 \cap$ ker $h = 0$. h is called *stochastic homomorphism* if $h(\Delta_1) \subset \Delta_2$. Clearly, h is stochastic if and only if it is nonnegative and baric. If a stochastic homomorphism h is bijective and its inverse map is stochastic as well then h is called a *stochastic isomorphism* . In this case $h\Delta_1 = \Delta_2$.

By identifying the reals as the stochastic space $(\mathbf{R}, 1)$ these definitions apply to the linear maps in E^*. Thus, a form f in E^* is nonnegative when $f(x) \geq 0$ for $x \in K$, i.e. when $f \in K^+$. By (3.7.5) the positive forms are those in Int K^+. Clearly, f is a positive form exactly when $f(x) > 0$ for all $x \in \Delta$. The weight form σ is the only stochastic form since σ is characterized by $\sigma(x) = 1$ for $x \in \Delta$.

By the Krein-Milman Theorem a compact, convex set Δ is the closed convex hull of its extremal points. The rays through these points are the *extremal rays* of K. Every extremal ray does not belong to a sum of other rays of K by definition. The stochastic space (E, Δ) is called *polyhedral* if Δ is a polyhedron, the convex hull of a finite set, or equivalently if K is a polyhedral cone. The stochastic space is polyhedral precisely when the set of extremal rays is finite. A list $\{u_1,\ldots,u_m\}$ such that $\{<u_1>,\ldots,<u_m>\}$ is the set of extremal rays, is called a *cone basis* for the polyhedral cone K. It is a *normalized* cone basis if $\sigma(u_i) = 1$, that is, $u_i \in \Delta$, for $i = 1,\ldots m$. As the cone K is solid a cone basis spans E. It is linearly independent as well, i.e. a basis for E, when $m = \dim E$, in which case Δ is a simplex and we call (E,Δ) a *simplicial stochastic space* and the cone K is called *simplicial* as well.

The motivating example of a stochastic space is *arithmetical n-space*: $(\mathbf{R}^n,\Delta)$ with Δ the standard unit simplex:

$$\Delta = \{x \in \mathbf{R}^n : \sum_{i=1}^{n} x_i = 1,\ x_i \geq 0 \text{ for } i = 1,\ldots,n\}.$$

The associated cone is $\mathbf{R}^n_+$, the set of vectors with nonnegative components, and the associated ordering is the usual one. The standard basis $\{e_1,\ldots,e_n\}$ is a normalized cone basis. The weight s, characterized by $s(e_i) = 1$ for $i = 1,\ldots m$ is given by

$$s = \sum_{i=1}^{n} e_i^*$$

where $\{e_1^*,\ldots,e_n^*\}$ is the dual basis in $\mathbf{R}^{n*}$ defined by $e_j^*(e_i) = \delta_{ij}$. So we have $s(x) = \sum_{i=1}^n x_i$.

Our primary interest is in simplicial spaces. However, the somewhat larger class of polyhedral spaces is closed under subspace and quotient space constructions while the simplicial family is not. For example, by mapping the standard basis onto a normalized cone basis we see that every polyhedral stochastic space is the image of an arithmetical stochastic space under a stochastic map. For a simplicial space there is a stochastic isomorphism with the arithmetical stochastic space of the same dimension.

The remainder of this section is taken up with procedures for recognizing whether a polyhedral space is in fact simplicial. For example, by (e) of Proposition 3.7.1 a closed cone K is polyhedral if and only if its dual K^+ is polyhedral. By using the isomorphism with arithmetical space we see that a simplicial cone K has a simplicial dual ($(\mathbf{R}^n_+)^+$ has $\{e_1^*,\ldots,e_n^*\}$ as cone basis) and so K is simplicial if and only if K^+ is.

For a simplicial space we will use the name *canonical basis* for the normalized cone basis listed in a fixed order, $\{e_1,\ldots,e_n\}$. As it is a basis any vector x in E can be written uniquely as $x = \sum_{i=1}^n \alpha_i e_i$ and $x \in K$ if and only if $\alpha_i \geq 0$ for $i = 1,\ldots,n$. For any such nonnegative vector x the *support* of x, denoted supp(x), is the set of basis vectors (or their corresponding indices) for which the coefficient is positive, i.e. $i \in \operatorname{supp}(x)$ if and only if $\alpha_i > 0$. We easily check the properties (for $x, y \in K$):

$$x = \sum \{\alpha_i e_i : i \in \operatorname{supp}(x)\}$$

$$\begin{aligned}
&\operatorname{supp}(x + y) = \operatorname{supp}(x) \cup \operatorname{supp}(y) \\
&\operatorname{supp}(\lambda x) = \operatorname{supp}(x) \qquad (\lambda > 0) \\
&\operatorname{supp}(x) = \emptyset \Leftrightarrow x = 0.
\end{aligned} \tag{3.7.13}$$

Using the canonical basis we can consider the dual space E^* as simplicial with canonical basis $\{e_1^*, \ldots, e_n^*\}$. Then

$$\Delta^* = \{f \in E^* : \sum_{k=1}^{n} f(e_k) = 1, \ f(e_k) \geq 0 \text{ for } k = 1, \ldots, n\}$$

and

$$K^* = K^+ = \{f \in E^* : f(e_k) \geq 0 \text{ for } k = 1, \ldots, n\}.$$

The weight on (E^*, Δ^*) is $\overline{e} = \sum_{k=1}^{n} e_k$ by duality ($E^{**} = E$).

With E_1 a subspace of a stochastic spase E we call E_1 a *stochastic subspace* if $\dim \Delta_1 = \dim E_1 - 1$ where Δ_1 is the intersection $E_1 \cap \Delta$. In that case (E_1, Δ_1) is a stochastic space in its own right and we will also call this pair a *stochastic subspace*, writing $(E_1, \Delta_1) \subset (E, \Delta)$ in that case. If s_1 is the weight for (E_1, Δ_1) then clearly $s_1 = s | E_1$ and so (E_1, s_1) is a baric subspace of (E, s). In general the cone K_1 generated by $E_1 \cap \Delta$ is $E_1 \cap K$ and so E_1 is a stochastic subspace if and only if $E_1 \cap K$ generates E_1. In particular, if $E_1 \cap \overset{\circ}{\Delta} \neq \emptyset$ then E_1 is a stochastic subspace. The embedding of the stochastic subspace E_1 into E is stochastic.

By (e) of Proposition 3.7.1 a stochastic subspace of a polyhedral space is polyhedral (K polyhedral implies $K_1 = E_1 \cap K$ polyhedral). We introduce ways to recognize when a stochastic subspace of a simplicial space is simplicial.

With respect to a canonical basis $\{e_1, \ldots e_n\}$ for a simplicial space (E, Δ), a list of vectors $\{u_1, \ldots u_m\}$ is called a *Gaussian system* if for $k = 1, \ldots, m$

$$u_k = \alpha_k e_{i_k} + \sum_{i \neq i_1, \ldots, i_m} \tau_{ik} e_i \qquad (\alpha_k \neq 0) \tag{3.7.14}$$

with $\{e_{i_1}, \ldots, e_{i_m}\}$ a subset of the basis. $\{u_1, \ldots u_m\}$ is called a *nonnegative Gaussian system* if the vectors $u_1, \ldots, u_m$ are nonnegative. It is called a *stochastic system* if the vectors $u_1, \ldots u_m$ are stochastic. We can always normalize a nonnegative system to get a stochastic system replacing u_k by $s(u_k)^{-1}u_k$. Clearly, every nonempty subset of the canonical basis is a stochastic Gaussian system. Every Gaussian system is obviously linear independent.

Because the basis is canonical u_k is a nonnegative vector if and only if the coefficients α_k and τ_{ik} are nonnegative.

Theorem 3.7.4 *Let (E, Δ) be a simplicial space and E_1 be a subspace of E. With $\Delta_1 = E_1 \cap \Delta$, (E_1, Δ_1) is a simplicial space (and, in particular, is a stochastic subspace of (E, Δ)) if and only if E_1 admits as a basis a nonnegative Gaussian system in E.*

The result is obviously valid for the coordinate subspace $E_1 = [e_{i_1}, \dots e_{i_m}]$ with basic body the face of Δ with vertices $e_{i_1}, \dots, e_{i_m}$. Note that for this stochastic subspace $E_1 \cap \overset{\circ}{\Delta} = \emptyset$.

Proof Assume that (E_1, Δ_1) is a simplicial subspase and let $\{u_1, \dots, u_m\}$ be its cone basis. The vectors $\{u_1, \dots, u_m\}$ are nonnegative and the set is linearly independent.

Now we prove that, as nonnegative vectors in E,

$$\operatorname{supp} u_k \not\subset \bigcup_{j \neq k} \operatorname{supp}(u_j) \qquad (1 \leq k \leq m). \qquad (3.7.15)$$

Were this not true then for some $\lambda > 0$ small enough

$$v = \sum_{j \neq k} u_j - \lambda u_k$$

would be nonnegative and so $v \in E_1 \cap K = K_1$. But the u_k-coefficient of v is negative and so v cannot be a nonnegative vector in the simplicial space (E_1, Δ_1) with canonical basis $\{u_1, \dots, u_m\}$.

So for $k = 1, \dots m$ we can choose

$$e_{i_k} \in \operatorname{supp}(u_k) \backslash \bigcup_{j \neq k} \operatorname{supp}(u_j).$$

By definition, the e_{i_k}-coordinate of u_k is positive but the e_{i_j}-coordinates of u_k are all 0 for $j \neq k$. So (3.7.14) holds and $\{u_1, \dots u_m\}$ is the required nonnegative Gaussian basis.

Conversely, if $\{u_1, \dots u_m\}$ is a nonnegative Gaussian basis for E_1 then, following (3.7.14) $\alpha_k > 0$ and $\tau_{ik} \geq 0$. As all u_k's lie in $K_1 = K \cap E_1$ it follows that K_1 generates E_1 and so E_1 is a stochastic subspace. To show that (E_1, Δ_1) is simplicial it is enough to prove that $\{u_1, \ldots, u_m\}$ generates the cone K_1.

We can expand any $x \in E_1$ with respect to the basis $\{u_1, \ldots, u_m\}$.

$$x = \sum_{k=1}^{m} \xi_k u_k.$$

What we must prove is that $x \in K_1$ implies $\xi_k \geq 0$ for $k = 1, \ldots, m$. By (3.7.14)

$$x = \sum_{k=1}^{m} \xi_k \alpha_k e_{i_k} + \sum{}^{*}$$

where $\sum^*$ is a linear combination of the e_i's with $i \neq i_1, \ldots, i_m$. Because $x \in K_1 \subset K$ its canonical coefficients in E are nonnegative. In particular, $\xi_k \alpha_k \geq 0$. $\alpha_k > 0$ then implies $\xi_k \geq 0$ as well. $\square$

Remark Because the normalized cone basis is unique up to ordering we see that up to normalization (to get the vectors stochastic) and up to numbering the nonnegative Gaussian basis is unique for a simplicial subspace.

There is also a dual space test whether a subspace is simplicial.

A list of linear forms $\{p_1, \ldots, p_l\}$ is a Gaussian system on E^* if for $h = 1, \ldots, l$

$$p_h = \beta_h e_{j_h}^* - \sum_{j \neq j_1, \ldots, j_l} \gamma_{jh} e_j^* \qquad (\beta_h \neq 0) \qquad (3.7.16)$$

with $\{e_{j_1}^*, \ldots, e_{j_l}^*\}$ l distinct elements of the dual basis. $\{p_1, \ldots p_l\}$ is called a *hyperbolic Gaussian system* of forms on a simplicial space (E, Δ) if the basis is the canonical one and the coefficients β_h and γ_{jh} are all nonnegative. Note that the subtraction in (3.7.16) means that the forms p_h are usually not nonnegative, i.e. they don't lie in K^+.

Theorem 3.7.5 *Let (E, Δ) be a simplicial space and E_1 be a subspace of E. E_1 is a simplicial subspace if and only if its annihilator $E_1^\perp$ in E^* admits a hyperbolic Gaussian basis.*

Equivalently, a subspace E_1 is a simplicial subspace exactly when it is the solution space of a system of equations, in canonical coordinates, of the form

$$x_{j_h} = \sum_{j \neq j_1, \ldots, j_l} \gamma_{jh} x_j \qquad (3.7.17)$$

with the coefficients γ_{jh} nonnegative.

Proof If E_1 is a simplicial subspace with nonnegative Gaussian basis of the form (3.7.14), normalize so that $\alpha_h = 1$ for $h = 1, \ldots m$ and let $\{j_1, \ldots j_l\}$ be the complement of the set $\{i_1, \ldots, i_m\}$ in $\{1, \ldots n\}$. Then define

$$p_h = e_{j_h}^* - \sum_{k=1}^{m} \tau_{j_h k} e_{i_k}^* \tag{3.7.18}$$

Clearly, $p_h(u_k) = 0$ for all h and k. Hence, $\{p_1, \ldots, p_l\}$ is a hyperbolic Gaussian basis for $E_1^\perp$, because $\dim E_1^\perp = n - \dim E_1 = l$.

The converse just reverses the argument. Starting from (3.7.16) normalized by $\beta_h = 1$ we let $\{i_1, \ldots, i_m\}$ be the complement of $\{j_1, \ldots, j_l\}$ in $\{1, \ldots, n\}$ and define

$$u_k = e_{i_k} + \sum_{h=1}^{l} \gamma_{i_k h} e_{j_h} \tag{3.7.19}$$

to get a nonnegative Gaussian basis of E_1. $\square$

Corollary 3.7.6 *Let $p = \sum_{i=1}^{n} \alpha_i e_i^*$ and E_1 be its annihilating hyperplane, the solution space of*

$$\sum_{i=1}^{n} \alpha_i x_i = 0. \tag{3.7.20}$$

Assuming $p \neq 0$ so that not all the α_i's vanish, E_1 is a simplicial subspace if and only if either exactly one $\alpha_i \neq 0$ in which case E_1 is a coordinate hyperplane, or exactly one α_i has sign opposite to the remaining nonvanishing coordinates. Assume that at least two of the α_i's are nonzero so that E_1 is not a coordinate hyperplane. Then E_1 is a stochastic subspace if and only if not all the coefficients have the same sign.

Proof First we can rewrite (3.7.20) in the form (3.7.17). Further, if all the nonzero coefficients have the same sign then $p(x) = 0$ and x nonnegative imply $\alpha_i x_i = 0$ for all i. So $K \cap E_1 = \{x : x \geq 0 \text{ and } x_i = 0 \text{ when } \alpha_i \neq 0\}$. This set spans a subspace of codimension greater than one and so $K \cap E_1$ does not generate E_1.

For the converse suppose that the basis has been ordered so that $\alpha_i > 0$ for $1 \leq i \leq r$, $\alpha_i < 0$ for $r + 1 \leq i \leq r + \rho$ and $\alpha_i = 0$ for $r + \rho < i$. Then define x by $x_i = (\sum_{j=1}^{r} \alpha_j)^{-1}$ for $1 \leq i \leq r$, $x_i = (\sum_{j=r+1}^{r+\rho} |\alpha_j|)^{-1} = -(\sum_{j=r+1}^{r+\rho} \alpha_j)^{-1}$ for $r + 1 \leq i \leq r + \rho$ and $x_i = 1$ for $r + \rho < i$. Clearly, x is a positive vector in E_1, i.e. $E_1 \cap \text{Int } K \neq \emptyset$. So E_1 is a stochastic subspace. $\square$

Using the ordering described in the above proof we can describe the cone basis for $K_1 = K \cap E_1$ in the general stochastic case. Define

$$v_{ij} = \frac{e_i}{\alpha_i} - \frac{e_j}{\alpha_j} = \frac{e_i}{\alpha_i} + \frac{e_j}{|\alpha_j|} \qquad (3.7.21)$$

for pairs ij with $1 \le i \le r$ and $r + 1 \le j \le r + \rho$. We claim that the set of such v_{ij}'s together with the basis vectors e_k for $k > r + \rho$ form a cone basis for K_1.

Note first that they are all nonnegative and satisfy (3.7.20), i.e. they do lie in $K_1 = K \cap E_1$. The vectors e_k are extremal in K_1 because they are even extremal in K. If $v_{ij} = x + y$ with $x, y \in E_1$ and $x, y \ge 0$ then as $\operatorname{supp}(x) \cup \operatorname{supp}(y) = \operatorname{supp}(v_{ij}) = \{e_i, e_j\}$ we can write $x = ae_i + be_j$ and $y = ce_i + de_j$ and from (3.7.20) we see that $a\alpha_i + b\alpha_j = 0 = c\alpha_i + d\alpha_j$. So $\{(a,b),(c,d)\}$ is a linearly dependent set in $\mathbf{R}^2$ and so x and y are proportional vectors. Thus, v_{ij} is extremal in K_1.

If we let K_2 be the cone generated by the set $\{v_{ij}, e_k : 1 \le i \le r < j \le r + \rho < k\}$ then we have shown $K_2 \subset K_1$. For the reverse inclusion it suffices to show $K_2^+ \subset K_1^+$, i.e. if a linear form f is nonnegative on the list of vectors then it is nonnegative on E_1. Write

$$f(x) = \sum_{i=1}^{r} \beta_i x_i + \sum_{j=r+1}^{r+\rho} \beta_j x_j + \sum_{k=r+\rho+1}^{n} \beta_k x_k.$$

Because $f \ge 0$ on v_{ij} and e_k we have

$$\frac{\beta_i}{\alpha_i} - \frac{\beta_j}{\alpha_j} \ge 0 \text{ and } \beta_k \ge 0.$$

As every $\beta_i/\alpha_i \ge$ every β_j/α_j we can find a number λ such that for all i, j:

$$\frac{\beta_i}{\alpha_i} \ge \lambda \ge \frac{\beta_j}{\alpha_j} \qquad (1 \le i \le r < j \le r + \rho).$$

Because $\alpha_i > 0 > \alpha_j$ we can rewrite this as

$$\beta_i \ge \lambda\alpha_i \text{ and } \beta_j \ge \lambda\alpha_j.$$

Hence, for $x \ge 0$ in E_1,

$$f(x) \ge \lambda p(x) = 0.$$

Thus, the order of K_1 is given by

$$\operatorname{ord} K_1 = r\rho + n - r - \rho = n - 1 + (r - 1)(\rho - 1). \qquad (3.7.22)$$

We recover the first part of Corollary 3.7.6 because ord $K_1 = n - 1$ if and only if $r = 1$ or $\rho = 1$.

This computation gives us a lot of stochastic subspaces which are not simplicial. The simplest is the solution space of

$$x_1 + x_2 - x_3 - x_4 = 0 \qquad (3.7.23)$$

in $\mathbf{R}^4$ with normalized cone basis given by

$$\{\frac{e_1 + e_3}{2}, \frac{e_1 + e_4}{2}, \frac{e_2 + e_3}{2}, \frac{e_2 + e_4}{2}\}. \qquad (3.7.24)$$

Our third approach characterizes simplicial subspaces as the image of a nonnegative projection. Recall that a linear operator A on a stochastic space (E, Δ) is called nonnegative (written $A \geq 0$) when $A(K) \subset K$. If the space is simplicial then $A \geq 0$ is equivalent to the nonnegativity of all the entries of the matrix of A with respect to the canonical basis.

Theorem 3.7.7 *Let (E, Δ) be a simplicial space and E_1 be a subspace of E. E_1 is a simplicial subspace if and only if there exists a nonnegative projection map P on E with $E_1 = \operatorname{Im} P$.*

Proof Assume that E_1 is a simplicial subspace with $\{u_1, \ldots u_m\}$ the nonnegative Gaussian basis satisfying (3.7.14) normalized by $\alpha_k = 1$ ($k = 1, \ldots m$). Define $P(e_{i_k}) = u_k$ ($k = 1, \ldots, m$) and $P(e_i) = 0$ ($i \neq i_1, \ldots i_m$). The definition extends to define a unique linear operator and clearly $P(u_k) = u_k$ from (3.7.14). So P is a projection with image $E_1 = [u_1, \ldots, u_m]$. Because $u_k \geq 0$ for all k, $P \geq 0$.

Conversely, suppose P is a nonnegative projection with image E_1. Observe first that $K_1(\equiv K \cap E_1)$ is the image $P(K)$. For by nonnegativity $P(K) \subset K$ and so $P(K) \subset K \cap \operatorname{Im}(P) = K_1$, while $K_1 \subset \operatorname{Im}(P)$ and $K_1 \subset K$ imply $K_1 = P(K_1) \subset P(K)$. Because K spans E, $K_1 = P(K)$ spans $E_1 = P(E)$.

Now let $\{u_1, \ldots, u_m\}$ be a cone basis for K_1 We prove that it is a Gaussian system. First, we show

$$\operatorname{supp}(u_i) \cap \operatorname{supp}(u_j) \subset \ker P \qquad (i \neq j). \qquad (3.7.25)$$

For if $e_k \in \operatorname{supp}(u_i)$ then $\operatorname{supp}(P(e_k)) \subset \operatorname{supp}(P(u_i)) = \operatorname{supp}(u_i)$ and so for small enough λ, $u_i \pm \lambda P(e_k)$ lies in K_1. But the midpoint of these two vectors is the extremal u_i. So $P(e_k)$ is a multiple of u_i. If $e_k \in \operatorname{supp}(u_j)$, too, then $P(e_k)$ is a multiple of u_j as well. But the two distinct extremal rays meet only at the origin, i.e. $P(e_k) = 0$.

On the other hand, $\text{supp}(u_i) \subset \ker P$ would imply $u_i = P(u_i) = 0$ which it is not. So (3.7.25) implies

$$\text{supp}(u_i) \not\subset \bigcup_{j \neq i} \text{supp}(u_j). \tag{3.7.26}$$

Just as in the proof of Theorem 3.7.4, using (3.7.15), this implies $\{u_1, \ldots u_m\}$ is a Gaussian system. $\square$

Now we can give a general description of a nonnegative projection.

Theorem 3.7.8 *A linear operator P on a simplicial space (E, Δ) is a nonnegative projection if and only if it can be written:*

$$P = \sum_{k=1}^{m} f_k \otimes u_k \tag{3.7.27}$$

with $m = \text{rank } P$, $\{u_1, \ldots, u_m\}$ a nonnegative Gaussian system and $\{f_1, \ldots, f_m\}$ a system of nonnegative linear forms biorthogonal to the $\{u_1, \ldots, u_m\}$, that is, $f_k(u_i) = \delta_{ki}$ $(i, k = 1, \ldots, m)$.

Proof If P satisfies (3.7.27) then clearly $P \geq 0$ (because each $u_k \geq 0$ and $f_k \geq 0$) and

$$P(u_i) = \sum_{k=1}^{m} f_k(u_i)u_k = u_i$$

implies P is a projection on the image $[u_1, \ldots, u_m]$.

Conversely, if P is a nonnegative projection we can choose a Gaussian basis $\{u_1, \ldots, u_m\}$ for its image E_1 which is also a cone basis for $P(K) = K_1 = K \cap E_1$ (see the proof of Theorem 3.7.7). We define $\{\overline{f}_1, \ldots, \overline{f}_m\}$ to be the coordinate functions on E_1 with respect to the basis $\{u_1, \ldots, u_m\}$, i.e. $\{\overline{f}_1, \ldots, \overline{f}_m\}$ is the basis of E_1^* dual to $\{u_1, \ldots, u_m\}$. Extend each $\overline{f}_k$ to E by $f_k(x) = \overline{f}_k(P(x))$. By definition of the coordinate functions

$$P(x) = \sum_{k=1}^{m} \overline{f}_k(P(x))u_k = \sum_{k=1}^{m} f_k(x)u_k,$$

i.e. (3.7.27) holds as does the biorthogonality condition. Finally, if $x \geq 0$ then $P(x) \geq 0$, i.e. $P(x) \in K_1$ and because E_1 is a simplicial subspace the coordinates $\overline{f}_k(P(x))$ are nonnegative for all k. Thus, $f_k \geq 0$ $(k = 1,\ldots,m)$ as well. $\square$

A nonnegative linear operator A on a simplicial space is called *indecomposable* if it has no nontrivial invariant coordinate subspace. For example, a positive operator is indecomposable.

Corollary 3.7.9 *An indecomposable nonnegative projection P has rank one and so is of the form*

$$P = f \otimes u \tag{3.7.28}$$

with $f >> 0$, $u >> 0$ and $f(u) = 1$. So P is positive.

Proof The coordinate subspaces spanned by $\text{supp}(u_k)$ are invariant from (3.7.27) ($e_i \in \text{supp}(u_k)$ implies $\text{supp}(P(e_i)) \subset \text{supp}(P(u_k)) = \text{supp}(u_k)$) and are distinct by (3.7.25). So $m = 1$ and $\text{supp}(u_1)$ is the entire basis for E, i.e. $u_1 >> 0$. Also the annihilator of the subspace spanned by $\text{supp}(f_1)$ (with respect to basis $\{e_1^*,\ldots,e_n^*\}$) is an invariant coordinate subspace as well. So $f >> 0$. $\square$

Recall that a linear operator A on a stochastic space (E, Δ) is called stochastic if $A(\Delta) \subset \Delta$, or equivalently, if A is nonnegative and $A^*s = s$ where s is the weight form on E. We now use Theorem 3.7.8 to characterize stochastic projections on a simplicial space.

It is convenient to introduce on the dual space E^* of a stochastic space the norm:

$$\| f \| = \max\{|f(x)| : x \in \Delta\}. \tag{3.7.29}$$

Since every vector in Δ is a convex combination of the extremal points we easily check that

$$\| f \| = \max_j\{|f(e_j)|\} \tag{3.7.30}$$

where $\{e_j\}$ is the set of extremal points of Δ, i.e. the cone basis for K normalized by $s(e_j) = 1$. Because this set spans E the functional $\| \, . \, \|$ is really a norm on E^*, that is, $\| f \| = 0$ only if $f = 0$. Clearly, $\| s \| = 1$ and for f_1, f_2 nonnegative forms we have

$$\| f_1 + f_2 \| \geq \max(\| f_1 \|, \| f_2 \|). \tag{3.7.31}$$

A system of nonnegative forms $\{p_1, \ldots p_m\}$ is called a *partition of the weight s* if

$$\sum_{i=1}^{m} p_i = s \text{ and } \| p_i \| = 1 \text{ for } i = 1,\ldots,m. \tag{3.7.32}$$

Lemma 3.7.10 *Let (E, Δ) be a simplicial space. $\{p_1, \ldots, p_m\}$ is a partition of the weight s if and only if*

$$p_i = e_{j_i}^* + \sum_{k \neq j_1, \ldots, j_m} \pi_{ik} e_k^* \qquad (i = 1, \ldots m) \qquad (3.7.33)$$

where $\{e_{j_1}^*, \ldots, e_{j_m}^*\}$ is a subset of the dual basis for the canonical basis and the matrix (π_{ik}) is stochastic, i.e.

$$\sum_{i=1}^{m} \pi_{ik} = 1 \text{ and } \pi_{ik} \geq 0 \ (i = 1, \ldots, m; k \neq j_1, \ldots, j_m). \qquad (3.7.34)$$

A fortiori, partition of the weight s is a nonnegative Gaussian system in E^* normalized by $\| p_i \| = 1$.

Proof Given (3.7.32) for each basis vector e_j

$$\sum_{l=1}^{m} p_l(e_j) = s(e_j) = 1 \qquad (3.7.35)$$

and for each $i = 1, \ldots m$

$$\max_j \{p_i(e_j)\} = \| p_i \| = 1. \qquad (3.7.36)$$

So for each i there is a basis vector e_{j_i} such that $p_i(e_{j_i}) = 1$. Because $p_l(e_{j_i}) \geq 0$ for each l, (3.7.35) implies $p_l(e_{j_i}) = 0$ if $l \neq j_i$. Formula (3.7.33) then follows. $p_i \geq 0$ then implies the coefficients $\pi_{ik} \geq 0$, while the equations of (3.7.34) follow from (3.7.32).

The converse is obvious. $\square$

Theorem 3.7.11 *A linear operator P on a simplicial space (E, Δ) is a stochastic projection if and only if it can be written*

$$P = \sum_{k=1}^{m} p_k \otimes u_k \qquad (3.7.37)$$

with $m = \text{rank } P$, $\{p_1, \ldots, p_m\}$ a partition of the weight s and $\{u_1, \ldots, u_m\}$ is a system of stochastic vectors with pairwise disjoint supports and biorthogonal to the $\{p_1, \ldots, p_m\}$, that is, $p_i(u_k) = \delta_{ik} \ (i, k = 1, \ldots, m)$.

Proof Just as in the proof of Theorem 3.7.8, P is a nonnegative projection onto E_1 spanned by $\{u_1,\ldots,u_m\}$. Furthermore, in this case $P^*(s) = s$ because

$$s(u_k) = \sum_{i=1}^{m} p_i(u_k) = 1 \qquad (k = 1,\ldots m)$$

and so

$$P^*(s) = \sum_{k=1}^{m} s(u_k)p_k = \sum_{k=1}^{m} p_k = s.$$

Now assume P is a stochastic projection. Then apply Theorem 3.7.8 to the dual projection P^* which is nonnegative on E^*. We obtain equation (3.7.37) for P with $\{p_1,\ldots,p_m\}$ a nonnegative Gaussian system biorthogonal to $\{u_1,\ldots u_m\}$. Hence,

$$p_i = \lambda_i e_{j_i}^* + \sum_{k \neq j_1\ldots j_m} \pi_{ik} e_k^*$$

with π_{ik}, λ_i nonnegative. By replacing each pair (p_k, u_k) by $(s(u_k)p_k, s(u_k)^{-1}u_k)$ we obtain that each u_k is stochastic. Because P is stochastic, i.e. $P^*(s) = s$ and each $s(u_k) = 1$ we see that

$$\sum_{i=1}^{m} p_i = s.$$

Applying this to each e_{j_i} we see that $\lambda_i = 1$ and so $1 = p_i(e_{j_i}) \leq \| p_i \| \leq \| s \| = 1$. Thus $\{p_1,\ldots,p_m\}$ is a partition of s.

Finally, for the supports of the u_i's we note that $p_k(u_i) = 0$ for $k \neq i$ and so $\operatorname{supp}(p_k) \cap \operatorname{supp}(u_i) = \emptyset$. (Here we are regarding the supports of p_k and u_i both as subsets of the index set $\{1,\ldots,n\}$ rather than as subsets of the canonical basis and its dual.) Hence

$$\operatorname{supp}(s - p_i) \cap \operatorname{supp}(u_i) = (\cup\{\operatorname{supp}(p_k) : k \neq i\}) \cap \operatorname{supp}(u_i) = \emptyset.$$

So if $i \neq j$, $\operatorname{supp}(u_i) \cap \operatorname{supp}(u_j)$ is disjoint from $\operatorname{supp}(s - p_i) \cup \operatorname{supp}(s - p_j) = \operatorname{supp}(s + (s - p_i - p_j))$. But the latter set is $\{1,\ldots,n\}$ because

$$s + (s - p_i - p_j) \geq s >> 0.$$

Hence, $\operatorname{supp}(u_i) \cap \operatorname{supp}(u_j) = \emptyset$ and the supports are pairwise disjoint. $\square$

Remark Note that the proof that P was a stochastic projection did not use $\| f_i \| = 1$ or the disjoint supports condition.

We can extend the notion "partition of s" to a general stochastic space (E, Δ). Let $\{p_i\}$ be a cone basis for the dual cone K^+ normalized by $\| p_i \| = 1$ (this is a finite set in the polyhedral case). Any element of K^+ is a nonnegative linear combination on the set $\{p_i\}$. In particular the weight s can be written as a finite nonnegative combination

$$s = \sum_i \lambda_i p_i. \tag{3.7.38}$$

which is called *the expansion of the weight* . In the simplicial case this expansion is unique but even in the general polyhedral case it need not be.

Lemma 3.7.12 *In every expansion of the weight s of the form (3.7.38) all the coefficients satisfy $\lambda_i \leq 1$. Furthermore, for each p_k there exists an expansion with $\lambda_k = 1$.*

Proof By (3.7.31):

$$\| s \| = \| \sum_i \lambda_i p_i \| \geq \lambda_k \| p_k \|$$

and so $\| s \| = \| p_k \| = 1$ imply $\lambda_k \leq 1$.

On the other hand, s is identically 1 on Δ while $\| p_k \| = 1$ implies $p_k \leq 1$ on Δ. Hence the form $s - p_k$ is nonnegative and so can be written as a nonnegative linear combination

$$s - p_k = \sum_i \gamma_i p_i,$$

i.e.

$$s = \sum_i \lambda_i p_i \text{ with } \lambda_k = 1 + \gamma_k.$$

With this expansion $\lambda_k = 1 + \gamma_k \geq 1$ and by the above general result $\lambda_k \leq 1$. $\square$

From this lemma we can test which subspaces F of the dual space E^* are simplicial subspaces when (E, Δ) is a simplicial space.

We are especially interested in subspaces which contain the weight s. Since s then lies in $F \cap \text{Int } \Delta$ it follows that such a subspace is always a stochastic, polyhedral subspace.

Theorem 3.7.13 *Let (E, Δ) be a simplicial space with F a linear subspace of E^*. F is a simplicial subspace and containing the weight form s as well if and only if it admits a basis $\{p_1, \ldots, p_m\}$ which provides a partition of s in E^*.*

Proof Assume F is simplicial with canonical basis $\{p_1, \ldots, p_m\}$ normalized by $\| p_i \| = 1$. Because $s \in F$ we can expand s as in (3.7.38) and the expansion is unique because F is simplicial. The lemma then implies that all λ_i's $= 1$. So $\{p_1, \ldots, p_m\}$ provides a partition of s.

Conversely, a partition of s is a nonnegative Gaussian system in E^* by Lemma 3.7.10. So $s \in F$ and the subspace F is simplicial by Theorem 3.7.4. $\square$

We now turn to quotient spaces. Recall that for a baric space (E, s) a subspace E_0 is an ideal subspace if $E_0 \subset \ker s$. So, in particular, this definition applies to a stochastic space (E, Δ). The quotient baric space $(E/E_0, s_0)$ with quotient map $h : E \to E/E_0$ and s_0 characterized by $s = h^* s_0$, is a stochastic space by defining $\Delta_0 = h(\Delta)$. Δ_0 clearly spans E/E_0 and s_0 is identically 1 on Δ_0. So $(E/E_0, \Delta_0)$ defines the quotient space $(E, \Delta)/E_0$ by Corollary 3.7.3, and the quotient map h is a stochastic linear operator. Clearly, the cone K_0 on Δ_0 is the image under h of the cone K on Δ. Also the quotient is polyhedral if (E, Δ) is. As usual extra conditions are required for the quotient to be simplicial.

Recall that the dual of the surjection h is the injection $h^* : (E/E_0)^* \to E^*$ mapping isomorphically onto the annihilator $E_0^\perp$.

Lemma 3.7.14 *The canonical isomorphism $h^* : (E/E_0)^* \to E_0^\perp$ identifies the dual cone K_0^+ with the intersection $K^+ \cap E_0^\perp$. In particular, this intersection generates $E_0^\perp$.*

Proof Because $s \in E_0^\perp \cap \text{Int } K^+$, $E_0^\perp \cap K^+$ generates $E_0^\perp$. If $f = h^*(f_0)$ then f_0 is nonnegative on $K_0 = h(K)$ if and only if f is nonnegative on K. Thus, $f_0 \in K_0^+$ if and only if $f \in K^+ \cap E_0^\perp$. $\square$

Theorem 3.7.15 *Let (E, Δ) be a simplicial space and E_0 be an ideal subspace. The quotient space $(E, \Delta)/E_0$ is simplicial if and only if E_0 admits a hyperbolic Gaussian basis. This is true, in turn, if and only if the annihilator $E_0^\perp$ admits a nonnegative Gaussian basis in E^*.*

Proof Because (E, Δ) is simplicial we can use the canonical dual basis to regard E^* as a simplicial space as well. The subspace $E_0^\perp$ is a simplicial subspace of E^* if and only if it admits a nonnegative Gaussian basis (Theorem 3.7.4) and if and only if its annihilator E_0 in $E^{**} = E$ admits a hyperbolic Gaussian basis (Theorem 3.7.5). Finally, the cone K_0 in $(E, \Delta)/E_0$ is simplicial if and only if its dual K_0^+ is simplicial in $(E/E_0)^*$, but by the lemma

h^* identifies K_0^+ with the intersection cone for the stochastic subspace $E_0^\perp$ of E^*. $\square$

We conclude with a general duality theorem between simplicial and ideal subspaces.

Theorem 3.7.16 *Let (E, Δ) be a stochastic space and E_0 be an ideal subspace. The quotient space $(E, \Delta)/E_0$ is simplicial if and only if there exists a simplicial subspace E_1 of E such that $E = E_0 \oplus E_1$, and such that $\Delta \subset E_0 + \Delta_1$ where $\Delta_1 = \Delta \cap E_1$ is the basic body of E_1.*

Proof Write $(\tilde{E}, \tilde{\Delta})$ for the quotient $(E, \Delta)/E_0$. If $\tilde{\Delta}$ is a simplex with vertices $\tilde{e}_1, \ldots, \tilde{e}_m$, choose $e_1, \ldots e_m$ in Δ so that $h(e_i) = \tilde{e}_i$, $i = 1, \ldots m$. Because $\{\tilde{e}_1, \ldots, \tilde{e}_m\}$ is a linearly independent set so is $\{e_1, \ldots, e_m\}$ and so these form the vertices of a simplex $\Delta_1 \subset \Delta$. Let E_1 be the subspace spanned by Δ_1. As $\{e_1, \ldots, e_m\}$ is a basis for E_1 and $\{\tilde{e}_1, \ldots, \tilde{e}_m\}$ is a basis for $\tilde{E}$ h maps E_1 isomorphically onto $\tilde{E}$. Hence, $E = E_0 \oplus E_1$. Also, $\Delta_1 \subset \Delta \cap E_1$. On the other hand, if $x \in \Delta \cap E_1$ then $x = \sum_{i=1}^m \lambda_i e_i$ with $1 = s(x) = \sum \lambda_i$ and $h(x) = \sum_{i=1}^m \lambda_i \tilde{e}_i$. Because $h(x) \in \tilde{\Delta}$ and $\tilde{\Delta}$ is a simplex with vertices $\tilde{e}_1, \ldots, \tilde{e}_m$ the coefficients λ_i are nonnegative and so $x \in \Delta_1$. Because $\Delta_1 = \Delta \cap E_1$, E_1 is a simplicial subspace. The condition $\Delta \subset E_0 + \Delta_1$ is equivalent to $h(\Delta) = h(\Delta_1)$ which is true because both are $\tilde{\Delta}$.

Conversely, suppose $E = E_0 \oplus E_1$ with $\Delta_1 = E_1 \cap \Delta$ a simplex on $\{e_1, \ldots, e_m\}$. Let $\tilde{e}_i = h(e_i)$ and note that $\{\tilde{e}_1, \ldots \tilde{e}_m\}$ is linearly independent because $\{e_1, \ldots, e_m\}$ is and h restricts to a linear isomorphism of E_1 onto $\tilde{E}$. Thus, $\{\tilde{e}_1, \ldots, \tilde{e}_m\}$ spans a simplex which is the image of Δ_1 under h. The additional condition $\Delta \subset \Delta_1 + E_0$ says that $\tilde{\Delta} = h(\Delta)$ is the same as $h(\Delta_1)$. Hence, $\tilde{\Delta}$ is a simplex. $\square$

Corollary 3.7.17 *Let (E, Δ) be a simplicial space and E_0 an ideal subspace with $h : (E, \Delta) \to (\tilde{E}, \tilde{\Delta}) \equiv (E, \Delta)/E_0$ the quotient map. $(\tilde{E}, \tilde{\Delta})$ is simplicial if and only if there exists a nonnegative operator $g \colon \tilde{E} \to E$ such that $hg = \mathrm{Id}_{\tilde{E}}$. g is then a stochastic operator. Furthermore, $P = gh$ is a stochastic projection operator defined on (E, Δ).*

Proof Assume g is a nonnegative homomorphism with $hg = id_{\tilde{E}}$. Because $h^*(\tilde{s}) = s$ (where $s, \tilde{s}$ are the weights) $g^*(s) = (hg)^*(\tilde{s}) = \tilde{s}$. As g is nonnegative and preserves weight it is stochastic. Hence, $P = gh$ is stochastic as well and $P^2 = ghgh = gh = P$. So P is a stochastic projection. By Theorem 3.7.8 the image E_1 of P is a simplicial subspace. Since

$hP = hgh = h$, ker $P = $ ker $h = E_0$. Hence, $E_0 \oplus E_1 = E$. With $\Delta_1 = E_1 \cap \Delta$ we have $\Delta_1 = P(\Delta_1) \subset P(\Delta) \subset E_1 \cap \Delta = \Delta_1$ because P is the identity on E_1 and is stochastic. So $\Delta_1 = P(\Delta)$ and $h(\Delta_1) = hP(\Delta) = h(\Delta)$. Applying the previous theorem we have that $(\tilde{E}, \tilde{\Delta})$ is simplicial.

Conversely, if $(\tilde{E}, \tilde{\Delta})$ is simplicial choose E_1 the splitting simplicial subspace of (E, Δ) given by the theorem. The restriction of h to E_1 is an isomorphism of E_1 on $\tilde{E}$ because $E_0 = $ ker h and $E_0 \oplus E_1 = E$. Define g to be the inverse map of this restriction. Clearly, $hg = \text{id}_{\tilde{E}}$. Because $h(\Delta_1) = h(\Delta) = \tilde{\Delta}$ it follows that $g(\tilde{\Delta}) = \Delta_1$ and so g is stochastic. $\square$

3.8   Stochastic Algebras

A *stochastic algebra* $(\mathcal{A}, \Delta)$ is a stochastic space with an algebra structure such that the basic body Δ is closed under multiplication. We can, of course, apply all the stochastic space concepts to such an algebra. For example, a *simplicial algebra* is a stochastic algebra whose underlying stochastic space is simplicial.

If K is the associated cone in a stochastic algebra it is clear, by normalization to Δ, that K is closed under multiplication, i.e. $x, y \geq 0$ implies $xy \geq 0$. Also, the weight form s is a character, that is, the equation

$$s(xy) = s(x)s(y) \tag{3.8.1}$$

follows for all x, y in $\mathcal{A}$ because K spans $\mathcal{A}$ and it clearly holds for x, y in Δ. Thus, a stochastic algebra is naturally a baric algebra.

Example An evolutionary algebra $\mathcal{A}_V$ for a population with unit simplex Δ and evolutionary operator V is the motivating example $(\mathcal{A}_V, \Delta)$ of a simplicial algebra. In fact, any simplicial algebra can be regarded as an evolutionary algebra by interpreting the structure constants of the multiplication with respect to the canonical basis as the inheritance coefficients. From now on we will use the phrase evolutionary algebra for any simplicial stochastic algebra.

Example Let A be a stochastic operator on a stochastic space $(\mathcal{A}, \Delta)$, with weight s. The A induced algebra structure:

$$xy = \frac{s(y)A(x) + s(x)A(y)}{2} \tag{3.8.2}$$

makes $(\mathcal{A}, \Delta)$ a stochastic algebra. If $(\mathcal{A}, \Delta)$ is simplicial and A has matrix (a_{ij}) with respect to the canonical basis then the structure constants are given by

$$p_{ik,j} = \frac{a_{ji} + a_{jk}}{2}. \tag{3.8.3}$$

In general, the evolutionary operator $V(x) = x^2$ coincides with A on the basic body Δ. In particular, we can get the dynamics of any finite Markov chain as the evolutionary operator of the A induced simplicial algebra where A is the matrix of the chain.

From any stochastic space we obtain the unit stochastic algebra by using the A induced algebra structure with A the identity.

A constant algebra is a stochastic algebra if and only if the basic body lies in the unit hyperplane and contains the defining vector.

If $(\mathcal{A}, \Delta)$ is a stochastic algebra then a subalgebra $\mathcal{A}_1 \subset \mathcal{A}$ is called a *stochastic subalgebra* if it is a stochastic subspace, i.e. $\Delta_1 = \mathcal{A}_1 \cap \Delta$ generates $\mathcal{A}_1$. Then $(\mathcal{A}_1, \Delta_1)$ is a stochastic algebra and s_1, the restriction of s to $\mathcal{A}_1$, is nonzero. In particular, $\mathcal{A}_1$ is automatically a baric subalgebra.

Example The subalgebra $\mathcal{A}^2$ is a stochastic subalgebra. To see that $\mathcal{A}^2 \cap K$ generates $\mathcal{A}^2$ we let $z = xy \in \mathcal{A}^2$. Because K generates $\mathcal{A}$, i.e. $K - K = \mathcal{A}$, we can write $x = x_1 - x_2$ and $y = y_1 - y_2$ with $x_1, x_2, y_1, y_2 \in K$. Then $z = z_1 - z_2$ with $z_1 = x_1 y_1 + x_2 y_2$, $z_2 = x_1 y_2 + x_2 y_1 \in \mathcal{A}^2 \cap K$.

Similarly, if $\mathcal{A}_1$ is a stochastic subspace of a stochastic algebra $(\mathcal{A}, \Delta)$ and $\Delta_1 = \mathcal{A}_1 \cap \Delta$ is closed under multiplication, then $\mathcal{A}_1$ is a stochastic subalgebra.

Example In an evolutionary algebra, a list $\{f_1, \ldots, f_m\}$ of invariant linear forms which is a hyperbolic Gaussian system defines a stochastic subalgebra $\cap_i \ker f_i$ which is simplicial by Theorem 3.7.5.

It can happen that in an evolutionary algebra no hyperbolic invariant linear forms exists (the constant algebra is such an example). So it is interesting, and useful, to have a condition for the existence of such forms.

An evolutionary algebra is called *J-irreducible* if the invariant linear forms distinguish the members of the canonical basis, that is, for $i \neq j$ there exists $f \in J$ such that $f(e_i) \neq f(e_j)$, or, equivalently, $e_i - e_j \notin J^\perp$ for $i \neq j$. For example, if there exists f in J whose coefficients with respect to the canonical basis are all distinct then the algebra is J-irreducible. The unit algebra is J- irreducible.

Lemma 3.8.1 *If $(\mathcal{A}, \Delta)$ is a J-irreducible evolutionary algebra then there exists a hyperbolic invariant linear form with all coefficients distinct.*

Proof Let $H_{ij} = \{f \in J : f(e_i) = f(e_j)\}$. By J- irreducibility each H_{ij} is a proper subspace of J ($i \neq j$) and so there exists $g \in J \backslash \cup_{i \neq j} H_{ij}$. Clearly the coefficients of g are all distinct. By rearranging the basis we can assume $g(e_1) > g(e_2) > \ldots > g(e_n)$. Choose λ in the interval $(g(e_2), g(e_1))$ and let $f = g - \lambda s$. Because $s(e_i) = 1$ for all i, f is a hyperbolic invariant form with all coefficients distinct. $\square$

Corollary 3.8.2 *In a J-irreducible evolutionary algebra there exists a simplicial subalgebra of codimension one.*

Proof If f is a hyperbolic invariant form the ker f is the required subalgebra. $\square$

Recall that a coordinate subspace $\mathcal{A}_1$ of a simplicial space is a simplicial subspace with $\mathcal{A}_1 \cap \Delta$ the face Γ whose vertices are the canonical vectors lying in $\mathcal{A}_1$. For an evolutionary algebra a face Γ of Δ is called an *invariant face* when $V(\Gamma) \subset \Gamma$. If the canonical basis $\{e_1, \ldots, e_n\}$ is so ordered that $e_i \in \Gamma$ exactly for $i \leq r$ then in terms of the structure constants $p_{ik,j}$ the criterion that Γ be invariant is:

$$p_{ik,j} = 0 \quad (i, k \leq r < j). \tag{3.8.4}$$

Thus, if a face Γ is invariant then the corresponding coordinate subspace is a simplicial subalgebra. It is called a *coordinate subalgebra* and is adequate to the biological notion of *subpopulation* .

Theorem 3.8.3 *Let Γ be a face of the simplex Δ in $(\mathcal{A}, \Delta)$. If there exists x in the* Int Γ *(interior with respect to the affine subspace Γ generates) such that $x^2 \in \Gamma$ then Γ is an invariant face. In particular, if* Int Γ *contains an idempotent then Γ is an invariant face.*

    Proof If Γ is spanned by $e_1, \ldots e_r$ and $x \in$ Int Γ then $x = \sum_{i=1}^{r} \xi_i e_i$ with $\xi_i > 0$. Because $x^2 \in \Gamma$ we have that for $j > r$ the e_j- coordinate of x^2:

$$\sum_{i,k=1}^{r} \xi_i \xi_k p_{ik,j} = 0.$$

But the terms of this sum are nonnegative and so must all vanish.

Because $\xi_i \xi_k > 0$, (3.8.4) follows. $\square$

    Remark: If for $x \in$ int Γ, $x^2 \in$ int Γ then int Γ itself is invariant with respect to V and so is closed under multiplication.

    When $\mathcal{A}_1$ is a subalgebra of a stochastic algebra $(\mathcal{A}, \Delta)$, the convex set $\Delta_1 = \mathcal{A}_1 \cap \Delta$, when nonempty, is closed under multiplication even when it does not generate $\mathcal{A}_1$ and so $\mathcal{A}_1$ is not a stochastic subspace. So in any case, $\tilde{\mathcal{A}}_1$, the subspace of $\mathcal{A}$ spanned by Δ_1, is a stochastic subalgebra and $\tilde{\mathcal{A}}_1 \subset \mathcal{A}_1$. $\tilde{\mathcal{A}}_1$ clearly includes any stochastic subalgebra of $\mathcal{A}$ which is contained in $\mathcal{A}_1$.

    On the other hand, if $\mathcal{A}_1$ is a subalgebra of an evolutionary algebra $(\mathcal{A}, \Delta)$ and $\Delta_1 = \mathcal{A}_1 \cap \Delta$ we define the *carrier* of Δ_1, $\Gamma(\Delta_1)$, to be the smallest face of Δ containing Δ_1, or equivalently the largest face Γ of Δ such that $\Delta_1 \cap$ Int $\Gamma \neq \emptyset$. Thus, the vertices $\{e\}$ of Γ are those in the support of some x in Δ_1. The equivalence between the two conditions follows from $\mathrm{supp}(\frac{x+y}{2}) = \mathrm{supp}(x) \cup \mathrm{supp}(y)$ for $x, y \in \Delta$. Because Δ_1 is invariant,

$\Delta_1 \cap \text{Int } \Gamma(\Delta_1) \neq \emptyset$ and $\Delta_1 \subset \Gamma(\Delta_1)$, it follows from Theorem 3.8.3 that $\Gamma(\Delta_1)$ is an invariant face and so generates a coordinate subalgebra.

We can combine these two constructions for any subalgebra $\mathcal{A}_1$ of an evolutionary algebra $(\mathcal{A}, \Delta)$ provided that $\mathcal{A}_1 \cap \Delta \neq \emptyset$.

An evolutionary algebra $(\mathcal{A}, \Delta)$ is called *indecomposable* if it contains no nontrivial coordinate subalgebras, or, equivalently, if the simplex Δ has no proper invariant face. In the case when the algebra structure is induced from a stochastic operator A this condition says that A has no invariant coordinate subspace, the usual indecomposability condition for nonnegative matrices described in the preceding section.

Theorem 3.8.4 *An evolutionary algebra is indecomposable if and only if every idempotent $x > 0$ is positive, i.e. $x > 0$ and $x^2 = x$ implies $x >> 0$.*

Proof $x^2 = x$ and $s(x) \neq 0$ imply $s(x) = 1$ and so $x > 0$ and $x^2 = x$ imply $x \in \Delta$. If such an x is not positive then the carrier, the face Γ such that $x \in \text{Int } \Gamma$, is a proper face of Δ and by Theorem 3.8.3 Γ is invariant.

Conversely, if Γ is an invariant face then by the Brouwer Fixed Point Theorem $V : \Gamma \to \Gamma$ has a fixed point x. $x > 0$ and $x^2 = x$ but as $x \in \Gamma$, a proper face of Δ, x is not positive. $\square$

Corollary 3.8.5 *If $\{p_{ik,j}\}$ is the family of structure constants for an evolutionary algebra and the nonnegative matrix $\{p_{ii,j} : i, j = 1, \ldots, n\}$ is indecomposable (in particular if $p_{ii,j} > 0$ for $i, j = 1, \ldots, n$) then the algebra is indecomposable.*

Proof If $x = \sum_{i=1}^{r} \xi_i c_i$ with $r < n$ (all ξ_i's ≥ 0) is an idempotent then

$$x = x^2 \geq \sum_{i=1}^{r} \xi_i^2 e_i^2$$

and so $p_{ii,k} = 0$ when $i \leq r < k$. $\square$

A different method for getting invariant faces comes from:

Theorem 3.8.6 *If f is an invariant linear form on an evolutionary algebra then*

$$\Gamma_f \equiv \{x : x \in \Delta \text{ and } f(x) = \max\{f(y); y \in \Delta\}\} \tag{3.8.5}$$

is an invariant face of Δ.

In other words, if a population admits a nontrivial linear conservation law then there exists a nontrivial subpopulation.

Proof Writing x in Δ as $\sum_{\text{supp } x} \xi_i e_i$ we have

$$f(x) = \sum_{\text{supp } x} \xi_i f(e_i) \leq \sum_{\text{supp} x} \xi_i M = M,$$

where M is the maximum value of f on Δ. As each $\xi_i > 0$ we see that $f(x) = M$ if and only if $f(e_i) = M$ for each e_i in the support of x, i.e. the maximum value of the linear form occurs at vertices. Thus, Γ_f is a face of Δ. By invariance of f, $f(x^2) = f(x)$ for x in Δ and so Γ_f is an invariant face. $\square$

Remark Note that if f is nonnegative then Γ_f is the set of points x of Δ at which $f(x) = \| f \|$.

Corollary 3.8.7 *If $(\mathcal{A}, \Delta)$ is an indecomposable evolutionary algebra (e.g. if $(p_{ii,j})$ is an indecomposable matrix for the structure constants $(p_{ik,j})$) then there is no nontrivial invariant linear form, i.e.* $\dim J = 1$.

Proof $\Gamma_f = \Delta$ exactly when f is a multiple of s. $\square$

If I is an ideal in a stochastic algebra $(\mathcal{A}, \Delta)$ we call I a *stochastic ideal* if $I \subset \ker s$. Thus I is a baric ideal for the associated baric algebra and an ideal subspace for the associated stochastic space. The quotient map $h : \mathcal{A} \to \tilde{\mathcal{A}} = \mathcal{A}/I$ is a baric algebra homomorphism and with $\tilde{\Delta} = h(\Delta)$, it is a stochastic homomorphism of stochastic spaces. Because Δ is closed under the multiplication in $\mathcal{A}$ and h is a homomorphism it follows that $\tilde{\Delta}$ is closed under multiplication in $\tilde{\mathcal{A}}$. Thus, $(\tilde{\mathcal{A}}, \tilde{\Delta}) = (\mathcal{A}, \Delta)/I$ is a stochastic algebra. Recall the usual examples: the barideal $\mathcal{B} = \ker s$, the annihilator $J^{\perp}$ of the space of invariant forms, and the annihilator ann $\mathcal{A}$ of the algebra.

The most important of these is $I = J^{\perp}$. We denote by $(\mathcal{A}_J, \Delta_J)$ the quotient $(\mathcal{A}, \Delta)/J^{\perp}$. We label by Q the associated cone in $\mathcal{A}_J$ and by C the cone of nonnegative invariant forms, i.e. $J \cap K^+$. By Lemma 3.7.14, C corresponds under the identification $h^* : (\mathcal{A}/J^{\perp})^* \to J$ with dual cone Q^+. The cone C will be a fundament tool in the study, in chapter 4, of stationary genetic structures. We will see that its cone basis (normalized by $\| \|$) can be interpreted as a complete list of conservation laws corresponding to genes.

If now $e_1, \ldots, e_n$ is the canonical basis of the evolutionary algebra $(\mathcal{A}, \Delta)$ we denote the image $h(e_i)$ in $\mathcal{A}_J$ by E_i for $i = 1, \ldots, n$. Δ_J is the convex hull of $\{E_1, \ldots, E_n\}$ and this set includes the extrema of Δ_J. If E is an extremal point of Δ_J then we define

$$\Gamma_E = \{x : x \in \Delta \text{ and } h(x) = E\}. \tag{3.8.6}$$

Γ_E is a face of Δ with vertices those e_i's such that $E_i = E$.

On the other hand we have defined for each $f \in J$ the invariant face Γ_f by (3.8.5). We will call such faces *J-induced*. Among these the most important are the minimal elements (with respect to inclusion).

Lemma 3.8.8 *The face Γ_f is minimal amongst the J-induced faces if and only if no two of its vertices are distinguished by any invariant linear form.*

Proof Given invariant linear forms f and g we can choose $\epsilon > 0$ small enough that $f + \epsilon g$ on each of the vertices of Γ_f takes on larger values than on any of the remaining vertices. Hence, $\Gamma_{f+\epsilon g}$ is a subset of Γ_f. It is a proper subset if g distinguishes two distinct vertices of Γ_f.

Conversely, if $\Gamma_g \overset{\subset}{\neq} \Gamma_f$ then g takes on larger values on the vertices of Γ_g than on the remaining vertices of Γ_f. $\square$

Theorem 3.8.9 *The correspondence $E \mapsto \Gamma_E$ yields a bijection between the extremal points of the basic body Δ_J of $(\mathcal{A}, \Delta)/J^\perp$ and the set of minimal J-induced faces of the basic simplex Δ.*

Proof If E is an extremal point for Δ_J we can apply the Separating Hyperplane Theorem to get a linear form f_0 in $\mathcal{A}_J^*$ such that $f_0(E_i) < f_0(E)$ for all $E_i \neq E$. $f \equiv f_0 \circ h$ is an invariant form because h^* maps $\mathcal{A}_J^*$ to J. Clearly, $\Gamma_f = \Gamma_E$. If e_1 and e_2 are vertices of Γ_E then $h(e_1) = h(e_2)$, i.e. $e_1 - e_2 \in J^\perp$ and so the invariant forms do not distinguish e_1 and e_2. By the lemma Γ_E is a minimal, J-induced invariant face.

Conversely, let Γ_f be a minimal, J-induced invariant face. By the lemma the vertices of Γ_f all map to the same E under h. In particular, f takes the vertices of Γ_f to the maximum value M and is less than M on the remaining vertices. As $f \in J$ it equals $h^*(f_0)$ for a form f_0 on $\mathcal{A}_J$. $f_0(E) = M$ when $f_0(E_i) < M$ for the remaining vertex images. In particular, E lies on the support hyperplane $\{f_0^{-1}(M)\}$ for Δ_J and so E is an extremal point. $\square$

We will call an extremal point E of Δ_J *isolated* if it is the image of a unique vertex e of the simplex Δ, or, equivalently, if $\dim \Gamma_E = 0$.

Corollary 3.8.10 *If E is an isolated extremal point of Δ_J with $\Gamma_E = \{e\}$ then the vertex e of Δ is an idempotent.*

Proof By Theorems 3.8.6, 3.8.9 Γ_E is invariant. So e^2 lies in Γ_E and must equal e. $\square$

In particular, for a J-irreducible evolutionary algebra the images of the canonical basis $E_i = h(e_i)$ are pairwise distinct and so each extremal point of Δ_J is isolated. As $\dim \Delta_J = n - 1 - \dim J^\perp = \dim J - 1$ there are at least $\dim J$ such extremal points. So we have

Corollary 3.8.11 *Let $(\mathcal{A}, \Delta)$ be a J-irreducible evolutionary algebra. Each vertex of Δ which maps to an extremal point of Δ_J is an idempotent. There are at least* $\dim J$ *such idempotent vertices.*

The vertex idempotent corresponds biologically to a type for which all offspring belong to the same type, such as homozygotes in a Mendel population. A vertex idempotent is called *nonsplitting type* (or *zygote*).

Corollary 3.8.12 *Distinct minimal J-induced faces are disjoint.*

Proof This is obvious from the theorem because $\Gamma_{E_1} \cap \Gamma_{E_2} = \emptyset$ for E_1, E_2 distinct extremals of Δ_J. It can also be proved directly: if two J-induced faces meet then their intersection is a J-induced face. For if $\Gamma_f \cap \Gamma_g \neq \emptyset$ then $\Gamma_f \cap \Gamma_g = \Gamma_{f+g}$. $\square$

To specify Corollary 3.8.10 we call a vertex e of the basic simplex Δ an *isolated idempotent* if $V(e) = e$ and there exists no face Γ of Δ containing e but with $\dim \Gamma > 0$ such that $V(\Gamma) = \{e\}$. Here V is the evolutionary operator $V(x) = x^2$. Note that if $x \in \Gamma$ and $V(\Gamma) = \{x\}$ then Γ is a very special example of an invariant face. In fact, $V(\Gamma)$ is a single point in Γ precisely when the coordinate subspace is a constant subalgebra.

Theorem 3.8.13 *If E is an isolated extremal point of Δ_J with $\Gamma_E = \{e\}$ then the vertex e is an isolated idempotent in the simplex Δ.*

Proof From Corollary 3.8.10 we already know e is an idempotent. Suppose e_1 is another vertex of Δ with $e_1^2 = e$. Then $E_1^2 = E$ in the factor algebra $\mathcal{A}/J^\perp$. But $\mathcal{A}/J^\perp$ is a unit algebra (see the discussion after Theorem 3.3.12) and so $E_1^2 = E_1$. Hence, $E_1 = E$ and so $e_1 \in \Gamma_E$. This contradicts $\Gamma_E = \{e\}$, the assumption that E is isolated. $\square$

The reversal of Theorem 3.8.13 requires us to replace the stochastic ideal $J^\perp$ by the smaller stochastic ideal $\mathrm{ann}\mathcal{A}$. We denote by h° the quotient map to $\mathcal{A}/\mathrm{ann}\mathcal{A}$ and by Δ_{ann} the basic body $h^\circ(\Delta)$ with associated cone Q°. Continuing this notation we denote by E_i° the image of the vertex e_i under h°. $E^\circ = h^\circ(e)$ is called an *isolated extremal point* of Δ_{ann} if it is an extremal point and $e - e_i \notin \mathrm{ann}\,\mathcal{A}$ for any vertex $e_i \neq e$ of Δ.

Theorem 3.8.14 *If e is an isolated idempotent vertex of the simplex Δ then its image $E^\circ = h^\circ(e)$ is an isolated extremal point of Δ_{ann}.*

Proof Note first that if $x \equiv e \bmod \mathrm{ann}\mathcal{A}$ then $x^2 = e^2 = e$. Thus, with $\Gamma_{E^\circ} = (h^\circ|\Delta)^{-1}(E)$ we have $V(\Gamma_{E^\circ}) = \{e\}$. If E° is an extremal point then Γ_{E° is a face of Δ and so because e is isolated it will follow that $\Gamma_{E^\circ} = \{e\}$. So it remains only to prove the E° is indeed an extremal point of Δ_{ann}.

Suppose $E^\circ = \sum \lambda_i E_i^\circ$ with $\lambda_i > 0$, $\sum \lambda_i = 1$ and $E_i^\circ \neq E^\circ$. Then $e \equiv \sum \lambda_i e_i \bmod \mathrm{ann}\, \mathcal{A}$ and, as above

$$e = e^2 = \sum \lambda_i \lambda_j e_i e_j.$$

But the vertex e is an extremal point of Δ and so the points $e_i e_j$ of Δ are all equal e. Also, we have

$$e = e^2 = \sum \lambda_i e_i e$$

and so $e_i e = e$ as well. Hence the face Γ of Δ with vertices e and these e_i's satisfies $V(\Gamma) = \{e\}$ which contradicts the assumption that e is isolated. $\square$

An evolutionary algebra is called *internally irreducible* if $e_i - e_j \notin \mathrm{ann}\, \mathcal{A}$ for distinct vertices e_i and e_j of the simplex Δ, or, equivalently, the set of image points $E_1^\circ, \ldots, E_n^\circ$ are all distinct in $\mathcal{A}/\mathrm{ann}\mathcal{A}$. In this case the extremal of Δ_{ann} are automatically isolated. Of course, J-irreducibility implies internal irreducibility.

Lemma 3.8.15 *An evolutionary algebra is internally irreducible if and only if for every pair of distinct vertices e_i, e_j there exists a vertex e_k such that*

$$e_i e_k \neq e_j e_k \tag{3.8.7}$$

Proof $e_i - e_j \in \mathrm{ann}\mathcal{A}$ if and only if $(e_i - e_j)x = 0$ for all x in $\mathcal{A}$ and so if and only if $(e_i - e_j)e_k = 0$ for all vertices e_k. $\square$

The gap between Theorems 3.8.13 and 3.8.14 does not occur when the algebra is conservative i.e. $\mathrm{ann}\mathcal{A} = J^\perp$. In that case, $\Delta_J = \Delta_{\mathrm{ann}}$, $E_i^\circ = E_i$, etc. and J-reducibility coincides with internal reducibility. So we get

Theorem 3.8.16 *Let (A, Δ) be a conservative evolutionary algebra. Let e be a vertex of Δ and $E = h(e)$ its image. The vertex e is an isolated idempotent if and only if E is an isolated extremal point of Δ_J. If the algebra is internally irreducible then the vertex e is an idempotent if and only if E is an extremal point of Δ_J. In this case all idempotents e_i's are isolated.*

Corollary 3.8.11 implies there are at least $\dim J$ such vertex idempotents in the canonical basis for $(\mathcal{A}, \Delta)$. It is important from a genetical point of view (see chapter 4).

3.9 The Kin of an Evolutionary Algebra. Normalization

For evolutionary algebras $(\mathcal{A}, \Delta)$ and $(\tilde{\mathcal{A}}, \tilde{\Delta})$ a stochastic algebra homomorphism $h : (\mathcal{A}, \Delta) \to (\tilde{\mathcal{A}}, \tilde{\Delta})$ is called an *internal kinship* if there exists a nonnegative linear homomorphism $g : \tilde{\mathcal{A}} \to \mathcal{A}$ such that: (1) g is a right inverse for h, i.e. $hg = \mathrm{id}_{\tilde{\mathcal{A}}}$, and (2) the operator $P = gh$ on $\mathcal{A}$ satisfies

$$(Px)^2 = x^2 \qquad (x \in \mathcal{A}). \qquad (3.9.1)$$

As usual, this condition is equivalent to

$$(Px)(Py) = xy \qquad (x, y \in \mathcal{A}). \qquad (3.9.2)$$

Theorem 3.9.1 *Let $h : (\mathcal{A}, \Delta) \to (\tilde{\mathcal{A}}, \tilde{\Delta})$ be an internal kinship of evolutionary algebras. The corresponding right inverse g is stochastic and the associated $P = gh$ is a stochastic projection. $\ker h = \ker P$ is a stochastic ideal of $(\mathcal{A}, \Delta)$ contained in $\mathrm{ann}\, \mathcal{A}$, and h induces a stochastic algebra isomorphism $\overline{h} : (\mathcal{A}, \Delta)/\ker h \to (\tilde{\mathcal{A}}, \tilde{\Delta})$. In particular, $\tilde{\Delta} = h(\Delta)$.*

Conversely, let I be a subspace in an evolutionary algebra $(\mathcal{A}, \Delta)$ with $I \subset \mathrm{ann}\, \mathcal{A}$ and with the quotient space $(\mathcal{A}, \Delta)/I$ simplicial. Then I is a stochastic ideal and the quotient map $q : (\mathcal{A}, \Delta) \to (\mathcal{A}, \Delta)/I$ is an internal kinship onto the simplicial algebra $(\mathcal{A}, \Delta)/I$.

Proof We know from Corollary 3.7.17 that g is stochastic and $P = gh$ is a stochastic projection, $\ker h = \ker P$. This time h is a stochastic algebra homomorphism and so $\ker h$ is stochastic ideal.

The identity (3.9.2) implies $\ker P \subset \mathrm{ann}\mathcal{A}$. Clearly, the induced map $\overline{h}$ is a stochastic algebra homomorphism and a bijection. But $\tilde{\Delta} = hg(\tilde{\Delta})$ which is contained in $h(\Delta)$ because g is stochastic. Thus, $\tilde{\Delta} = h(\Delta)$ because h is stochastic. Hence, the algebra homomorphism $\overline{h}^{-1}$ takes $\tilde{\Delta}$ to Δ and so $\overline{h}^{-1}$ is stochastic.

On the other hand, if I is a stochastic ideal with $(\mathcal{A}, \Delta)/I$ simplicial then the quotient map q admits a nonnegative right inverse g by Corollary 3.7.17.

It remains to show that $I \subset \mathrm{ann}\, \mathcal{A}$ implies (3.9.1) for $P = gq$. We have $x - Px \in \ker P = \ker q = I$. Consequently,

$$x^2 - (Px)^2 = (x - Px)(x + Px) = 0.$$

□

Note that any linear subspace of ann $\mathcal{A}$ is an ideal.

Remark Observe that g need not be an algebra homomorphism and so its image, the subspace Im P, need not be a subalgebra. So it will usually not be true that $(Px)(Py) = P(xy)$, in other words, the product $(Px)(Py)$, which equals xy by (3.9.2), need not lie in the subspace Im P.

Because h is surjective, the existence of an internal kinship $h : (\mathcal{A}, \Delta) \to (\tilde{\mathcal{A}}, \tilde{\Delta})$ implies $\dim \mathcal{A} \geq \dim \tilde{\mathcal{A}}$ with equality only when h is a stochastic algebra isomorphism. Evolutionary algebras $(\mathcal{A}, \Delta)$ and $(\tilde{\mathcal{A}}, \tilde{\Delta})$ with $\dim \mathcal{A} \geq \dim \tilde{\mathcal{A}}$ are called *internal kin* when such an internal kinship $h : (\mathcal{A}, \Delta) \to (\tilde{\mathcal{A}}, \tilde{\Delta})$ exists.

We can restate condition (3.9.1) in terms of the evolutionary operator $V(x) = x^2$ on $\mathcal{A}$:

$$VP = V. \tag{3.9.3}$$

This coincides with (3.9.1) for $x \in \Delta$. If (3.9.1) is valid for $x \in \Delta$ then it is valid for $x > 0$. It implies (3.9.1) for all x because a quadratic form is 0 when it is 0 for $x > 0$. Because h is an algebra homomorphism we have $\tilde{V} h = hV$ where $\tilde{V}$ is the evolutionary operator on $\tilde{\Delta}$. Composing with the right inverse g we get:

$$\tilde{V} = hVg. \tag{3.9.4}$$

The operators V and $\tilde{V}$ are then called *internal kin* as well.

We now provide an alternative, and useful for some applications, way of describing an internal kinship.

Theorem 3.9.2 *An internal kinship* $h : (\mathcal{A}, \Delta) \to (\tilde{\mathcal{A}}, \tilde{\Delta})$ *can be written as*

$$h = \sum_{i=1}^{m} p_i \otimes \tilde{e}_i \tag{3.9.5}$$

where $m = \dim \tilde{\mathcal{A}}$, $\{\tilde{e}_1, \ldots, \tilde{e}_m\}$ *is a canonical basis for* $(\tilde{\mathcal{A}}, \tilde{\Delta})$ *and* $\{p_1, \ldots, p_m\}$ *is a partition of the weight* s *on* $(\mathcal{A}, \Delta)$. *For* $x \in \mathcal{A}$, x^2 *depends only on the values* $\{p_1(x), \ldots, p_m(x)\}$.

Conversely, if $\{p_1, \ldots p_m\}$ *is a partition of* s *and for all* $x \in \mathcal{A}$, x^2 *depends only on the values* $\{p_1(x), \ldots p_m(x)\}$ *then with* $I = \cap \{\ker p_i\}$ *and with* h *the quotient map onto* $(\mathcal{A}, \Delta)/I$, h *is an internal kinship of the form (3.9.5) using the given partition* $\{p_i\}$.

Proof With $\{\tilde{e}_1^*,\ldots,\tilde{e}_m^*\}$ the dual of the canonical basis any homomorphism h to $(\tilde{\mathcal{A}},\tilde{\Delta})$ is of the form (3.9.5) with $p_i = h^*\tilde{e}_i^*$. Each p_i is nonnegative and satisfies $\parallel p_i \parallel \le 1$ because h is stochastic, which also implies $\sum p_i = h^*(\sum \tilde{e}_i^*) = h^*\tilde{s} = s$. Finally, because $h(\Delta) = \tilde{\Delta}$, $\tilde{e}_i$ is in the image of h and so p_i takes on the value 1 of Δ, i.e. $\parallel p_i \parallel = 1$ for $i = 1,\ldots,m$. (3.9.3) follows that x^2 depends only on the congruence class of x mod ker P and because ker P = ker h = $\cap_i\{\text{ker } p_i\}$ this class depends only on the values $\{p_1(x),\ldots,p_m(x)\}$.

For the converse let $p : \mathcal{A} \to \mathbf{R}^m$ be the map given by $p(x) = (p_1(x),\ldots,p_m(x))$. Because each $p_i \ge 0$ and $\sum p_i = s$ which is 1 on Δ, it follows that p is a stochastic space homomorphism of $(\mathcal{A},\Delta)$ onto arithmetical space $(\mathbf{R}^m,\Delta^{m-1})$. Because $\parallel p_i \parallel = 1$ there exists $u_i \in \Delta$ such that $p_i(u_i) = 1$ and so $p_j(u_i) = 0$ for $j \ne i$. Hence p maps u_i to $\overline{e}_i$ in the canonical basis $\{\overline{e}_1,\ldots,\overline{e}_m\}$ of $\mathbf{R}^m$. Thus, p maps Δ in $\mathcal{A}$ onto the unit simplex in $\mathbf{R}^m$ and so induces a stochastic space isomorphism $\overline{p}$ between the quotient $(\mathcal{A},\Delta)/\text{ker } p$ and $(\mathbf{R}^m,\Delta^{m-1})$. Obviously, ker p = $\cap$ker p_i. With $\tilde{e}_i = \overline{p}^{-1}(\overline{e}_i)$ it follows that the quotient map h is of the form (3.9.5), ker h = ker p. Finally, x^2 depends only on the values $\{p_1(x),\ldots,p_m(x)\}$ and so x^2 depends just on the congruence class of x mod ker h. As usual this means xy depends only on the congruence classes of x and y. This implies ker $h \subset$ ann $\mathcal{A}$. So h is an internal kinship by the previous theorem. $\square$

Example Let $(\mathcal{A},\Delta)$ be an evolutionary algebra which is internally reducible. This means that in the quotient algebra $(\mathcal{A},\Delta)/\text{ann}\mathcal{A}$ we have some repetitions among the images $E_1^0,\ldots,E_n^0$ of the canonical basis vectors $e_1,\ldots e_n$. To be specific suppose $E_1^0 = \ldots = E_{n_1}^0$, $E_{n_1+1}^0 = \ldots = E_{n_2}^0,\ldots,E_{n_{m-1}+1}^0 = \ldots = E_{n_m}^0$ $(n_m = n)$ and $E_{n_1}^0,\ldots,E_{n_m}^0$ are pairwise distinct. Then $\{e_j - e_{n_i} : n_{i-1} < j < n_i, i = 1,\ldots,m\}$ all lie in ann $\mathcal{A}$ (by convention $n_0 \equiv 0$). Together they form a basis for an $n - m$ dimensional subspace $I \subset$ ann $\mathcal{A}$. To see that $(\mathcal{A},\Delta) \to (\mathcal{A},\Delta)/I$ is an internal kinship we observe that

$$s_i = \sum_{n_{i-1} < j \le n_i} e_j^* \quad (i = 1,\ldots,m) \tag{3.9.6}$$

is a partition of the weight and x^2 depends only on the values $s_1(x),\ldots,s_m(x)$. The canonical basis $\{\tilde{e}_1,\ldots,\tilde{e}_m\}$ in the quotient is obtained by taking $\tilde{e}_i$ to be the common image of the e_j's with $n_{i-1} < e_j \le n_i$. the quotient map is $\sum_{i=1}^m s_i \otimes \tilde{e}_i$. The assumptions defining $n_1,\ldots,n_m$ are equivalent to saying of the structure constants $\{p_{rt,j} : r,t,j = 1,\ldots,n\}$ for $(\mathcal{A},\Delta)$

$$p_{rt,j} = p_{n_in_k,j} \quad (n_{i-1} < r \le n_i;\; n_{k-1} < t \le n_k) \tag{3.9.7}$$

and the associated constants for $(\tilde{A}, \tilde{\Delta}) = (A, \Delta)/I$ are given by

$$\tilde{p}_{ik,l} = \sum_{n_{l-1} < j \leq n_l} p_{n_i n_k, j}. \tag{3.9.8}$$

Thus, the biological meaning of internal reducibility is the existence of a partition of the original population types $\{1, \ldots, n\}$ into a nontrivial family of classes $\{1, \ldots, n_1\}$, $\{n_1 + 1, \ldots, n_2\}, \ldots$ in such a way that the inheritance coefficients of offspring depend only on the classes of the parents. Two types belonging to the same class are called *internally kin*. A type which is not internally kin to any other is called *internally isolated*. These classes are the types of a new reduced population (3.9.8). The reduction described above is called the *internal union of types*. This procedure is in accordance with the addition of probabilities. In such a situation the heredity is determined by the reduced types. In this sense the original differentiation is excessive. A population is internally reducible if and only if its evolutionary operator V depends only on the sums of coordinates according to a nontrivial partition of set $\{1, \ldots, n\}$.

Theorem 3.9.3 *If $h : (A, \Delta) \to (\tilde{A}, \tilde{\Delta})$ is an internal kinship then the dual map $h^* : \tilde{A}^* \to A^*$ is an isometric embedding. It is a nonnegative map taking the cone $\tilde{K}^+$ of nonnegative forms on $\tilde{A}$ onto $K^+ \cap (\ker h)^\perp$, the cone of nonnegative forms on A which vanish on $\ker h$. Also, h^* maps $\tilde{J}$ bijectively onto J, where J and $\tilde{J}$ are the subspaces of invariant forms on A and $\tilde{A}$, with $h^*(\tilde{C}) = C$, where $\tilde{C} = \tilde{K}^+ \cap \tilde{J}$ and $C = K^+ \cap J$.*

Proof The identification of $\tilde{K}^+$ with $K^+ \cap (\ker h)^\perp$ follows from Lemma 3.7.14 and the isomorphism of $(\tilde{A}, \tilde{\Delta})$ with $(A, \Delta)/\ker h$.

Because $\tilde{\Delta} = h(\Delta)$ we have

$$\| \tilde{f} \| = \max_{\tilde{\Delta}} |\tilde{f}| = \max_{\Delta} |f| = \| f \|$$

when $f = \tilde{f} \circ h = h^*(\tilde{f})$. So h^* is an isometry.

Because $\ker h \subset \text{ann} A \subset J^\perp$ Theorem 3.3.12 implies that h^* maps $\tilde{J}$ bijectively to J, and by nonnegativity $h^*(\tilde{C}) \subset C$. On the other hand if $f = \tilde{f} \circ h$ with f in C and $\tilde{f}$ in $\tilde{J}$ then f nonnegative on Δ implies $\tilde{f}$ is nonnegative on $\tilde{\Delta} = h(\Delta)$. So $\tilde{f} \in \tilde{C}$. $\square$

Dual to the concept of internal kinship is external kinship. For evolutionary algebras (A, Δ) and $(\tilde{A}, \tilde{\Delta})$ a stochastic algebra homomorphism

$h : (\tilde{A}, \tilde{\Delta}) \to (A, \Delta)$ is called an *external kinship* if there exists a nonnegative linear operator $g : A \to \tilde{A}$ such that (1) g is a left inverse for h, i.e. $gh = \mathrm{id}_{\tilde{A}}$, and (2) the operator $P = hg$ on A satisfies

$$P(x^2) = x^2 \quad (x \in A) \tag{3.9.9}$$

or, equivalently, if

$$P(xy) = xy \quad (x, y \in A). \tag{3.9.10}$$

Theorem 3.9.4 *Let $h : (\tilde{A}, \tilde{\Delta}) \to (A, \Delta)$ be an external kinship of evolutionary algebras. The operator $P = hg$ associated with the nonnegative left inverse g is a nonnegative projection.* $\mathrm{Im}\, h = \mathrm{Im}\, P$ *is a simplicial subalgebra of (A, Δ) containing A^2 and h induces a stochastic algebra isomorphism onto its image. In particular $h(\tilde{\Delta}) = \Delta \cap \mathrm{Im}\, h$.*

Conversely, let $(\tilde{A}, \tilde{\Delta})$ be a simplicial subspace of (A, Δ) which contains A^2. Then, $(\tilde{A}, \tilde{\Delta})$ is a simplicial subalgebra of (A, Δ) and the embedding is an external kinship.

Proof As in the dual case $P^2 = hghg = hg$ and so P is a projection. Similarly, $Ph = h$ and so $\mathrm{Im}\, h = \mathrm{Im} P$. Also, $\mathrm{Im} P \supset A^2$ due to (3.9.10). P is nonnegative because h and g are.

Because h is stochastic $h(\tilde{\Delta}) \subset \Delta \cap \mathrm{Im}\, h$. On the other hand if $x \in \Delta \cap \mathrm{Im}\, h$ then $x = h(y)$ for y in $\tilde{A}$. $y = (gh)(y) = g(x)$ is nonnegative because x and g are. Also, $\tilde{s}(y) = (h^*s)(y) = s(x) = 1$. So y lies in $\tilde{\Delta}$ and $x \in h(\tilde{\Delta})$. Hence, h induces a stochastic isomorphism of $(\tilde{A}, \tilde{\Delta})$ onto $(\mathrm{Im}\, h, \Delta \cap \mathrm{Im}\, h)$ and so $\mathrm{Im}\, h$ is a simplicial subalgebra.

On the other hand, if $\tilde{A}$ is a simplicial subspace of (A, Δ) then by Theorem 3.7.7 there exists a nonnegative projection P on A with image equal to $\tilde{A}$. With $\tilde{\Delta} = \Delta \cap \tilde{A}$ let h denote the embedding and g the map from A to $\tilde{A}$ induced by P. Since $\tilde{A} \supset A^2$, h is an external kinship with above g and P. $\square$

Notice that a linear subspace which contains A^2 is a subalgebra (even an ideal though not a baric ideal, of course).

Remark g need not be an algebra homomorphism and g and P need not be stochastic. However, because $gh = \mathrm{id}_{\tilde{A}}$ we do have:

$$g(\Delta \cap \mathrm{Im}\, h) = \tilde{\Delta}. \tag{3.9.11}$$

It means that the restriction $g|\mathrm{Im}\, h$ is stochastic. An external kinship $h : (\tilde{A}, \tilde{\Delta}) \to (A, \Delta)$ is injective and so we have $\dim \tilde{A} \le \dim A$, with equality only when h is a stochastic algebra isomorphism. Evolutionary algebras $(\tilde{\mathcal{A}}, \tilde{\Delta})$ and $(\mathcal{A}, \Delta)$ with $\dim \tilde{\mathcal{A}} \leq \dim \mathcal{A}$ are called *external kin* when an external kinship $h : (\tilde{\mathcal{A}}, \tilde{\Delta}) \to (\mathcal{A}, \Delta)$ exists.

In terms of the evolutionary operator V on $\mathcal{A}$ (3.9.19) is equivalent to

$$PV = V. \tag{3.9.12}$$

Also, $h\tilde{V} = Vh$ (h is an algebra homomorphism) implies

$$\tilde{V} = gVh \tag{3.9.13}$$

where $\tilde{V}$ is the evolutionary operator on $\tilde{\mathcal{A}}$. $\tilde{V}$ and V are then called *external kin* as well.

Theorem 3.9.5 *An external kinship $h : (\tilde{\mathcal{A}}, \tilde{\Delta}) \to (\mathcal{A}, \Delta)$ can be written as*

$$h = \sum_{i=1}^{m} \tilde{p}_i \otimes u_i \tag{3.9.14}$$

where $m = \dim \tilde{\mathcal{A}}$, $\{\tilde{p}_1, \ldots, \tilde{p}_m\}$ is a partition of the weight $\tilde{s}$ on $(\tilde{\mathcal{A}}, \tilde{\Delta})$ and $\{u_1, \ldots, u_m\}$ is a stochastic Gaussian system of vectors in $\mathcal{A}$. For $x \in \mathcal{A}$, x^2 is a linear combination of $u_1, \ldots u_m$.

Conversely, if $\{u_1, \ldots, u_m\}$ is a stochastic Gaussian system of vectors which generate by linear combination all the x^2's for x in $\mathcal{A}$ then with the simplex $\tilde{\Delta}$ generated by $\{u_1, \ldots, u_m\}$, subspace $\tilde{\mathcal{A}} = [\tilde{\Delta}] \supset \mathcal{A}^2$, $\tilde{\Delta} = \Delta \cap \tilde{\mathcal{A}}$ and h the embedding of $(\tilde{\mathcal{A}}, \tilde{\Delta})$ into $(\mathcal{A}, \Delta)$, h is an external kinship of the form (3.9.14).

Proof By Theorem 3.7.8 we can write

$$P = \sum_{i=1}^{m} p_i \otimes u_i \tag{3.9.15}$$

where $m = \mathrm{rank}\, P = \dim \tilde{\mathcal{A}}$, $\{u_1, \ldots u_m\}$ is a nonnegative Gaussian system (which we can normalize to assume stochastic) and $\{p_1, \ldots p_m\}$ a nonnegative family of forms biorthogonal to the u_i's.

Writing $\tilde{p}_i = h^* p_i$ we obtain (3.9.14) because $Ph = h$. Because h is a stochastic and $p_i \geq 0$ we have $\tilde{p}_i \geq 0$. Also

$$\tilde{s} = h^* s = h^* P^* s = \sum_i s(u_i)\tilde{p}_i = \sum_i \tilde{p}_i.$$

In particular, $\| \tilde{p}_i \| \leq 1$. Because $h(\tilde{\Delta}) = \Delta \cap \operatorname{Im} P$ and $u_i \in \Delta \cap \operatorname{Im} P$ imply that for some $\tilde{e}_i \in \tilde{\Delta}$, $h(\tilde{e}_i) = u_i$ and so $\| \tilde{p}_i \| \geq \tilde{p}_i(\tilde{e}_i) = p_i(u_i) = 1$. Thus, $\|\tilde{p}_i\| = 1$ and $\{\tilde{p}_1, \ldots, \tilde{p}_m\}$ is a partition of $\tilde{s}$. Finally,

$$V(x) = x^2 = P(x^2) = \sum_i p_i(x^2) u_i.$$

Conversely, if $\{u_1, \ldots, u_m\}$ is a stochastic Gaussian system generating a subspace $\tilde{A}$ which contains all $V(x)$'s then $\tilde{A} \supset A^2$. Also, by Theorem 3.7.4, $\tilde{A}$ is a simplicial subspace and the proof of that result shows that the vertices of $\tilde{\Delta} = \Delta \cap \tilde{A}$ are $u_1, \ldots u_m$. By the previous theorem the embedding h is an external kinship.

Because $\{u_1, \ldots u_m\}$ is the canonical basis for $\tilde{A}$ we can write P in the form (3.9.15) and so from $h = Ph$ we get (3.9.14) with the given system $\{u_1, \ldots u_m\}$. $\square$

Theorem 3.9.6 *If $h : (\tilde{A}, \tilde{\Delta}) \to (A, \Delta)$ is an external kinship then the dual map $h^* : A^* \to \tilde{A}^*$ is a nonnegative surjective map of norm $\|h\| \leq 1$. Also, h^* maps J injectively (and even isometrically) into $\tilde{J}$, where J and $\tilde{J}$ are the subspaces of invariant forms on A and $\tilde{A}$. In particular, $\dim J \leq \dim \tilde{J}$. If $h^*(J) = \tilde{J}$, i.e. h^* is an isomorphism of J onto $\tilde{J}$, then $h^*(C) = \tilde{C}$, where $C = K^+ \cap J$ and $\tilde{C} = \tilde{K}^+ \cap \tilde{J}$.*

Proof h^* is nonnegative because h is and

$$\| h^* f \| = \max_{\tilde{\Delta}} |f \circ h| = \max_{h(\tilde{\Delta})} |f| \leq \max_{\Delta} |f| = \| f \|.$$

That h^* maps J injectively into $\tilde{J}$ follows from $\operatorname{Im} h \supset A^2$ and Theorem 3.3.12. Moreover, if $f \in J$ then

$$\|f\| = \max_{\Delta \cap A^2} |f| \leq \max_{h(\tilde{\Delta})} |f| = \|h^* f\|.$$

Finally, we must show that for $\tilde{f} = h^* f$, $\tilde{f} \in \tilde{K}^+$ and $f \in J$ imply $f \in K^+$. But by invariance $x \in \Delta$ implies

$$f(x) = f(x^2) = f(P(x^2)) = \tilde{f}(g(x^2)) \geq 0.$$

Hence, f is nonnegative. $\square$

We now consider two examples having biological interpretation.

Example A type j is called *disappearing* in the population if its probability in the next generation is identically 0. This means the coordinate function e_j^* is a disappearing form, $e_j^* \in N = (\mathcal{A}^2)^\perp$, or equivalently for an evolutionary algebra $e_j^*(V(x)) = 0$ for all x in Δ. A population with at least one disappearing type is called *degenerate*. Degeneracy means that V maps Δ into some proper face $\tilde{\Delta}$ (clearly $\tilde{\Delta}$ is then an invariant face). Equivalently, $\mathcal{A}^2$ is contained in a proper coordinate subspace. If $(\mathcal{A}, \Delta)$ is degenerate we let $\tilde{\mathcal{A}}$ be the smallest coordinate subspace containing $\mathcal{A}^2$. Clearly, $\tilde{\Delta} = \Delta \cap \tilde{\mathcal{A}}$ is the smallest face of Δ containing $V(\Delta)$ and the embedding of $(\tilde{\mathcal{A}}, \tilde{\Delta})$ into $(\mathcal{A}, \Delta)$ is an external kinship.

In terms of structure constants $\{p_{ik,j}\}$ j is a disappearing type if and only if

$$p_{ik,j} = 0 \qquad\qquad (i, k = 1, \ldots n). \qquad\qquad (3.9.16)$$

Elimination of the disappearing types is by definition the restriction to the subalgebra $\tilde{\mathcal{A}}$ where all the disappearing coordinates e_j^* equal zero.

Example Nondisappearing types i, j are called *externally kin* if their probabilities are in a constant ratio in the next generation. This means that for some $\alpha > 0$ the coordinate functions satisfy $e_i^*(V(x)) = \alpha e_j^*(V(x))$ for all x in Δ. Equivalently, the difference $e_i^* - \alpha e_j^*$ is a disappearing form. When two distinct externally kin types occur we call the algebra or the population *externally reducible*.

For any externally reducible evolutionary algebra, separate the canonical basis into external kin classes: $\{e_1, \ldots, e_{n_1}\}, \{e_{n_1+1}, \ldots, e_{n_2}\}, \ldots \{e_{n_{m-1}+1}, \ldots, e_{n_m}\}$ (with $n_m = n$) so that

$$f_j = e_j^* - \alpha_j e_{n_i}^* \qquad\qquad (n_{i-1} < j < n_i, n_0 = 0) \qquad\qquad (3.9.17)$$

is a disappearing form with the $\alpha_j's > 0$. Because the forms are disappearing $\mathcal{A}^2 \subset \tilde{\mathcal{A}} \equiv \cap_j \ker f_j$. Because the family of f_j's is clearly a hyperbolic Gaussian system, $\tilde{\mathcal{A}}$ is a simplicial subspace by Theorem 3.7.5, and so with $\tilde{\Delta} = \Delta \cap \tilde{\mathcal{A}}$, the embedding h of $(\tilde{\mathcal{A}}, \tilde{\Delta})$ into $(\mathcal{A}, \Delta)$ is an external kinship by Theorem 3.9.4.

The vertices of $\tilde{\Delta}$ are the stochastic Gaussian system $\{u_1, \ldots u_m\}$:

$$u_i = \frac{1}{N_i}\left(e_{n_i} + \sum_{n_{i-1} < j < n_i} \alpha_j e_j\right) \qquad\qquad (3.9.18)$$

with N_i the normalization constant

$$N_i = \sum_{n_{i-1} < j \leq n_i} \alpha_j \qquad\qquad (\text{with } \alpha_{n_i} \equiv 1). \qquad\qquad (3.9.19)$$

In terms of the structure constants, to say that f_j is a disappearing form says:

$$p_{kl,j}/\alpha_j = p_{kl,n_i} \qquad n_{i-1} < j \leq n_i$$

and the structure constant $\tilde{p}_{\alpha\beta,i}$ on $\tilde{A}$ ($=$ the u_i-coefficient of $u_\alpha u_\beta$) is given by

$$\tilde{p}_{\alpha\beta,i} \;=\; \frac{N_i}{N_\alpha N_\beta} \sum_{\substack{n_{\alpha-1}<k\leq n_\alpha \\ n_{\beta-1}<l\leq n_\beta}} p_{kl,n_i}\alpha_k\alpha_l$$

$$= \frac{1}{N_\alpha N_\beta} \sum_{\substack{n_{\alpha-1} < k \leq n_\alpha \\ n_{\beta-1} < l \leq n_\beta \\ n_{i-1} < j \leq n_i}} p_{kl,j}\alpha_k\alpha_l$$

$$(3.9.20)$$

The reduction to $(\tilde{A}, \tilde{\Delta})$ described above is called *external union of types.*

A canonical basis vector which is neither disappearing nor externally kin with any other is called *externally isolated.*

A *kinship chain* is a sequence of evolutionary algebras

$$(A_1, \Delta_1), (A_2, \Delta_2), \ldots, (A_r, \Delta_r) \qquad (3.9.21)$$

such that each pair $\{(A_k, \Delta_k), (A_{k+1}, \Delta_{k+1})\}$ are either externally or internally kin. We call the kinship chain *normal* if $\dim A_k \geq \dim A_{k+1}$ for each k. (A, Δ) and $(\tilde{A}, \tilde{\Delta})$ are called *kin* if they can be connected by a kinship chain. They are *normal kin* (with $\dim A \geq \dim \tilde{A}$) if they can be connected by a normal kinship chain.

We now consider some kinship invariants.

Theorem 3.9.7 *If algebras (A, Δ) and $(\tilde{A}, \tilde{\Delta})$ are kin then there is a bijection (defined by the corresponding chain of homomorphisms) between the varieties of idempotents* Id A *and* Id $\tilde{A}$*. In particular,* $\dim \mathrm{Id}\, A = \dim \mathrm{Id}\, \tilde{A}$*.*

Proof Apply Theorems 3.9.1, 3.9.4 and 3.1.12. $\square$

Suppose that a class of baric algebras is defined by some identities, possibly involving the weight .For example, conservative algebras are characterized by (3.3.16) and Bernstein algebras are defined (3.3.17). If (A, Δ) belongs to such a class then so does any kin to (A, Δ) because the defining identities are preserved by the passage to subalgebra, quotient algebra and isomorphic image. In particular, this applies to the class of conservative evolutionary algebras and Bernstein evolutionary algebras.

Theorem 3.9.8 *If conservative evolutionary algebras (A, Δ) and $(\tilde{A}, \tilde{\Delta})$ are kin then there is an isometric and cone isomorphism (defined by the dual of the corresponding chain of homomorphism) between the spaces J and $\tilde{J}$ of invariant forms. In particular* $\dim J = \dim \tilde{J}$.

Proof All the algebras in the chain are conservative. The internal kinship maps induce isomorphisms of invariant form spaces by Theorem 3.9.3. For external kin the result follows from Theorem 3.9.6 together with Corollary 3.4.9, Theorem 3.4.17 and Theorem 3.9.7 which imply that in the conservative case we have $\dim J = \dim \tilde{J}$. $\square$

Remark Without the conservative assumption we can at least conclude that $\dim J \leq \dim \tilde{J}$.

An evolutionary algebra without degeneracies, internally and externally irreducible is called *normal*. If (A, Δ) is not normal we can apply one of the three reductions: elimination of the disappearing types, internal or external union of types. By successive applications we can build a normal kinship chain from any evolutionary algebra to one which is normal. The process must terminate because the dimensions are finite and strictly decreasing. We call this *normalization*.

Chapter 4

Stationary Gene Structure

4.1   Statement of Problem. Examples

The Mendel mechanism of heredity at an autosomal locus determines a two-level population with an identically equilibrious gamete (gene) pool. Correspondingly, the multiallelic Hardy-Weinberg operator is parametrized by the invariant linear forms. It is natural to raise the problem of an explicit description of all evolutionary operators with that property. This problem, whose solution will be set forth in the present chapter forms the genetically interpretable part of the broader Bernstein problem of explicitly describing all stationary evolutionary operators, $V^2 = V$. The general Bernstein problem is studied in the next chapter. The populations treated in chapters 4 to 6 are assumed to be free.

A free population with an evolutionary operator V has the *stationary gene structure (s.g.s)* if $V(x)$ depends only on the values of invariant linear forms of x, i.e. if $x_1 \equiv x_2 (\mathrm{mod} J^{\perp})$ implies $V(x_1) = V(x_2)$ (in this sense there exist sufficiently many linear conservation laws).

The equivalent algebraic definition of s.g.s is given by the following propositions.

Lemma 4.1.1 *For a population to have s.g.s it is necessary and sufficient that the corresponding evolutionary algebra $\mathcal{A}_V$ would be conservative.*

Proof The sufficiency immediately follows from Theorem 3.3.4, since $V(x) = x^2$. To prove the necessity let us consider the basic simplex Δ^{n-1}. Let $x \in \mathrm{Int}\Delta^{n-1}$ and $y \in J^{\perp}$. Then $x + \lambda y \in \Delta^{n-1}$ for sufficiently small $|\lambda|$. Therefore, $(x + \lambda y)^2 = x^2$ and it follows that $xy = 0$. It means that

$y \in \text{ann} \mathcal{A}_V$. We have the inclusion $J^{\perp} \subset \text{ann } \mathcal{A}_V$ which proves that the algebra is conservative. $\square$

Now it is easy to show that in the presence of s.g.s. the evolutionary operator is parametrized by the invariant linear forms.

Let us consider the cone C of nonnegative invariant linear forms. It is finitely generated. Its basis

$$p_i = \sum_{j=1}^{n} \pi_{ij} x_j \qquad (1 \le i \le r) \qquad (4.1.1)$$

is uniquely determined by the norming condition $\| p_i \| = 1$ for $i = 1, \ldots, r$, i.e.

$$\max_{j} \pi_{ij} = 1 \qquad (1 \le i \le r). \qquad (4.1.2)$$

In the presence of s.g.s. we will denote by μ map (4.1.1) of Δ^{n-1} into $\mathbf{R}^r$ and call it the *canonical meiosis operator*. In what follows the attribute "canonical" will be usually omitted, since other meiosis operators in s.g.s. will not be introduced (it would be permissible but not expedient, since only the canonical meiosis operator allows the genetic interpretation).

Lemma 4.1.2 *If a population has s.g.s., then it is two-level with the meiosis operator μ, quadratic fertilization operator ϕ:*

$$x'_j = \sum_{i,k=1}^{r} c_{ik,j} p_i p_k \qquad (4.1.3)$$

$(c_{ki,j} = c_{ik,j})$ *and identically equilibrious gamete pool* $(\mu\phi = \text{id})$.

Proof Choose some basis $f_1, \ldots, f_m$ in J where m (here and to the end of Chapter 4) denotes dim J. According to Corollary 3.3.5

$$V(x) = \sum_{j,l=1}^{m} f_j(x) f_l(x) u_{jl}. \qquad (4.1.4)$$

But the cone C generates J. Hence, every form f_j is a linear combination of the form $p_1, \ldots, p_r$ and

$$V(x) = \sum_{i,k=1}^{r} p_i(x) p_k(x) v_{ik} \qquad (4.1.5)$$

where v_{ik} are some linear combinations of the vectors u_{jl}. Let

$$v_{ik} = \sum_{j=1}^{n} c_{ik,j} e_j. \qquad (4.1.6)$$

Introducing the quadratic operator ϕ of the form (4.1.3) with the coefficients $c_{ik,j}$ defined by (4.1.6), we obtain $V = \phi\mu$ according to (4.1.5). On the other hand, $\mu V = \mu$ since the forms p_i $(1 \le i \le r)$ are invariant. Hence, $\mu\phi\mu = \mu$. Restricting the operator ϕ to Im μ (in a manner suiting a two-level population) we have $\mu\phi = \mathrm{id}$. $\square$

According to Theorem 1.3.4 the operator ϕ should be quadratic, but the reference to this theorem is justified only when the population is two-level.

Formula (4.1.5) (or the preceding formula (4.1.4)) gives a parametrization of an evolutionary operator via the invariant linear forms. It is obvious that *if, on the contrary, the evolutionary operator is parametrized via some set of invariant linear forms, then the population has s.g.s.* For example, *the Mendel multiallelic population has s.g.s.* Its (canonical) meiosis operator coincides with that described in section 2.1. Indeed, by the Hardy-Weinberg Law (2.1.11) $V(x)$ depends on $p_1, \ldots, p_m$ and $\sum_i p_i = s$. Thus, any invariant linear form depends on $p_1, \ldots, p_m$ only, since $f(x) = f(V(x))$. Therefore, the linear forms f, $p_1, \ldots, p_m$ are functionally dependent (in the domain $s(x) \ne 0$) which implies that they are linearly dependent. Since $p_1, \ldots, p_m$ are linearly independent by virtue of (2.1.11) they form the basis of the space J. If, at the same time, $f \in C$, i.e. $f \ge 0$, then all $\lambda_i \ge 0$ in the expansion $f = \sum_i \lambda_i p_i$ since a coordinate x_{ii} enters only p_i (with the coefficient 1). Thus $\{p_i\}_1^m$, is the normalized base for the cone C.

Lemma 4.1.2 may be inverted: *if a population is two-level with the identically equilibrious gamete pool, then it has s.g.s.* Indeed, if $V = \phi\mu$ and $\mu\phi = \mathrm{id}$, then $\mu V = \mu$, i.e. the coordinate form of the operator μ is given by the invariant linear forms, and as $V = \phi\mu$, these forms parametrize the evolutionary operator V.

As in Section 2.1, the fact that the evolutionary operator on the gamete level is unit implies the stationarity.

Corollary 4.1.3 *In the presence of s.g.s. the evolutionary operator is Bernstein:* $V^2 = V$.

Proof $V^2 = \phi\mu\phi\mu = \phi\mu = V$. $\square$

Thus, the Stationarity Principle is valid in the presence of s.g.s., but it is a more general property .

Example Let us consider the evolutionary operator in Δ^2:

$$x_1' = x_1(s + w)$$

$$x_2' = \frac{1}{2}((x_2 + x_3)s - x_1 w) \qquad (4.1.7)$$

$$x_3' = \frac{1}{2}((x_2 + x_3)s - x_1 w)$$

where, as usual, $s = x_1 + x_2 + x_3$ (i.e. $s|\Delta^2 = 1$) and $w = x_3 - x_2$. It is evident that $w' = 0$, i.e. the linear form is disappearing. As a result, $x_i'' = x_i'$ ($i = 1, 2, 3$), i.e. operator (4.1.7) is Bernstein. Further, if a linear form $f(x) = \alpha_1 x_1 + \alpha_2 x_2 + \alpha_3 x_3$ is invariant, i.e. $f' = sf$ then it follows from (4.1.7) that $(\alpha_1 - \frac{1}{2}(\alpha_2 + \alpha_3))x_1 w$ is divisible by s. This is possible for $\alpha_1 = \frac{1}{2}(\alpha_2 + \alpha_3)$ only. But in this case the identity $f' = sf$ takes on the form $f(x) = \alpha_1(x_1 + x_2 + x_3) = \alpha_1 s(x)$ (on dividing by s). Thus, there is no nontrivial invariant linear form here and, hence, a parametrization of operator (4.1.7) by such forms is impossible, i.e. there is no s.g.s.

The meiosis operator coefficients in s.g.s. satisfy the inequalities $0 \leq \pi_{ij} \leq 1$. This makes possible their interpretation in probabilistic terms related to genes. Let us introduce the symbols $A_1, \ldots, A_r$ which will be formally called *genes*. On the other hand, we have the zygotes (types) $e_1, \ldots, e_n$. The quantity π_{ij} is said to be *the probability that the gene A_i is generated by the zygote e_j* (via meiosis). This interpretation requires that some norming conditions should be fulfilled, for example,

$$\sum_{i=1}^{r} \pi_{ij} = 1 \qquad (1 \leq j \leq n). \qquad (4.1.8)$$

Accordingly the variable p_i should be called the probability of the gene A_i. In such a fashion, we identify the gametes with the genes and the gamete pool with *the gene pool*. This will be justified by an interpretation below in which the genes will turn out to be allelic.

If the population is in a state x, then its gene pool is in the state $\mu(x) = (p_1(x), \ldots, p_r(x))$ arising in the meiosis. The probabilities $p_i(x)$ are invariant, i.e. the Gene Conservation Law is valid: $p_i' = p_i$ ($1 \leq i \leq r$). If (4.1.8) is fulfilled then

$$\sum_{i=1}^{r} p_i = s, \qquad (4.1.9)$$

i.e. the linear forms $\{p_i\}_1^r$ generate a partition of the weight (and, conversely, this obviously implies (4.1.8)). But the validity of (4.1.9) is not necessary for s.g.s. If (4.1.9) is fulfilled, then s.g.s. is called *an elementary gene structure*, briefly, *e.g.s.* *The Mendel multiallelic population has e.g.s.* by virtue of (2.1.11). The simplest example of a nonelementary s.g.s. is following.

Example (Quadrille Law). Let us consider the evolutionary operator in Δ^3:

$$x_1' = (x_1 + x_2)(x_1 + x_3), \quad x_2' = (x_1 + x_2)(x_2 + x_4)$$

$$x_3' = (x_3 + x_4)(x_1 + x_3), \quad x_4' = (x_3 + x_4)(x_2 + x_4). \tag{4.1.10}$$

The forms

$$p_1 = x_1 + x_2, \quad p_2 = x_3 + x_4$$

$$q_1 = x_1 + x_3, \quad q_2 = x_2 + x_4 \tag{4.1.11}$$

are invariant and at the same time

$$x_1' = p_1 q_1, \quad x_2' = q_1 q_3, \quad x_3' = p_2 q_1, \quad x_4' = p_2 q_2. \tag{4.1.12}$$

Hence, s.g.s is present here.

Since $\operatorname{rank}(p_1, q_1, p_2, q_2) = 3$, then $\dim J \geq 3$. On the other hand, $\dim J \leq 4$, trivially. But not all linear forms are invariant (for example, x_1). Hence, $\dim J = 3$. Any three out of the four forms p_1, p_2, q_1, q_2 provide a basis for J. Let us show that together they form the basis of the cone (evidently, normalized). This means that (4.1.11) is *the meiosis operator* and we note at once that, instead of (4.1.9) there are two expansions of the weight,

$$p_1 + p_2 = q_1 + q_2 = s \tag{4.1.13}$$

i.e. the given s.g.s is nonelementary.

Let $f(x) = \alpha_1 x_1 + \alpha_2 x_2 + \alpha_3 x_3 + \alpha_4 x_4 \in C$. Representing f as a linear combination, $f = a_1 p_1 + a_2 p_2 + b_1 q_1 + b_2 q_2$ we obtain by comparing the coefficients at x_i ($1 \leq i \leq 4$): $a_1 + b_1 = \alpha_1$, $a_1 + b_2 = \alpha_2$, $a_2 + b_1 = \alpha_3$, $a_2 + b_2 = \alpha_4$. Therefore, $\alpha_1 + \alpha_4 = \alpha_2 + \alpha_3$.

Let $\alpha_1 \geq \alpha_2$. Put $\rho = \alpha_1 - \alpha_2 = \alpha_3 - \alpha_4$. Then $\alpha_1 = \alpha_2 + \rho$, $\alpha_3 = \alpha_4 + \rho$ and $f = \alpha_2 p_1 + \alpha_4 p_2 + \rho q_1$. At the same time $\alpha_2, \alpha_4, \rho \geq 0$ which means that p_1, p_2, q_1, q_2 generate the cone C. They are its extreme points, since if, for example, $p_1 = \lambda p_2 + \omega_1 q_1 + \omega_2 q_2$ then $\lambda + \omega_1 = 0$ (the coefficient at x_3), $\lambda + \omega_2 = 0$ (the coefficient at x_4), and one of the coefficients $\lambda, \omega_1, \omega_2$ would be negative.

The natural genetic interpretation of the Quadrille Law is the following. Let us relate the genes A_1, A_2, B_1, B_2 to the forms p_1, p_2, q_1, q_2 respectively, and call A_1 and A_2 *male genes* and B_1 and B_2 *female genes*. Thus, at the gene level the sex differentiation will be observed, as at the same time, the zygotes e_1, e_2, e_3, e_4 are bisexual or hermaphroditic. Let us define their genotypes to be $A_1B_1, A_1B_2, A_2B_1, A_2B_2$ in accordance with (4.1.12). In the process of meiosis every zygote generates the two genes constituting it, and these genes enter the male and female parts of the gene pool (with the probability 1). This is in exact correspondence with the meiosis operator (4.1.11) and normalizing relations (4.1.13). The *fertilization operator* (4.1.12) describes the formation of zygotes of the next generation by the union of the corresponding genes chosen from the male or female parts of one gene pool at random and independently.

4.2 Zygote Genotypes. Principle of Gene Inheritance

Starting with the present section and to the end of this chapter, we assume that each population has s.g.s. It means that the heredity is controlled by Gene Conservation Law.

Let us call a set of genes G_j *the reduced genotype of the zygote* e_j if the probability of their generation by e_j in meiosis is positive: $G_j = \{A_i : \pi_{ij} > 0\}$. For example, in the Mendel population the reduced genotype of the homozygote AA is $\{A\}$ and the reduced genotype of the heterozygote Aa is $\{A, a\}$.

Lemma 4.2.1 *The reduced genotype of every zygote is nonvoid, i.e. every zygote generates at least one gene via meiosis.*

Proof Let us consider some weight expansion $s = \sum_i \lambda_i p_i$. Then $\sum_i \lambda_i \pi_{ij} = 1$ for $j = 1, \ldots, n$, which implies that for every j there exists such an i ($1 \leq i \leq r$) that $\pi_{ij} \neq 0$, i.e. $\pi_{ij} > 0$ since $\pi_{ij} \geq 0$. $\square$

The Principle of Gene Inheritance, proved below, may be formulated in genetic language as follows: *if a zygote generates some gene in meiosis, then this gene participates in all fertilization acts, leading to the origination of the given zygotes*, i.e. this zygote cannot originate without this gene's participation. In this sense, s.g.s. rules out self-originating of genes.

The mathematical formulation of the Principle of Gene Inheritance requires the additional assumption of the nonnegativity of the coefficients $c_{ik,j}$

in a quadratic representation of the fertilization operator. As stated in section 1.3, this representation, generally speaking, is not unique. For uniqueness it is necessary and sufficient that the linear forms $p_1, \ldots, p_r$ be linearly independent, which is equivalent to e.g.s. as we will see below. In the case of e.g.s. it will be proved (Section 4.3) that $c_{ik,j} \geq 0$, in the general case of s.g.s. it will be proved (section 4.4) that there exists such a quadratic representation of the fertilization operator for which $c_{ik,j} \geq 0$. Thus, finally, the Principle of Gene Inheritance for s.g.s does not need additional assumptions, but for now we cannot dismiss them and we will call a s.g.s. *nonnegative* if the fertilization operator admits a quadratic representation with the coefficients $c_{ik,j} \geq 0$. If $c_{ik,j} > 0$ (or $c_{ik,j} = 0$) for some i, k, j, then the fertilization act uniting the genes A_i and A_k leads (respectively, does not lead) to arising of the zygote e_j. Such is the natural interpretation of formulas (4.1.3) for $c_{ik,j} \geq 0$, since $p_i p_k$ is the probability of the combination of the genes A_i and A_k, represented in the gene pool with the probabilities p_i and p_k respectively.

Theorem 4.2.2 (Principle of Gene Inheritance) . *Let all $c_{ik,j} \geq 0$ in a population which has s.g.s. If $\pi_{ij} > 0$ for some i, j then $c_{\alpha\beta,j} = 0$ when both α and β are different from i.*

Proof Let us represent the evolutionary operator in the form

$$x'_\gamma = \sum_{\alpha=1}^{r} p_\alpha(x) P_{\alpha\gamma}(x) \qquad (1 \leq \gamma \leq n)$$

where

$$P_{\alpha\gamma}(x) = \sum_{\beta=1}^{r} c_{\alpha\beta,\gamma} p_\beta(x). \qquad (4.2.1)$$

By the invariance of the form $p_i(x)$ we have (in the simplex Δ^{n-1})

$$p_i(x) = p_i(V(x)) = \sum_{\gamma=1}^{n} \pi_{i\gamma} x'_\gamma = \sum_{\alpha=1}^{r} p_\alpha(x) P_\alpha(x). \qquad (4.2.2)$$

Here

$$P_\alpha(x) = \sum_{\gamma=1}^{n} \pi_{i\gamma} P_{\alpha\gamma}(x) \qquad (1 \leq \alpha \leq r) \qquad (4.2.3)$$

Let us restrict ourselves in (4.2.2) to such $x \in \Delta^{n-1}$ for which $\operatorname{supp} x \cap \operatorname{supp} p_i = \emptyset$, i.e. $p_i(x) = 0$. Since all linear forms p_α, P_α are nonnegative by virtue of nonnegativity of the coefficients $c_{\alpha\beta,\gamma}$ and $\pi_{i\gamma}$ it follows that

$$p_\alpha(x)P_\alpha(x) = 0 \qquad (1 \le \alpha \le r) \qquad (4.2.4)$$

on the entire coordinate subspace L generated by the associated face of Δ^{n-1}. The identities (4.2.4) are meaningful for $\alpha \ne i$ as $p_\alpha|L \ne 0$. Indeed

$$\operatorname{supp} p_\alpha \not\subset \operatorname{supp} p_i \qquad (\alpha \ne i) \qquad (4.2.5)$$

since in the opposite case the combinations $p_i \pm \lambda p_\alpha$ for small $|\lambda|$ belong to the cone C, and their half-sum is equal to p_i in contradiction to the fact that p_i is an extreme point in C.

Now it follows from (4.2.4) that $P_\alpha(x) = 0$ for $x \in L$, $\alpha \ne i$. Hence, $\operatorname{supp} P_\alpha \subset \operatorname{supp} p_i (\alpha \ne i)$ which by (4.2.3) and nonnegativity of the coefficients $\pi_{i\gamma}$ implies $\operatorname{supp} P_{\alpha\gamma} \subset \operatorname{supp} p_i$ if $\alpha \ne i$ and $\pi_{i\gamma} > 0$. In particular, $\operatorname{supp} P_{\alpha j} \subset \operatorname{supp} p_i$ $(\alpha \ne i)$ as, by the assumption, $\pi_{ij} > 0$. Because the coefficients $c_{\alpha\beta,j}$ are nonnegative it follows now from (4.2.1) for $\gamma = j$ that $\operatorname{supp} p_\beta \subset \operatorname{supp} p_i$ if $c_{\alpha\beta,j} > 0$ and $\alpha \ne i$. For $\beta \ne i$ it contradicts (4.2.5). So if $\alpha \ne i, \beta \ne i$ then $c_{\alpha\beta,j} = 0$. $\square$

Corollary 4.2.3 *If $\pi_{ij} > 0$, then the quadratic form x'_j (i.e. the j-th coordinate of $V(x)$) is divisible by $p_i(x)$ and the quotient is a nonnegative invariant linear form.*

Proof Indeed, the Principle of Gene Heredity implies

$$x'_j = c_{ii,j}p_i^2 + 2p_i \sum_{\alpha \ne i} c_{\alpha i}p_\alpha. \qquad (4.2.6)$$

This formula, related to the fertilization operator, is transformed by the substitution $p_i = p_i(x)$ into the i-th evolutionary equation. $\square$

Corollary 4.2.3 is a working formulation, and we will also call it the Principle of Gene Heredity.

Because a quadratic form which is not identically equal to zero, cannot have more than two linear divisors, Corollary 4.2.3 implies

Corollary 4.2.4 *Any nondisappearing zygote generates not more than two genes in the meiosis.*

If a zygote generates one gene only, for example, A_i, then it is called *a homozygote* and the combination A_iA_i—its *genotype*. If a zygote generates two genes, for example $A_i, A_k (i \neq k)$ then it is called *a heterozygote* and either combination A_iA_k, A_kA_i—its *genotype*.Thus, the genotype of any nondisappearing zygote contains two genes, which may be identical. In this sense the zygotes are *diploid*.

For the above-described examples of s.g.s., i.e. Mendel and quadrille, the general notion of a genotype which we have just introduced coincides with that in the previous sense. By the way, a quadrille population has no homozygotes. This example shows that not all pairwise gene combinations are necessarily zygote genotypes.

Different zygotes may have the same genotype.

Example Let us consider the s.g.s. (even e.g.s.) with fertilization operator

$$x_1' = p_1^2, \; x_2' = p_2^2, \; x_3' = p_1p_2, \; x_4' = p_1p_2$$

and meiosis operator

$$p_1 = x_1 + \frac{1}{2}x_3 + \frac{1}{2}x_4, \quad p_2 = x_2 + \frac{1}{2}x_3 + \frac{1}{2}x_4$$

(it is easy to check that the latter is canonical). There are two genes, A_1 and A_2 here corresponding to the forms p_1 and p_2. The zygotes e_1, e_2, e_3, e_4 have the genotypes $A_1A_1, A_2A_2, A_1A_2, A_1A_2$.

Note that the population in this example is internally reducible: $V(x)$ depends on $x_1, x_1, x_3 + x_4$ only. After the internal union of the zygotes e_3, e_4 the population is transformed to the Mendel diallelic one, i.e. we get the Hardy-Weinberg Law $x_1' = p_1^2, \; x_2' = p_2^2, \; \tilde{x}_3' = 2p_1p_2$ where $\tilde{x}_3 = x_3 + x_4$, $p_1 = x_1 + \frac{1}{2}\tilde{x}_3, \; p_2 = x_2 + \frac{1}{2}\tilde{x}_3$.

The coincidence of genotypes of the different zygotes is an artifact. We want to eliminate all such anomalous differentiations. The presence of disappearing zygotes (population degeneracy) is also anomalous, since the genotypes of such zygotes are not defined. The following theorem clears up these situations.

Theorem 4.2.5 *Let nondegenerate population have a nonnegative s.g.s. Its different zygotes have distinct genotypes if and only if the population is normal.*

Note, once more, that the a priori condition of nonnegativity of s.g.s. is temporary.

Proof If the population is internally reducible, then there exist internally kin zygotes, for example, e_1 and e_2. This means that $e_1 \equiv e_2 (\mathrm{mod}\, J^\perp)$. But then $p_i(e_1) = p_i(e_2)$ i.e. $\pi_{i1} = \pi_{i2}$ $(1 \le i \le r)$, and hence the genotypes of e_1 and e_2 coincide.

If the population is externally reducible, then there exist externally kin zygotes, for example e_1 and e_2. It means that,

$$x_1' = cx_2' \qquad\qquad (c = \mathrm{const} > 0) \qquad\qquad (4.2.7)$$

in the evolutionary equations. Let the zygote e_1 generate the gene A_1. Then by the Principle of Gene Inheritance we have $x_1' = p_1 f$ where f is a non-negative invariant linear form. If the zygotes e_1 or e_2 generate a gene different from A_1, for example A_2, then by the Principle Gene Inheritance p_2 divides x_1' or p_2 divides x_2' and then by (4.2.7) p_2 again divides x_1', i.e. in any case f is a multiple of p_2. Thus either e_1 and e_2 are of the same genotype $A_1 A_1$ or f is a multiple of some form $p_2, \ldots, p_r$, for example, p_2. It remains to consider the latter possibility. We have

$$x_1' = c_1 p_1 p_2, \quad x_2' = c_2 p_1 p_2 \qquad\qquad (c_1 > 0, c_2 > 0). \qquad (4.2.8)$$

Under these conditions each zygote e_1, e_2 may have the genotype $A_1 A_1$ or $A_2 A_2$ or $A_1 A_2$. Suppose that the genotypes of these zygotes are different. Then at least one of them is a homozygote. Let e_1, say, have genotype $A_1 A_1$. This means that x_1 is present in p_1, but is not present in $p_1, \ldots, p_r$. Then it follows from (4.2.8) that the quadratic forms x_1', x_2' do not contain x_1^2. But $\sum_j x_j' = s^2 = (\sum_j x_j)^2 = x_1^2 + \cdots$. So $n > 2$ and for some $j > 2$ the form x_j' contains x_1^2. Let it be x_3', for example. But x_1^2 must then be present in x_3' through the fertilization operator due to p_1^2, i.e. $x_3' = c_{11,3} p_1^2 + \ldots$ where $c_{11,3} > 0$. It follows now from the Principle of Gene Inheritance that the zygote e_3 has the genotype $A_1 A_1$, i.e. the same as the zygote e_1.

Thus, it is proved that if the different zygotes have different genotypes, then the population is internally and externally irreducible. Besides, we have assumed that it is nondegenerate. Hence, the population is normal. To prove the converse (and for later applications) we will need the following lemma (which does not require any condition related to normality).

Lemma 4.2.6 *If e_j is a homozygote and $A_i A_i$ is its genotype, then*

$$\pi_{ij} = 1 \qquad\qquad (4.2.9)$$

and every expansion of the weight contains p_i with the coefficient 1.

Proof According to Lemma 3.7.12 there exists an expansion of the weight containing p_i with the coefficient 1:

$$s = p_i + \sum_{k \neq i} \lambda_k p_k.$$

But by the definition of a homozygote the linear forms p_k $(k \neq i)$ do not contain x_j. So x_j is present in p_i with the same coefficient as in s, i.e. $\pi_{ij} = 1$.

Let $s = \sum_k \alpha_k p_k$ be some expansion of the weight. Comparing the coefficients at x_j, we obtain $1 = \alpha_i \pi_{ij} = \alpha_i$ since x_j is present in p_i only. $\square$

Thus, *every homozygote generates the corresponding gene with probability* 1.

Going back to the proof of Theorem 4.2.5, suppose that the population is normal. If it contains two homozygotes say e_1 and e_2 with the same genotype, say $A_1 A_1$, then by Lemma 4.2.6 $\pi_{11} = \pi_{12} = 1$, i.e. $p_1 = x_1 + x_2 + \cdots$ and $p_2, \ldots, p_r$ do not depend on x_1 and x_2. This contradicts to the internal irreducibility. If the population contains two heterozygotes, say e_1 and e_2 with the same genotype, say, $A_1 A_2$, then by the Principle of Gene Heredity $x'_1 = c_1 p_1 p_2$, $x'_2 = c_2 p_1 p_2$ where $c_1 > 0, c_2 > 0$ owing to the nondegeneracy. Therefore, $x'_1 = c x'_2$ $(c > 0)$ which contradicts external irreducibility. $\square$

Theorem 4.2.5 states that the population differentiation with s.g.s. is genotypical if and only if the population is normal. For this reason it is possible to restrict oneself to normal populations for the problem of an explicit description of s.g.s. At the same time the general case is reduced to the normal one by normalization. It is convenient to start this process by eliminating the disappearing zygotes. Afterwards the population becomes and remains nondegenerate. The eliminating step, as well as the next one (union of the internally or externally kin zygotes) decreases the number of zygotes.

Let us point out the main invariants of the normalization process.

Theorem 4.2.7 *The normalization process 1) preserves s.g.s.; 2) the space of invariant linear forms is subjected to an isometric cone isomorphism; 3) the process preserves the state space of the gene pool.*

By assertion 2) the gene set does not change in the process of normalization.

Proof Assertion 1) follows from Lemma 4.1.1 and invariance of the class of conservative basic algebras under normalization. Assertion 2) is contained in Theorem 3.9.8. To prove 3) let us consider one step of the normalization procedure. The internal kinship $h : (\mathcal{A}, \Delta) \to (\tilde{\mathcal{A}}, \tilde{\Delta})$ (or external kinship $h : (\tilde{\mathcal{A}}, \tilde{\Delta}) \to (\mathcal{A}, \Delta)$) is related to this step. According to Theorems 3.9.3 and 3.9.6 h^* determines the isometric cone isomorphism $\tilde{J} \to J$ (or $J \to \tilde{J}$).

Let $\mu, \tilde{\mu}$ be the corresponding meiosis operators. Because h^* bijectively maps a normalized basis of the cone $\tilde{C}$ (or C) onto the normalized basis of the cone C (or $\tilde{C}$), these bases may be identified. Then $\mu = \tilde{\mu} h | \Delta$ for the internal kinship and $\tilde{\mu} = \mu h | \tilde{\Delta}$ for the external kinship. In the former case, according to Theorem 3.9.1, $h(\Delta) = \tilde{\Delta}$ and, hence, Im $\tilde{\mu} =$ Im μ. In the latter case, according to Theorem 3.9.4, $h(\tilde{\Delta}) = \Delta \cap$ Im h and Im $\tilde{\mu} =$ Im $\mu | (\Delta \cap$ Im $h) =$ Im μ since any invariant linear form takes on the equal values at the points $x \in \Delta$ and x^2, and $x^2 \in$ Im h for the external kinship. $\square$

In the normalization process identification of internally or externally kin zygotes occurs. This identification, generally speaking need not agree with the genotypical differentiation.

Example Let us consider e.g.s. with the fertilization operator

$$x_1' = p_1^2 + \frac{1}{4}p_1 p_2, \ x_2' = p_2^2 + \frac{1}{2}p_1 p_2, \ x_3' = \frac{1}{4}p_1 p_2, \ x_4' = p_1 p_2$$

and the meiosis operator

$$p_1 = x_1 + x_3 + \frac{1}{2}x_4, \ p_2 = x_2 + \frac{1}{2}x_4.$$

The zygotes e_1, e_2, e_3, e_4 have the genotypes $A_1 A_1, A_2 A_2, A_1 A_1, A_1 A_2$ respectively. The genotype of the zygotes e_3, e_4 are different but, nevertheless, they are externally kin. Their union results in the heterozygote $A_1 A_2$, i.e. the genotype is preserved for e_4 and is changed for e_3.

Thus, the notion of a zygote genotype has an invariant sense in a normal population only. By the way, the same example shows that e_3 is internally kin to e_1 and their genotypes coincide. If e_3 is identified not with e_4 but with e_1, then it will agree with the genotypical differentiation.

The normalization process is called *genotypically correct* if the zygote identification agrees with the genotypical differentiation at each step.

Theorem 4.2.8 *Any population having s.g.s admits a genotypically correct normalization.*

Proof Let us start the normalization by eliminating disappearing zygotes. Genotypically, this step is correct, since it is not accompanied with any identification of zygotes. After that, the population becomes and remains nondegenerate, which we will forth assume.

The internal union of zygotes is always genotypically correct as was seen in the beginning of proof of Theorem 4.2.5. As a result of the internal union of zygotes, the population (having s.g.s) becomes internally irreducible. Indeed, every invariant linear form $\tilde{f}$ on the quotient algebra $(\mathcal{A}, \Delta)/J^{\perp} \equiv (\tilde{\mathcal{A}}, \tilde{\Delta})$ is defined by some invariant linear form f on the algebra $(\mathcal{A}, \Delta)$. If $E_1, \ldots, E_n$ are the images of the elements $e_1, \ldots, e_n$ and, for example, $E_1 \neq E_2$ then E_1 and E_2 are not internally kin in the algebra $(\tilde{\mathcal{A}}, \tilde{\Delta})$ since if $\tilde{f}(E_1) = \tilde{f}(E_2)$ for all $\tilde{f} \in \tilde{J}$ then $f(e_1) = f(e_2)$ for all $f \in J$ which implies $E_1 = E_2$.

If a population is internally irreducible, then the internal union of zygotes is genotypically correct. To be more exact, in this case all externally kin zygotes have the same heterozygote genotype. It also follows from the proof of Theorem 4.2.5. Indeed, as was shown, an internally irreducible population contains no two homozygotes with the same genotype, and in this situation any two externally kin zygotes have the same heterozygote genotype.

Now define the genotypically correct normalization process by induction. The initial step is elimination of disappearing zygotes. If, after that, the population is internally reducible, then the next step is the internal union of zygotes. If it is internally irreducible but externally reducible, then the next step is the external union of zygotes. If it is both internally and externally irreducible, then it is normal and the process of over. $\square$

For a genetically correct normalization the genotypes of all (nondisappearing) zygotes are preserved. Indeed, if, for example, $e_1, \ldots, e_\nu$ are internally united in E_1, then in $p_1, \ldots, p_r$ the sum $x_1 + \cdots + x_\nu$ is substituted by some variable s_1. The latter, evidently, enters those and only those forms which do not contain x_1 (and simultaneously $x_2, \ldots, x_\nu$). This means that E_1 is of the same genotype as e_1 (and simultaneously $e_2, \ldots, e_\nu$). If $e_1, \ldots, e_\nu$ are externally united in ϵ_1 (and the population is internally irreducible), then $p_1, \ldots, p_r$ undergo the change of variables $x_i = \alpha_i \xi_1$ $(1 \leq i \leq \nu)$ where all $\alpha_i > 0$, $\sum_i \alpha_i = 1$. At the same time, two of these forms, for example, p_1 and p_2 contain all the coordinates and the remaining forms $p_3, \ldots, p_r$ are independent of these coordinates. Hence, ξ_1 enters p_1, p_2 and does not enter $p_3, \ldots, p_r$. The zygote ϵ_1 is of the same genotype as e_1 (and simultaneously $e_2, \ldots, e_\nu$). In general, *a heterozygote type is preserved under any normalization* but they can "swallow" the homozygote genotypes associated with them.

Let us consider how the normalization affects splitting. Recall a non-splitting zygote is such a type for which all offspring belong to the same type, i.e. it is a basis idempotent.

Theorem 4.2.9 *Nonsplitting is preserved under normalization.*

A splitting however, may fail. In the aforesaid example e_3 is a splitting homozygote. After the internal union with e_1 (which is genotypically correct) one obtains a nonsplitting homozygote.

Proof In the process of the internal union the image of an idempotent in the corresponding quotient algebra remains idempotent. External union is ruled out, since any basis idempotent is externally isolated. Indeed, if, say, $e_1^2 = e_1$ then $x_1' = x_1^2 + \cdots$ and so x_1^2 does not enter x_j' $(j > 1)$ i.e. x_j' for $j > 1$ cannot be a multiple of x_1'. $\square$

4.3 Elementary Gene Structure (E.G.S.)

An elementary gene structure (e.g.s.), according to the definition given in section 4.1, is an s.g.s. such that $\sum_i p_i = s$, i.e. $\sum_i p_i(x) = 1$ for the states $x \in \Delta^{n-1}$,. It can be interpreted as saying the genes $A_1, \ldots, A_r$ belong to the same, a fortiori unique, part of the gene pool. In turn, this may be described as allelism in the absence of gene sex differentiation.

Theorem 4.2.7 implies the preservation of e.g.s. under normalization.

In the case of e.g.s. the problem of the explicit description will be solved in the present section.

Lemma 4.3.1 *For a s.g.s. to be elementary it is necessary and sufficient that the forms p_i ($1 \le i \le r$) be linearly independent, i.e. the cone C of nonnegative invariant linear forms be symplicial.*

Proof In the case of e.g.s. $\{p_i\}_1^r$ is a partition of the weight and hence, is a Gaussian and moreover, linearly independent, system. Conversely, if the system $\{p_i\}_1^r$ is linearly independent, then the expansion $s = \sum_i \lambda_i p_i$ is unique. But by Lemma 3.7.12 for every i there exists such a decomposition of the weight which contains p_i with the coefficient 1. Therefore, all $\lambda_i = 1$ $(1 \le i \le r)$. $\square$

Corollary 4.3.2 *For a s.g.s. to be elementary it is necessary and sufficient that $r = m$, i.e. ord $C = \dim J$.*

In the following discussion of e.g.s. we will use m instead of r.

Corollary 4.3.3 *If $m \le 2$ or $n \le 3$, then a s.g.s. is elementary.*

Proof Indeed, any cone in a space of a dimension not exceeding 2 is simplicial. It remains to note that for $m = n = 3$ the evolutionary algebra is unit. $\square$

Corollary 4.3.4 *The coefficients $c_{ik,j}$ in the quadratic representation of the fertilization operator are uniquely determined if and only if s.g.s is elementary.*

Lemma 4.3.5 *Every e.g.s. is nonnegative.*

Proof The system $\{p_i\}_1^m$ is Gaussian. In particular, we can reorder the canonical basis of $\mathcal{A}$ and $p_1, \ldots, p_m$ so that

$$p_i(e_i) = 1 \qquad\qquad (i = 1, \ldots, m)$$

$$p_j(e_i) = 0 \qquad\qquad (i, j = 1, \ldots, m;\ i \neq j). \qquad (4.3.1)$$

Therefore, $c_{ik,j}$ must be equal to the inheritance coefficient $p_{ik,j}$. But this is nonnegative. $\square$

Thus, if the population has elementary gene structure the Principle of Gene Inheritance of the previous section applies.

Theorem 4.3.6 *If the population has elementary gene structure we can choose the ordering of $\{e_1, \ldots, e_n\}$ and $\{p_1, \ldots p_m\}$ so that the evolutionary operator has the following form:*

$$x_i' = p_i P_i \qquad\qquad (i = 1, \ldots, m)$$

$$x_j' = p_{i_j} P_j \qquad\qquad (j = m+1, \ldots, n) \qquad (4.3.2)$$

where the invariant forms $\{p_1, \ldots, p_m\}$ and $\{P_1, \ldots, P_n\}$ are given by:

$$p_i(x) = x_i + \sum_{j=m+1}^{n} \pi_{ij} x_j, \qquad\qquad (i = 1, \ldots, m) \qquad (4.3.3)$$

and

$$P_l = \sum_{k=1}^{m} \tau_{lk} p_k, \qquad\qquad (l = 1, \ldots, n). \qquad (4.3.4)$$

Here (π_{ij}) is a column stochastic $m \times (n-m)$ matrix with $i_j \equiv \min\{i : \pi_{ij} > 0\}$ for each $j = m+1, \ldots, n$; (τ_{lk}) is a nonnegative $n \times m$ matrix which is related to (π_{ij}) by the equations:

$$\tau_{ii} + \sum_{j:i_j=i} \pi_{ij} \tau_{ji} = 1$$

$$\tau_{ik} + \sum_{j:i_j=i} \pi_{ij} \tau_{jk} + \sum_{j:i_j=k} \pi_{ij} \tau_{ji} = 1 \qquad (4.3.5)$$

$$\sum_{j:i_j \neq i} \pi_{ij} \tau_{jk} = 0$$

where $1 \leq i, k \leq m$, $m+1 \leq j \leq n$ and $i \neq k$.

Conversely, if the evolutionary operator of a population can be written in the above form then the population has e.g.s. with meiosis operator (4.3.3) and fertilization operator (4.3.2).

Proof (4.3.3) and stochasticity of the matrix (π_{ij}) follow because $p_1, \ldots, p_m$ are nonnegative forms satisfying (4.3.1) and

$$\sum_{i=1}^{m} p_i = s. \tag{4.3.6}$$

The equations (4.3.2) follow from the Principle of Gene Inheritance. The coefficients for quotient forms P_l are nonnegative because so are the coefficients of the fertilization operator. Then the equations (4.3.5) say exactly that the forms $p_1, \ldots p_m$ are invariant. For invariance says $p_i' = sp_i$ which says, using (4.3.6)

$$p_i P_i + \sum_{j=m+1}^{n} \pi_{ij} p_{i_j} P_j = p_i (\sum_{k=1}^{m} p_k) \tag{4.3.7}$$

or equivalently by (4.3.4):

$$p_i \sum_{k=1}^{m} \tau_{ik} p_k + \sum_{j=m+1}^{n} \pi_{ij} p_{i_j} \sum_{k=1}^{m} \tau_{jk} p_k = p_i (\sum_{k=1}^{m} p_k). \tag{4.3.8}$$

As $p_1, \ldots p_m$ are linearly independent forms we can equate the coefficients of $p_i p_k$ in (4.3.8) and get the equations (4.3.5).

For the converse, we have, so far, that $p_1, \ldots p_m$ are linear independent nonnegative invariant forms satisfying (4.3.6). We see that the population has s.g.s. and that the forms $p_1, \ldots p_m$ form a basis for J. In particular, if $f \in J$ we can write f as the linear combination $\sum_{i=1}^{m} \alpha_i p_i$, and if x is in the subspace with $x_{m+1} = \cdots = x_n = 0$ then by (4.3.3) $f(x) = \sum_{i=1}^{m} \alpha_i x_i$. In particular, if f is nonnegative the coefficients $\alpha_1, \ldots \alpha_m$ are nonnegative and so $p_1, \ldots, p_m$ generate the cone C. By linear independence of $p_1, \ldots p_m$ we see that the population has e.g.s. $\square$

From now on we will assume the ordering is chosen so that equations (4.3.2)-(4.3.5) hold.

Now we see the state space of the gene pool for a population admitting e.g.s. is the simplex Δ^{m-1}. The latter is the image of simplex Δ^{n-1} with respect to meiosis operator (4.3.3). Δ^{n-1} is the state space on the zygote level.

Consider e.g.s genotypically. We see from (4.3.3) that $e_1, \ldots, e_m$ are homozygotes. Zygote e_j for $j > m$ is a homozygote if and only if it is nondisappearing and there exists i such that $\pi_{ij} = 1$. This is possible only when the population is internally reducible.

Example Define a population with $n = 5$ having e.g.s. with meiosis operator

$$p_1 = x_1 + x_3 + \frac{1}{2}x_5, \; p_2 = x_2 + x_4 + \frac{1}{2}x_5$$

and with fertilization operator

$$x_1' = x_3' = \frac{1}{2}p_1^2, \; x_2' = x_4' = \frac{1}{2}p_2^2$$

$$x_5' = 2p_1p_2.$$

So the matrix (π_{ij}) is:

$$\begin{pmatrix} 1 & 0 & \frac{1}{2} \\ 0 & 1 & \frac{1}{2} \end{pmatrix}$$

Thus, not only e_1 and e_2 but e_3 and e_4 are homozygotes. Notice that all of the zygotes are splitting.

Theorem 4.3.7 *In a population with e.g.s. all of the heterozygotes are splitting. If a population with s.g.s. is internally irreducible then it has e.g.s. if and only if there are m nonsplitting zygotes, in which case these are the homozygotes $e_1, \ldots, e_m$.*

Proof If e_j has genotype $A_1 A_2$ then by the Principle of Gene Inheritance $x_j' = 2bp_1p_2$ with $b \geq 0$. By (4.3.1) the coefficients of $x_1 x_2$ in x_j' is $2b$, but

$$\sum_{k=1}^{n} x_k' = s(x') = s(x)^2 = (\sum_{k=1}^{n} x_k)^2 \tag{4.3.9}$$

and so the coefficient of $x_1 x_2$ in x_j' is at most 2, i.e. $b \leq 1$.

Now (4.3.6) implies $\pi_{1j} + \pi_{2j} \leq 1$. Consequently, $\pi_{1j}\pi_{2j} \leq \frac{1}{4}$ and so the coefficient of x_j^2 in x_j', $2b\pi_{1j}\pi_{2j}$, is at most $\frac{1}{2}$. But e_j is idempotent if and only if the coefficient of x_j^2 is x_j' is 1 (and so the coefficient of x_j^2 in x_k' with $k \neq j$ is zero). Hence, e_j is splitting.

If the population has s.g.s. then it is conservative by Lemma 4.1.1. If it is also internally irreducible we can apply Theorem 3.8.16 to observe that e_j is an idempotent vertex, i.e. a nonsplitting zygote, if and only if its image is an extremal of the cone Q in $\mathcal{A}/J^{\perp}$. Because Q is a proper cone in a space of dimension m its cone basis consists of at least m extrema with exactly m precisely when the cone is simplicial. This is equivalent to e.g.s. by Corollary 4.3.1. In sum a s.g.s. population which is internally irreducible has at least m nonsplitting zygotes and exactly m precisely when e.g.s. holds. $\square$

Corollary 4.3.8 *If an internally irreducible population has e.g.s then all the homozygotes are nonsplitting.*

In principle Theorem 4.3.6 gives a complete description of the evolutionary operator of an arbitrary population which has e.g.s. However, the parameters π_{ij} and τ_{li} are not independent because of the relations given in the equations of (4.3.5). When the population satisfies the biologically reasonable condition of normality, obtainable by kin reductions, then we can give a sharper description.

Theorem 4.3.9 *If a normal population has elementary gene structure one can write the evolutionary operator as follows:*

$$x_i' = p_i^2 + 2 \sum_{k \neq i:1}^{m} \theta_{ik} p_i p_k \qquad (i = 1, \ldots, m)$$

$$x_j' = 2\theta_j p_{i_j} p_{k_j}, \qquad (j = m+1, \ldots, n) \qquad (4.3.10)$$

where the invariant forms $\{p_1, \ldots, p_m\}$ are given by:

$$p_i(x) = x_i + \sum_{j=m+1}^{n} \pi_{ij} x_j, \qquad (i = 1, \ldots, m). \qquad (4.3.11)$$

Here (π_{ij}) is a column stochastic $m \times (n-m)$ matrix with exactly two nonzero entries in every column: In column j, we define $\pi_j = \pi_{i_j j}$ and $\overline{\pi}_j = \pi_{k_j j}$ with $i_j < k_j$, so that

$$\pi_j + \overline{\pi}_j = 1 \qquad j = m+1, \ldots, n. \qquad (4.3.12)$$

For distinct values of $j = m+1, \ldots, n$ the pairs (i_j, k_j) are different.
All $\theta_{ik} \geq 0$ and $\theta_j > 0$. They are related to the π_{ij}'s by the relations:

$$\theta_{i_j k_j} + \pi_j \theta_j = \frac{1}{2}, \quad \theta_{k_j i_j} + \overline{\pi}_j \theta_j = \frac{1}{2} \qquad (4.3.13)$$

for $j = m+1, \ldots n$, and for all pairs (i,k) with $i \neq k$ different from (i_j, k_j) and (k_j, i_j) for all $j = m+1, \ldots n$:

$$\theta_{ik} = \theta_{ki} = \frac{1}{2}. \qquad (4.3.14)$$

Conversely, if the evolutionary operator of the population has this form then it is normal and has e.g.s.

Proof Ordering according to Theorem 4.3.6, Corollary 4.3.8 implies that the homozygotes $e_1, \ldots e_m$ are nonsplitting, and so the coefficient of x_i^2 in x_i' is 1 for $i = 1, \ldots, m$. So (4.3.2)–(4.3.4) says that $\tau_{ii} = 1$ for $1 \le i \le m$. Thus, with $\theta_{ik} \equiv \frac{1}{2}\tau_{ik}$ ($1 \le i, k \le m$) we have the first equation of (4.3.10).

Because the population is internally irreducible for $j = m + 1, \ldots, n$, e_j is a heterozygote and so $\pi_{ij} < 1$ for all i by stochasticity. Notice that the population is nondegenerate and so every e_j is either a homo- or heterozygote. Also from nondegeneracy, the Principle of Gene Inheritance says that $\pi_{ij} > 0$ for two values of i. We label these two values i_j and k_j with $i_j < k_j$ and thence define π_j and $\overline{\pi}_j$ as in the statement. (4.3.12) follows from stochasticity. From the third equation of (4.3.5) and nonnegativity of all the terms we have $\pi_{ij}\tau_{jk} = 0$ if $i \ne k$ or i_j. If $i = k_j$ then $\pi_{ij} > 0$ and so

$$\tau_{jk} = 0 \qquad\qquad (k \ne k_j,\ j = m + 1, \ldots, n). \qquad\qquad (4.3.15)$$

So for $j > m$, P_j defined by (4.3.4) is $\tau_{jk_j}p_{k_j}$ and so with $\theta_j \equiv \frac{1}{2}\tau_{jk_j}$ we have (4.3.10) from (4.3.2). Because the population is nondegenerate the nonnegative number θ_j is, in fact, positive. Observe that the equations for x_j' follow directly from the Principle of Gene Inheritance applied to the heterozygote e_j.

Distinct values of j yield different pairs (i_j, k_j) for otherwise the population would be externally reducible.

From the second equation in (4.3.5) we have for $1 \le i, k \le m$ with $i \ne k$:

$$\tau_{ik} + \sum_{l : i_l = i} \pi_{il}\tau_{lk} + \sum_{l : i_l = k} \pi_{il}\tau_{li} = 1 \qquad\qquad (4.3.16)$$

First, set $i = i_j$ and $k = k_j$. Notice that if $l \ne j$ then $i_l = i = i_j$ implies $k_l \ne k_j = k$. So from (4.3.15) the only nonzero term in the first sum occurs with $l = j$ and all of the terms in the second sum equal zero. We are left with the first equation of (4.3.13) and the second follows similarly from the case $i = k_j$ and $k = i_j$. Finally, if the pair i, k with $i \ne k$ is none of the pairs (i_j, k_j) nor (k_j, i_j) then both sums in (4.3.16) vanish and so we get (4.3.14).

These arguments are reversible and so given (4.3.10)–(4.3.14) we can recover (4.3.5) and so by Theorem 4.3.6 the population has e.g.s. It is nondegenerate because all θ_j's are positive, externally irreducible because the pairs (i_j, k_j) are distinct, and internally irreducible because all the $\pi_{ij} < 1$. Thus, the population is normal. $\square$

Let us try to explain in genetic language the pattern of heredity for a normal population with e.g.s. We interpret the forms $p_1, \ldots, p_m$ as the probabilities in the gene pool of the "alleles" $A_1, \ldots, A_m$. Fertilization creates, in addition to the homozygotes $A_1 A_1, \ldots A_m A_m$, the heterozygotes $A_{i_{m+1}} A_{k_{m+1}}, \ldots, A_{i_n} A_{k_n}$. There are no more genotypes in the population and so

$$n \leq \frac{1}{2} m(m+1). \tag{4.3.17}$$

In meiosis the homozygotes and heterozygotes split to produce gametes but we have *meiotic drive* or perhaps *gamete choice* so that alleles may be introduced into the gene pool with unequal probabilities $\pi_j, \overline{\pi}_j$:

$$A_i A_i \mapsto A_i$$

$$A_{i_j} A_{k_j} \mapsto \pi_j A_{i_j} + \overline{\pi}_j A_{k_j} \tag{4.3.18}$$

represents the meiosis operator on the basis vectors.

In fertilization the meeting of two genes A_{i_j} and A_{k_j} yields the heterozygote $A_{i_j} A_{k_j}$ with probability θ_j but yields the homozygote $A_{i_j} A_{i_j}$ and $A_{k_j} A_{k_j}$ with probabilities $\theta_{i_j k_j}$ and $\theta_{k_j i_j}$ respectively. According to this

$$\theta_j + \theta_{i_j k_j} + \theta_{k_j i_j} = 1 \tag{4.3.19}$$

formally follows from (4.3.12) and (4.3.13). This nonMendel way of reaching homozygotes can be interpreted as a death of one of the genes with a duplication of its complement.

If the heterozygote $A_i A_k$ is forbidden, i.e. $i < k$ but (i, k) is not one of the pairs (i_j, k_j) then the meeting of gametes A_i and A_k yield homozygotes $A_i A_i$ and $A_k A_k$ with probability $\theta_{ik} = \theta_{ki} = \frac{1}{2}$.

The union of A_i with A_i yields the homozygote $A_i A_i$ with probability one.

The deviations from the usual Mendel model are balanced so that the gene probabilities remain stationary, i.e. $p_i' = p_i$ for $i = 1, \ldots, m$, which follows from (4.3.13). According to these *equations of gene balance* the deficit of an allele due to the meiotic drive is compensated by the duplication in fertilization. The expression $\theta_{i_j k_j} + \pi_j \theta_j$ gives the probability of the gene A_{i_j} in the next generation due to meeting with a gene of type A_{k_j} while A_{k_j}'s are produced with probability $\theta_{k_j i_j} + \overline{\pi}_j \theta_j$ and by (4.3.13) these probabilities are equal. Consequently, the relative frequencies of A_{i_j} versus A_{k_j} are not disturbed in the following generation.

In the simplest case $n = 3$ and $m = 2$. There are two alleles A_1 and A_2 and three genotypes $A_1 A_1, A_2 A_2$ and $A_1 A_2$. If the last is not forbidden then we get the evolutionary operator (cf. Section 2.1)

$$x'_1 = p_1^2 + 2\theta_{12} p_1 p_2$$
$$x'_2 = p_2^2 + 2\theta_{21} p_1 p_2 \qquad (4.3.20)$$
$$x'_3 = 2\theta_3 p_1 p_2.$$

Here we have

$$p_1 = x_1 + \pi_3 x_3, \; p_2 = x_2 + \overline{\pi}_3 x_3. \qquad (4.3.21)$$

The parameters $\pi_3, \overline{\pi}_3$ and θ_3 are positive, θ_{12}, θ_{21} are nonnegative and

$$\pi_3 + \overline{\pi}_3 = 1, \; \theta_{12} + \pi_3 \theta_3 = \frac{1}{2}, \; \theta_{21} + \overline{\pi}_3 \theta_3 = \frac{1}{2}. \qquad (4.3.22)$$

We can use as independent parameters π_3, θ_3 with $0 < \pi_3 < 1$ and $0 < \theta_3 < \min(\frac{1}{2\pi_3}, , \frac{1}{2(1-\pi_3)})$. So we get a two parameter family of normal e.g.s populations. The Mendel special case occurs when $\pi_3 = \frac{1}{2}$ and $\theta_3 = 1$ so that $\overline{\pi}_3 = \frac{1}{2}$ and $\theta_{12} = \theta_{21} = 0$.

In general, we get the Mendel population when $n = \frac{1}{2} m(m + 1)$ so that there are no forbidden pairs, alternate genes have the same probability in meiosis ($\pi_j = \overline{\pi}_j = \frac{1}{2}$) and only heterozygotes are formed at the meeting of different gametes ($\theta_j = 1$). It then follows that $\theta_{ik} = \frac{1}{2}$ for all pairs (i, k) with $i \neq k$.

We will refer to the evolutionary operator of Theorem 4.3.9 as the *Extended Hardy-Weinberg operator or Law*. Given an allele number m and a choice of marked pairs (i_j, k_j) $(j = m + 1, \ldots, n)$ the family of such operators is $2(n - m)$ dimensional with independent parameters $0 < \pi_j < 1$ and $\theta_j < \min(\frac{1}{2\pi_j}, \frac{1}{2\overline{\pi}_j})$ (with $\overline{\pi}_j = 1 - \pi_j$).

4.4   Nonnegativity for an Arbitrary S.G.S.

The purpose of this section is to prove:

Theorem 4.4.1 *Every population with s.g.s. admits a nonnegative extended fertilization operator. So writing $x' = \phi(p)$ as*

$$x'_j = \sum_{\alpha,\beta=1}^{r} c_{\alpha\beta,j} p_\alpha p_\beta \qquad (1 \leq j \leq n) \qquad (4.4.1)$$

we can assume that coefficients $c_{\alpha\beta,j}$ are nonnegative.

In particular, the Principle of Gene Inheritance and its corollary (Theorem 4.2.2 and Corollary 4.2.3) apply to every population with s.g.s.

Proof (1) Recall that an evolutionary algebra $(\mathcal{A}, \Delta)$ with vertices $\{e_1, \ldots, e_n\}$ is internally irreducible if for any pair e_i, e_j with $i \neq j$ the images E_i and E_j are distinct in $\mathcal{A}/\text{ann}\,\mathcal{A}$ i.e. $\{E_i, E_j\}$ is linearly independent. We now sharpen this condition calling $(\mathcal{A}, \Delta)$ *doubly internally irreducible* if every three distinct vertices e_i, e_j and e_k are linearly independent mod ann $\mathcal{A}$, i.e. $\{E_i, E_j, E_k\}$ is a linearly independent list in $\mathcal{A}/\text{ann}\,\mathcal{A}$. For the first part of the proof we will assume that $(\mathcal{A}, \Delta)$ has s.g.s. and is doubly internally irreducible. In particular, because $J^{\perp} = \text{ann}\,\mathcal{A}$ by Lemma 4.1.1(c) the algebra is J- irreducible.

By Corollary 3.8.2. $(\mathcal{A}, \Delta)$ admits a simplicial subalgebra $(\overline{\mathcal{A}}, \overline{\Delta})$ of codimension 1 (with $\overline{\Delta} = \Delta \cap \overline{\mathcal{A}}$). To be precise $\overline{\mathcal{A}} = \ker \psi$ where ψ is a hyperbolic, invariant linear form with all coefficients distinct. We can write:

$$\psi(x) = x_1 - \sum_{i=2}^{n} a_i x_i \qquad (a_i > 0). \qquad (4.4.2)$$

According to (3.7.21) the canonical basis $\{v_2, \ldots, v_n\}$ for $(\overline{\mathcal{A}}, \overline{\Delta})$ is given by

$$v_i = \frac{a_i e_1 + e_i}{a_i + 1}. \qquad (4.4.3)$$

Dually, the canonical coordinates $\xi_2, \ldots, \xi_n$ for a vector in $\overline{\mathcal{A}}$ are given in terms of its canonical coordinates in $\mathcal{A}$ by:

$$\xi_i = (a_i + 1)x_i \qquad (2 \leq i \leq n). \qquad (4.4.4)$$

Because conservatism is inherited by subalgebras and because $(\overline{\mathcal{A}}, \overline{\Delta})$ is simplicial, Lemma 4.1.1(c) implies that $(\overline{\mathcal{A}}, \overline{\Delta})$ itself is an evolutionary algebra with s.g.s. We will show, in fact, that it has e.g.s. We first prove that it is internally irreducible.

Let $v_i \equiv v_j (\text{mod ann } \overline{\mathcal{A}})$ with $2 \leq i \neq j$. Then by Theorem 3.3.13, $v_i \equiv v_j (\text{mod} J^{\perp})$ and so $v_i - v_j$ lies in ann $\mathcal{A}$. By (4.4.3) the three vectors e_1, e_i and e_j are then linearly dependent (mod ann $\mathcal{A}$) contradicting double internal irreducibility of $\mathcal{A}$.

Because $\overline{\mathcal{A}}$ is internally irreducible we can apply Theorem 4.3.7 to prove the e.g.s. property by counting the idempotents among the canonical basis. Because $\psi \in J$ and $\overline{\mathcal{A}} = \ker \psi$ Theorem 3.3.13 implies that for the space $\overline{J}$ of invariant forms on $\overline{\mathcal{A}}$

$$\dim \overline{J} = m - 1. \qquad (4.4.5)$$

By Corollary 3.8.11 there are at least $m-1$ idempotents among $\{v_2, \ldots, v_n\}$. To show that there are no more and hence that the e.g.s. property holds we assume, instead, that the subspace spanned by such idempotents has dimension at least m and so, since $\overline{J}^{\perp} = \operatorname{ann} \overline{A}$, there exists $v \neq 0$ in ann $\overline{A}$ which is a linear combination of vertex idempotents.

Let $v_i^2 = v_i$. According to Theorem 3.8.3 and formula (4.4.3) the subspace $A = [e_1, e_i]$ defines a simplicial subalgebra of $\mathcal{A}$. Because dim $A = 2$ there are only two possibilities: either there is no nontrivial invariant linear form on A or all linear forms on A are invariant. In the first case the barideal $= \ker s|A$ is annihilated by every invariant form on A and so by the restriction to A of every invariant form on $\mathcal{A}$. In particular, $e_1 - e_i$ lies in $J^{\perp}$ and so, by conservatism, in ann $\mathcal{A}$. This contradicts the assumption that $\mathcal{A}$ is internally irreducible. The first possibility is thus excluded and the second says that the algebra A is a unit algebra. In particular, $v_i e_1 = \frac{1}{2}(v_i + e_1)$. So for any linear combination v of such idempotents v_i we have $v e_1 = \frac{1}{2}(v + s(v)e_1)$. In particular, were v in ann $\mathcal{A}$ we would have $v e_1 = 0$ and $s(v) = 0$. Consequently, $v = 0$ contradicting the construction of the previous paragraph.

So the codimension 1 subalgebra $\overline{A}$ has e.g.s.

Consequently, by Theorem 4.3.6 the evolutionary operator of the population $\overline{A}$ has the form:

$$\xi_j' = q_{i_j}(\xi) Q_j(\xi) \qquad\qquad (2 \le j \le n) \qquad\qquad (4.4.6)$$

where

$$q_i(\xi) = \xi_i + \sum_{k=m+1}^{n} \kappa_{ik} \xi_k \qquad\qquad (i \le 2 \le m) \qquad\qquad (4.4.7)$$

are linear forms invariant on $\overline{A}$, (κ_{ik}) is a stochastic matrix, and the Q_j's are nonnegative linear combinations of the q_i's.

The formulae (4.4.4) can be interpreted as a nonnegative operator $\overline{A}^* \rightarrow A^*$, providing an extension of linear forms which preserves nonnegativity. This extension is uniquely defined by the absence of coordinate x_1. We denote by $\hat{g}$ the extension of a form g in $\overline{A}^*$. By Theorem 3.3.13 $\hat{g} \in J$ if $g \in \overline{J}$.

So from (4.4.6) we have in the original population:

$$x_j' = \alpha_j \hat{q}_{i_j}(x) \hat{Q}_j(x) + \psi(x)\theta_j(x) \ (2 \le j \le n). \qquad\qquad (4.4.8)$$

Here $\alpha_j = (a_j + 1)^{-1}$.

The θ_j's are linear forms on $\mathcal{A}$ and we show they are invariant by using stationarity of the evolutionary operator $(V^2 = V)$. This implies that on the simplex $x_j'' = x_j'$ for all j, i.e. the x_j''s are invariant quadratic forms. Now the $\hat{q}_{i_j}$'s, $\hat{Q}_j$'s and ψ are invariant forms and so from (4.4.8) it follows that $\theta_j(x') = \theta_j(x)$ on the set Δ. Because this set spans $\mathcal{A}$, the forms θ_j are invariant.

The forms q_i $(2 \leq i \leq m)$ make up a basis for $\overline{J}$ and so the extensions $\hat{q}_i$ $(2 \leq i \leq m)$ together with ψ provide a basis for J. Expand θ_j via this basis to get

$$\theta_j = \sum_{i=2}^{m} \beta_{ji}\hat{q}_i + \gamma_j\psi \qquad (2 \leq j \leq n).$$

Substituting, we can rewrite (4.4.8) as

$$x_j' = \alpha_j\hat{q}_{i_j}(x)\hat{Q}_j(x) + \sum_{i=2}^{m} \beta_{ji}\hat{q}_i(x)\psi(x) + \gamma_j\psi^2(x). \qquad (4.4.9)$$

Notice that

$$\hat{q}_i(x) = \frac{x_i}{\alpha_i} + \sum_{k=m+1}^{n} \frac{\kappa_{ik}}{\alpha_k}x_k \qquad (2 \leq i \leq m)$$

implies that $\hat{q}_i(x)$ does not depend on x_l for $l \neq i$, $2 \leq l \leq m$. For $l \neq i_j$ the coefficient of x_l^2 in x_j' is:

$$\frac{-\beta_{jl}}{\alpha_l}a_l + \gamma_j a_l^2 \geq 0. \qquad (4.4.10)$$

This coefficient is nonnegative, as indicated, because it is the structure constant $p_{ll,j} \geq 0$ for the multiplication of the simplicial algebra $\mathcal{A}$. Here the a_l is the positive coefficient in ψ, c.f. (4.4.2).

Similarly, in (4.4.9) the coefficient of $x_1 x_l$ is:

$$\frac{\beta_{jl}}{\alpha_l} - 2\gamma_j a_l = 2p_{1l,j} \geq 0. \qquad (4.4.11)$$

Finally, the coefficient of x_1^2 in (4.4.9) is:

$$\gamma_j = p_{11,j} \geq 0.$$

On the other hand, multiplying (4.4.11) by a_l and adding to (4.4.10) we see that $-\gamma_j a_l^2 \geq 0$ and so $\gamma_j \leq 0$. So (4.4.12) implies $\gamma_j = 0$ and so from

(4.4.10) and (4.4.11) $\beta_{jl} = 0$ $(2 \leq l \leq m, l \neq i_j)$. The sum defining θ_j reduces to

$$\theta_j = \beta_{ji_j}\hat{q}_{i_j} \qquad (2 \leq j \leq n) \qquad (4.4.12)$$

and so (4.4.8) becomes

$$x'_j = \hat{q}_{i_j}(x)R_j(x) \qquad (2 \leq j \leq n) \qquad (4.4.13)$$

with $R_j = \alpha_j\hat{Q}_j + \beta_{ji_j}\psi$ invariant.

When $x > 0$ the quadratic forms x'_j and the linear forms q_{i_j} are positive and so the linear forms R_j have the same property. Consequently the coefficients in the R_j's are nonnegative, i.e. $R_j \geq 0$. Expanding the $\hat{q}_{i_j}$'s and R_j's as nonnegative linear combinations of the cone basis $p_1, \ldots, p_r$ for C we get the required representation (4.4.1) except for the first coordinate x'_1. To close this gap it suffices to replace ψ by the form $-(\psi + \lambda s)$ where λ lies between the largest and next to largest among the coefficients a_i $(2 \leq i \leq l)$.

(2) Now we prove the theorem in general by induction on n. When $n \leq 3$ we have e.g.s. by Corollary 4.3.3. So we assume $n > 3$ and that the theorem holds for $n-1$ dimensional populations. If $\mathcal{A}$ is doubly internally irreducible then the result follows by (1) above. So among $E_1, \ldots, E_n$, the images of $e_1, \ldots, e_n$ in $\mathcal{A}/J^\perp$, we can assume that there is a linearly dependent set of three, which we can take to be $\{E_1, E_2, E_3\}$.

Let us assume, provisionally, that the population is internally irreducible. Then (because $s(E_i) = 1$ for $1 \leq i \leq n$) each pair amongst $E_1, \ldots, E_n$ is linearly independent. So for a nontrivial liner combination $\lambda_1 E_1 + \lambda_2 E_2 + \lambda_3 E_3 = 0$ none of the coefficients vanish. We can assume $\lambda_1 > 0$, $\lambda_2 > 0$, and $\lambda_3 < 0$. Consequently, we can write $E_3 = c_1 E_1 + c_2 E_2$ with $c_1, c_2 > 0$, and $c_1 + c_2 = 1$ (again by $s(E_i) = 1$).

This means that for every invariant linear form f : $f(e_3) = c_1 f(e_1) + c_2 f(e_2)$. Thus, every f in J can be written in terms of the variables:

$$\xi_1 = x_1 + c_1 x_3, \xi_2 = x_2 + c_2 x_3, \xi_4 = x_4, \ldots, \xi_n = x_n \qquad (4.4.14)$$

and so $V(x)$ depends only on these variables. This list of variables comprise a partition of s so the map described by (4.4.14) to $\mathbf{R}^{n-1}$ is an internal kinship (by Theorem 3.9.2) of the original algebra $(\mathcal{A}, \Delta)$ onto an $n-1$ dimensional evolutionary algebra $(\tilde{\mathcal{A}}, \tilde{\Delta})$. The latter is conservative as the quotient algebra of a conservative algebra and so has s.g.s. By the inductive hypothesis:

$$\xi'_j = \sum_{\alpha,\beta=1}^{r} c_{\alpha\beta,j}q_\alpha(\xi)q_\beta(\xi) \qquad (1 \leq j \leq n-1). \qquad (4.4.15)$$

Here $\{q_\alpha\}$ is the normalized basis for the cone $\tilde{C}$ and $c_{\alpha\beta,j} \geq 0$. By Theorem 3.9.3 the change of variables (4.4.14) converts $q_\alpha(\xi)$ into $p_\alpha(x)$ with $\{p_\alpha\}$ a normalized cone basis for C. From (4.4.15) we immediately have

$$x'_j = \sum_{\alpha,\beta=1}^{r} c_{\alpha\beta,j} p_\alpha(x) p_\beta(x) \qquad (4 \leq j \leq n) \qquad (4.4.16)$$

and it suffices to consider x'_1, x'_2 and x'_3.

By Lemma 4.2.1 the variable ξ_1 appears with a positive coefficient in at least one form q_α. We suppose it is q_1. Then by the Principle of Gene Inheritance, Corollary 4.2.2, we have

$$\xi'_1 = q_1(\xi) Q(\xi) \qquad (4.4.17)$$

with Q some nonnegative invariant form on the algebra $\tilde{A}$.

Using $\xi_1 = x_1 + c_1 x_3$, $c_1 > 0$ we will prove that both x'_1 and x'_3 are divisible by $p_1(x)$ $(= q_1(\xi)$ via (4.4.14)$)$. The quotients will be nonnegative invariant forms. Expanding these by using $p_1,\ldots,p_r$ we get the required representation for x'_1 and x'_3. Analogously we use ξ'_2 to get a representation of x'_2 (and again for x'_3).

First, rewrite (4.4.17) as

$$x'_1 + c_1 x'_3 = p_1(x) P(x) \qquad (4.4.18)$$

with $P(x) \equiv Q(\xi)$ via (4.4.14).

Consider the face Γ of Δ on all vertices complementary to $\mathrm{supp}(p_1)$. On this face $p_1(x)$ is identically zero and so from (4.4.18) we have $x'_1 = 0$, $x'_3 = 0$ (recall x'_1 and x'_3 are nonnegative on Δ). Consequently these equations hold on the coordinate subspace, $[\Gamma]$, spanned by Γ.

Choose from among $\{p_\alpha\}_1^r$ a basis for J which includes p_1. After reordering we can suppose it is $p_1, p_2, \ldots p_m$. Because of s.g.s., x'_1 and x'_3 are quadratic forms of $p_1, p_3, \ldots p_m$. Let us denote by $\pi_2, \pi_3, \ldots \pi_m$ the restrictions of the forms $p_2, \ldots p_m$ to the subspace $[\Gamma]$. It is enough to show that $\{\pi_2, \ldots \pi_m\}$ is linearly independent. For then $x'_1 = 0$ and $x'_3 = 0$ on $[\Gamma]$ will imply that every term in the expansions of x'_1 and x'_3 will contain p_1, i.e. they are divisible by p_1.

Let $\sum_{i=2}^{m} a_i \pi_i = 0$ with some coefficients a_i. Then the invariant linear form

$$f = \sum_{i=2}^{m} a_i p_i \qquad (4.4.19)$$

depends only on coordinates x_j with e_j in the support of p_1. Hence, $p_1 + \lambda f$ remains nonnegative provided $|\lambda|$ is small enough. Let us expand using the cone basis to get

$$p_1 + \lambda f = \sum_{k=1}^{r} b_k p_k \qquad\qquad (b_k \geq 0).$$

But $\mathrm{supp}(p_1 + \lambda f) = \mathrm{supp}\,(p_1)$ and so from (4.2.5) all the b_k's $= 0$ for $k \geq 2$. Thus, f is proportional to p_1. Now from (4.4.19) and linear independence of $\{p_1,\ldots,p_m\}$ we have that $a_i = 0$ for $i \geq 2$.

In the final remaining case, the internally reducible population, we can assume $E_1 = E_2$. Just as above all forms f are functions of $\xi_1 = x_1 + x_2, \xi_3 = x_3,\ldots,\xi_n = x_n$. This is just the extremal case of the above situation with $c_1 = 1$ and $c_2 = 0$. Proceeding as above we get (4.4.16) for $j \geq 3$ from the analogue of (4.4.15). As above we get the representation of x_1' and x_2' from that of ξ_1'. $\square$

4.5 Genetic Criterion for E.G.S.

Elementary gene structure can be characterized from the genetic point of view by the presence of homozygotes in the population.

Theorem 4.5.1 *In order that an s.g.s. population be elementary it is necessary and sufficient that there be at least one homozygote in the population.*

By contrast, heterozygotes need not occur.

Example Only the homozygotes $A_1 A_1, \ldots, A_n A_n$ occur when the meiosis operator is given by $p_i = x_i$ $(1 \leq i \leq n)$ and the fertilization operator by $x_i' = p_i(\sum_{j=1}^{n} p_j)$ $(1 \leq i \leq n)$, i.e. for the identity evolutionary operator.

There is only one homozygote, $A_1 A_1$, when the meiosis operator is $p_1 = \sum_{i=1}^{n} x_i$ and the fertilization operator is $x_1' = p_1^2$, $x_j' = 0$ $(2 \leq j \leq n)$. Here the zygotes other than 1 are disappearing and the evolutionary operator is constant.

Necessity in Theorem 4.5.1 follows from Theorem 4.3.3. In particular, formula (4.3.3) shows that $e_1, \ldots, e_m$ are homozygotes. We will prove sufficiency by using Lemma 4.2.6 and some related ideas which we will now develop. Throughout the remainder of the section we will assume s.g.s. without further remark.

Lemma 4.5.2 *All internally isolated homozygotes are nonsplitting.*

Proof Let e_1 be an internally isolated homozygote and E_1 be its image in the quotient algebra $\mathcal{A}/\text{ann } \mathcal{A} = \mathcal{A}/J^{\perp}$. The isolation assumption means $E_1 \neq E_i$ for $i \neq 1$. We will prove that E_1 is an extremal point for the basic body of the quotient. It then follows from Corollary 3.8.10 that e_1 is idempotent, i.e. the homozygote is nonsplitting.

Suppose instead that

$$E_1 = \sum a_i E_i \tag{4.5.1}$$

where the index i varies in a subset S of $\{2, \ldots, n\}$, the coefficients a_i are positive and $\sum a_i = 1$.

If $A_1 A_1$ is the genotype of zygote e_1 then by Lemma 4.2.6, $\pi_{11} = 1$, i.e. $p_1(e_1) = 1$. But by (4.5.1) $e_1 \equiv \sum a_i e_i \pmod{J^{\perp}}$ and as $p_1 \in J$ we then have

$$1 = p_1(e_1) = \sum a_i p_1(e_i) = \sum a_i \pi_{1i}.$$

Consequently, $\pi_{1i} = 1$ for all i in S. We also have that $\pi_{\alpha 1} = 0$ for $\alpha > 1$ because e_1 is a homozygote. So we get from (4.5.1) $\pi_{\alpha i} = 0$ for all i in S

and $\alpha > 1$. Thus, $e_1 \equiv e_i \pmod{J^\perp}$ for all i in S and so $E_1 = E_i$ contrary to the assumption. $\square$

Following the definition of a normal population we will call a zygote a *normal zygote* if it is not disappearing and it's internally and externally isolated.

Lemma 4.5.3 *In a nondegenerate population, if e_j is a normal, splitting heterozygote with genotype $A_\alpha A_\beta$ then*

$$\pi_{\alpha j} + \pi_{\beta j} = 1 \tag{4.5.2}$$

and each expansion of the weight contains p_α and p_β with coefficients equal to 1.

Proof Consider the image E_j of e_j in the quotient algebra $\mathcal{A}_J = \mathcal{A}/J^\perp = \mathcal{A}/\operatorname{ann}\mathcal{A}$. Because $E_j \neq E_l$ for $l \neq j$ but e_j is not idempotent, the point E_j is not an extremal one in the basic body Δ_J:

$$E_j = \sum_{l \neq j} a_l E_l \qquad (a_l \geq 0, \sum_{l \neq j} a_l = 1).$$

From this we get

$$\pi_{\alpha j} = p_\alpha(e_j) = \sum_{l \neq j} a_l p_\alpha(e_l) = \sum_{l \neq j} a_l \pi_{\alpha l}$$

and, analogously,

$$\pi_{\beta j} = \sum_{l \neq j} a_l \pi_{\beta l}.$$

But if $\pi_{\alpha l} \neq 0$ for some $l \neq j$ then $\pi_{\beta l} = 0$. For otherwise e_l would be disappearing or it would have genotype $A_\alpha A_\beta$ and so e_l would be externally kin with e_j by the Principle of Gene Inheritance. These possibilities are excluded by the assumptions of the lemma. Consequently, we have the inequality

$$\pi_{\alpha j} + \pi_{\beta j} = \sum_{l \neq j} a_l(\pi_{\alpha l} + \pi_{\beta l}) \leq \sum_{l \neq j} a_l = 1. \tag{4.5.3}$$

Consider now any expansion of the weight:

$$s = \sum_{\gamma=1}^{r} \lambda_\gamma p_\gamma.$$

By Lemma 3.7.1 $0 \leq \lambda_\gamma \leq 1$. Comparing the coefficients of x_j we have

$$1 = \lambda_\alpha \pi_{\alpha j} + \lambda_\beta \pi_{\beta j}.$$

From this and from (4.5.3) it follows that $\lambda_\alpha = \lambda_\beta = 1$ and so that (4.5.2) holds as well. $\square$

Corollary 4.5.4 *In a normal population, if some gene is created (via meiosis) by a splitting heterozygote then it is created by some homozygote as well.*

Proof Suppose gene A_α is created by some splitting heterozygote, say e_1. Because the form p_α is normalized there exists some zygote e_k which creates A_α with probability 1, i.e. such that $\pi_{\alpha k} = 1$. For each other gene A_β there exists, by Lemma 3.7.12, some expansion of the weight with coefficient 1 on p_β. By the above lemma the coefficient of p_α is also 1:

$$s = p_\alpha + p_\beta + \sum_{\gamma \neq \alpha, \beta} \lambda_\gamma p_\gamma.$$

Applying this equation to e_k we have

$$1 = \pi_{\alpha k} + \pi_{\beta k} + \sum_{\gamma \neq \alpha, \beta} \lambda_\gamma \pi_{\gamma k} \geq 1 + \pi_{\beta k}.$$

So $\pi_{\beta k}$ cannot be positive. Because $\pi_{\beta k} = 0$ for all $\beta \neq \alpha$ it follows that $A_\alpha A_\alpha$ is the genotype of e_k. $\square$

The analogue of Lemma 4.5.3 for nonsplitting heterozygotes does not require the assumptions associated with normality.

Lemma 4.5.5 *If e_j is a nonsplitting heterozygote with genotype $A_\alpha A_\beta$ then*

$$\pi_{\alpha j} = \pi_{\beta j} = 1, \tag{4.5.4}$$

and there exists an expansion of the weight in which that the coefficients of p_α and p_β are 0 and 1 respectively.

Proof For simplicity we will label $\pi_{\alpha j} = a$ and $\pi_{\beta j} = b$. By assumption $a, b > 0$ and

$$p_\alpha(x) = a x_j + \sum_{l \neq j} \pi_{\alpha l} x_l$$

$$\tag{4.5.5}$$

$$p_\beta(x) = bx_j + \sum_{l \neq j} \pi_{\beta l} x_l.$$

By the Principle of Gene Inheritance $x'_j = cp_\alpha p_\beta$. From this and from (4.5.5) we see that $x'_j = cabx_j^2 + \ldots$. But e_j is idempotent and so the coefficient $p_{jj,j}$ of x_j^2 in x'_j is 1, i.e. $cab = 1$ and so

$$x'_j = \frac{p_\alpha p_\beta}{ab}. \tag{4.5.6}$$

Now suppose that $\pi_{\alpha l} \neq 0$ for some $l \neq j$. By the Principle of Gene Inheritance

$$x'_l = p_\alpha \left(\sum_{\gamma=1}^{r} c_{l\gamma} p_\gamma \right)$$

with $c_{l\gamma} \geq 0$. But x'_l does not contain x_j^2 (because the structure equation $p_{jj,j} = 1$ implies $p_{jj,l} = 0$). So from (4.5.5) $c_{l\alpha} = c_{l\beta} = 0$, and

$$x'_l = p_\alpha \left(\sum_{\gamma \neq \alpha, \beta} c_{l\gamma} p_\gamma \right) \tag{4.5.7}$$

for all $l \neq j$ such that $\pi_{\alpha l} \neq 0$. Substituting (4.5.6) and (4.5.7) into (4.5.5) we have

$$p'_\alpha = ax'_j + \sum_{l \neq j} \pi_{\alpha l} x'_l = p_\alpha \left(\frac{p_\beta}{b} + \sum_{\gamma \neq \alpha, \beta} \theta_\gamma p_\gamma \right)$$

with $\theta_\gamma \geq 0$.

On the other hand, invariance implies $p'_\alpha = p_\alpha s$ and so

$$\frac{p_\beta}{b} + \sum_{\gamma \neq \alpha, \beta} \theta_\gamma p_\gamma = s. \tag{4.5.8}$$

From this we have $b \geq 1$ as the form p_β is normalized but from (4.5.5) (and the same reason) $b \leq 1$. Thus, $1 = b = \pi_{\beta j}$ and (4.5.8) is the required expansion of the weight s. Interchanging the roles of α and β we get that $\pi_{\alpha j} = 1$ and that s admits an alternative expansion with coefficients of p_α and p_β, 1 and 0 respectively. $\square$

We now prove sufficiency in Theorem 4.5.1.

We can assume that the population is normal. For by Theorem 4.2.8 we can always do a genotype preserving normalization. So if the original population contains a homozygote there will be one in the normalized population as well. If then the normalized population has e.g.s. so, too does the original because the e.g.s. property is an invariant with respect to normalization.

Let $\{e_1,\ldots,e_m\}$ be the set of homozygotes in the normal population with corresponding genes $A_1,\ldots A_m$. These genes are pairwise distinct by Theorem 4.2.5. The set $\{p_1,\ldots,p_m\}$ is linearly independent because coordinate x_i occurs only in p_i. We will prove that no other genes exist. The e.g.s. property then follows from of Lemma 4.3.1.

Suppose instead that genes $A_{m+1},\ldots,A_r$ exist and are created from no homozygote. So they are created by heterozygotes and by Corollary 4.5.4 these heterozygotes are all nonsplitting. Thus, genes $A_1,\ldots,A_m$ are created by homozygotes $e_1,\ldots,e_m$ and splitting heterozygotes $e_{m+1},\ldots,e_\nu$ while genes $A_{m+1},\ldots,A_r$ are created by nonsplitting heterozygotes $e_{\nu+1},\ldots,e_n$. We derive a contradiction.

For a vector x we denote by u and v the projections on $[e_1,\ldots e_\nu]$ and $[e_{\nu+1},\ldots,e_n]$, respectively. Then $p_1,\ldots,p_m$ depend only on u with

$$p_i = x_i + \sum_{j=m+1}^{\nu} \pi_{ij}x_j \qquad (1 \le i \le m), \qquad (4.5.9)$$

but $p_{m+1},\ldots,p_r$ depend only on v.

We apply the identity:

$$\sum_{j=1}^{n} x'_j = s^2(x) = s^2(u) + s^2(v) + 2s(u)s(v) \qquad (4.5.10)$$

and compare bilinear terms in u and v.

By the Principle of Gene Inheritance $x'_{m+1},\ldots,x'_\nu$ depend only on u while $x'_{\nu+1},\ldots,x'_n$ depend only on v. Finally, for the homozygotes

$$x'_i = p_i(u)[P_i(u) + 2Q_i(v)] \qquad (1 \le i \le m) \qquad (4.5.11)$$

where P_i and Q_i are nonnegative linear forms. So from (4.5.10)

$$\sum_{i=1}^{m} p_i(u)Q_i(v) = s(u)s(v). \qquad (4.5.12)$$

But from (4.5.9) and Lemma 4.5.3:

$$\sum_{i=1}^{m} p_i(u) = s(u). \qquad (4.5.13)$$

Comparing (4.5.12) with (4.5.13) we get from linear independence of $\{p_1,\ldots,p_m\}$ that

$$Q_i(v) = s(v) \qquad (1 \le i \le m). \qquad (4.5.14)$$

On the other hand from (4.5.9) and invariance of the forms p_i we have for $1 \leq i \leq m$:

$$p'_i = x'_i + \sum_{j=m+1}^{\nu} \pi_{ij} x'_j = p_i(u)(s(u) + s(v)). \qquad (4.5.15)$$

Again comparing the uv cross terms, this time from (4.5.11) and (4.5.15) we have

$$2Q_i(v) = s(v) \qquad\qquad (1 \leq i \leq m). \qquad (4.5.16)$$

But (4.5.14) and (4.5.16) contradict one another.

4.6 Nonelementary S.G.S.

In this paragraph we will complete the description of evolutionary operators for populations with s.g.s. Theorems 4.3.6 and 4.3.9 describe the elementary case. By Theorem 4.5.1 in the remaining, nonelementary, case the population consists entirely of heterozygotes. If the population is normal we can further conclude by Corollary 4.5.4. that these heterozygotes are all nonsplitting.

Let us now look at a normal, nonelementary s.g.s. population and consider any gene A_1. Suppose that it is created by zygotes $e_1, \ldots, e_\beta$. By Lemma 4.5.5:

$$p_1 = x_1 + \cdots + x_\beta \qquad (4.6.1)$$

Now each zygote $e_1, \ldots e_\beta$ creates a gene different from $A_1 : B_1, \ldots, B_\beta$. These genes are all distinct by normality, see Theorem 4.2.5. By the Principle of Gene Inheritance we have:

$$x'_j = c_j p_1 q_j \qquad (1 \leq j \leq \beta)$$

where form q_j corresponds to gene B_j and the coefficient c_j is positive.

Just as in (4.6.1) the nonzero coefficients in each q_j are all 1's. The heterozygotes are nonsplitting and so the coefficient of x_j^2 in x'_j is 1. Thus, each c_j is 1, as well, and we have:

$$x'_j = p_1 q_j \qquad (1 \leq j \leq \beta). \qquad (4.6.2)$$

Furthermore, by invariance of the form p_1 we have

$$p_1 s = p'_1 = \sum_{j=1}^{\beta} x'_j = p_1 \left(\sum_{j=1}^{\beta} q_j \right)$$

and so we have an expansion of the weight:

$$s = \sum_{j=1}^{\beta} q_j. \tag{4.6.3}$$

From this and the analogue of (4.6.1) for the q_j's we see that $q_1, \ldots, q_\beta$ provides a partition of the set of zygotes $\{e_1, \ldots, e_n\}$ into β disjoint sets. The jth set of coordinates then sums to yield q_j. Thus, each zygote $e_1, \ldots, e_n$ creates exactly one of the genes $B_1, \ldots B_\beta$ with probability 1. We refer to this as the set of *female genes*.

Now the genotypes of $e_1, \ldots, e_\beta$ are known, namely $A_1 B_1, \ldots A_1 B_\beta$. Just as formula (4.6.1) lead to the partition of $\{e_1, \ldots, e_n\}$ into β sets each creating a female gene, similarly, we can start with q_1 and obtain an expansion of the weight similar to (4.6.3).

$$s = \sum_{i=1}^{\alpha} p_i, \tag{4.6.4}$$

and so get another partition of $\{e_1, \ldots, e_n\}$ this time into α distinct sets each of which creates a *male gene* $A_1, \ldots, A_\alpha$ in addition to some female gene B_j.

Let us check that $A_i \neq B_j$ for any pair ij. When $i = 1$ this is obvious from the heterozygote definition of $B_1, \ldots B_\beta$. With $i \neq 1$, supp $p_1 \cap$ supp $p_i = \emptyset$. But as there exists a zygote with genotype $A_1 B_j$ we have supp $p_1 \cap$ supp $q_j \neq \emptyset$. Therefore, $p_i \neq q_j$.

Thus, each zygote among $\{e_1, \ldots, e_n\}$ creates a unique male gene A_i and a unique female B_j and so the association from zygote to genotype is a correspondence from the set $\{e_1, \ldots, e_n\}$ to the Cartesian product $\{A_1, \ldots, A_\alpha\} \times \{B_1, \ldots, B_\beta\}$. This association is one-to-one by the genotypic characterization of zygotes (Theorem 4.2.5, again). It is also onto, i.e. each pair $A_i B_j$ is the genotype of a zygote. For in general the set of genotypes involving B_j is $A_{i_1} B_j, \ldots A_{i_\nu} B_j$ with $1 = i_1 < \ldots < i_\nu \leq \alpha$ and just as before we have the expansion of the weight

$$s = \sum_{k=1}^{\nu} p_{i_k}.$$

Comparing this with (4.6.4) we see that no terms can be omitted and so the subset $\{i_1, \ldots i_n\}$ consists of all of $\{1, \ldots \alpha\}$.

We have thus proved one direction of the following:

Theorem 4.6.1 *If a normal population has stationary gene structure but is not elementary then the coordinates can be labeled* $\{x_{ij} : 1 \leq i \leq \alpha, 1 \leq j \leq$

$\beta\}$ *with* $\alpha > 1$, $\beta > 1$ *and* $n = \alpha\beta$, *so that its evolutionary operator is given by:*

$$x'_{ij} = p_i q_j \qquad (1 \leq i \leq \alpha,\, 1 \leq j \leq \beta) \qquad (4.6.5)$$

with

$$p_i = \sum_{j=1}^{\beta} x_{ij} \qquad (1 \leq i \leq \alpha)$$

$$q_j = \sum_{i=1}^{\alpha} x_{ij} \qquad (1 \leq j \leq \beta). \qquad (4.6.6)$$

Conversely, if the evolutionary operator is of this form the population is normal with nonelementary s.g.s.

Notice that in contrast with the elementary case there are no continuously varying parameters in (4.6.5) and (4.6.6). This family of evolutionary operators is discrete.

The conditions $\alpha > 1$ and $\beta > 1$ follow from the fact that if $\alpha = 1$ and $\beta = 1$ the population is a unit, and so has e.g.s. In particular, we get

Corollary 4.6.2 *If a normal population has s.g.s and its dimension, n, is a prime number then it has e.g.s.*

Notice that when $n = 4$ and $\alpha = \beta = 2$ we get the Quadrille Law (see section 4.1). So we will call the operator of (4.6.5) the *Extended Quadrille Law* (or *operator*).

By Corollary 4.6.2 dimension 4 is minimal for the occurrence of a nonelementary s.g.s. population. Notice that the process of normalization preserves the nonelementary s.g.s. property and reduces dimension to get normality. So with $n \leq 4$ a nonelementary s.g.s population had to be normal to begin with and so have $n = 4$.

Corollary 4.6.3 *With dimension $n \leq 4$ the Quadrille Law is the only nonelementary s.g.s. population.*

We now complete the proof of Theorem 4.6.1 by proving the reverse direction. Note first that the forms are invariant:

$$p'_i = \sum_{j=1}^{\beta} x'_{ij} = \sum_{j=1}^{\beta} p_i q_j = p_i \sum_{j=1}^{\beta} q_j = p_i s,$$

and similarly for q'_j. Consequently, the population has s.g.s. with meiosis operator (4.6.6) and fertilization operator (4.6.5). Clearly, the forms p_i and q_j are nonnegative and normalized. We check that they are extremal for the cone C.

From (4.6.5) and (4.6.6) $\{p_1, \ldots, p_\alpha, q_1, \ldots, p_\beta\}$ span J. Now for $f \in C$ we can write

$$f = \sum_{i=1}^{\alpha} \lambda_i p_i + \sum_{j=1}^{\beta} \omega_j q_j.$$

Comparing x_{ij} coefficients we see that $\lambda_i + \omega_j \geq 0$ for all $1 \leq i \leq \alpha,, 1 \leq j \leq \beta$. From which $\min_i(\lambda_i) \geq \max_j(-\omega_j)$. So there exists a λ such that $\min_i(\lambda_i) \geq \lambda \geq \max_j(-\omega_j)$. That is, $\lambda_i - \lambda \geq 0$ and $\omega_j + \lambda \geq 0$ for all i, j. Now use the obvious linear relations

$$\sum_{i=1}^{\alpha} p_i = \sum_{j=1}^{\beta} q_j = s \tag{4.6.7}$$

to represent f as

$$f = \sum_{i=1}^{\alpha} (\lambda_i - \lambda) p_i + \sum_{j-1}^{\beta} (\omega_j + \lambda) q_j$$

and here all the coefficients are nonnegative. Hence, the cone C is generated by $\{\hat{p}_1, \ldots p_\alpha, q_1, \ldots, p_\beta\}$

Also, the linear relation (4.6.7) is the only relation (up to constant multiple) among these forms for if $\sum \lambda_i p_i + \sum \omega_j q_j = 0$ we use the same argument as above. First, $\lambda_i + \omega_j = 0$ for all pairs i, j and so there is a common $\lambda = \lambda_i = -\omega_j$, yielding (4.6.7) times λ. From this it easily follows that the forms are extremal. For if

$$p_1 = \sum_{i=2}^{\alpha} \lambda_i p_i + \sum_{j=1}^{\beta} \omega_j q_j$$

we must have $\lambda_i = -1$ for $i = 2, \ldots, \alpha$ (and $\omega_j = 1$ for all j) and so the coefficients cannot be all nonnegative.

The relation (4.6.7) shows that the population is not elementary.

Finally, the population is clearly nondegenerate and externally irreducible. It is also internally irreducible because coordinate x_{ij} is included in both p_i and q_j and any other x_{kl} is missing in either p_i or q_j. So zygotes e_{ij} and e_{kl} are not internally kin. Thus, the population is normal. $\square$

Notice that the Extended Quadrille Law can be written in matrix form. Write the state vectors $x = (x_{ij})$ as an $\alpha \times \beta$ matrix and then the meiosis operator is given by

$$p = x\epsilon_f, \quad q = \epsilon_m x \tag{4.6.8}$$

where ϵ_f is the $\beta \times 1$ column of 1's and ϵ_m is the $1 \times \alpha$ row of 1's. For the fertilization operator we multiply the male $\alpha \times 1$ column vector p with the female $1 \times \beta$ row vector q:

$$x' = pq. \tag{4.6.9}$$

In matrix form the Gene Conservation Law follows from:

$$p' = x'\epsilon_f = pq\epsilon_f = p(q\epsilon_f) = sp$$

because

$$q\epsilon_f = \epsilon_m x\epsilon_f = s(x).$$

Similarly $q' = sq$.

There are some direct consequences of Theorem 4.6.1.

Corollary 4.6.4 *The space of genotypes for a nonelementary s.g.s. population is the Cartesian product of two simplices: $\Delta^{\alpha-1} \times \Delta^{\beta-1}$.*

Proof In the normal case the meiosis operator maps Δ^{n-1} onto $\Delta^{\alpha-1} \times \Delta^{\beta-1}$ by the theorem. The assumption of normality can then be eliminated by Theorem 4.2.7. $\square$

From this, we get another criterion for e.g.s.

Corollary 4.6.5 *In order that an s.g.s. population be elementary it is necessary and sufficient that the space of genotypes be a simplex.* $\square$

We can observe the difference between elementary and nonelementary s.g.s. by comparing the number r of genotypes with m, the dimension of the space of invariant forms J.

Corollary 4.6.6 *For any population with s.g.s. we have*

$$m = r - l + 1 \tag{4.6.10}$$

where l is the number of sexes at the gene level. So $l = 1$ in the e.g.s. case and $l = 2$ in the nonelementary case. $\square$

The mechanism associated with the Extended Quadrille Law is similar to the simplest Quadrille Law. There are α male genes $A_1, \ldots A_\alpha$ and β female genes $B_1, \ldots, B_\beta$. They yield $n = \alpha\beta$ bisexual zygotes A_iB_j ($1 \le i \le \alpha$; $1 \le j \le \beta$). In meiosis an A_iB_j individual produces A_i type male and B_j type female gametes. Fertilization consists of union of a male and a female gamete selected randomly from the two gene pools.

In the nonnormal case continuous parameters do appear. These are correlations among zygotes with the same genotype. Let us consider this more general situation but still assume the population is nondegenerate.

Nonelementary s.g.s. means the nonoccurrence of homozygotes (see Theorem 4.5.1). The heterozygotes with the same genotypes (but only those) are externally kin by the Principle of Gene Inheritance. By uniting such externally kin sets of zygotes we obtain a nonelementary s.g.s. population with distinct genotypes for different zygotes. Consequently, by Theorem 4.2.5 the resulting population is normal and Theorem 4.6.1 applies to it. So in the reduced population, and the initial population as well, we have genes $A_1, \ldots A_\alpha$; $B_1, \ldots B_\beta$ and genotypes A_iB_j ($1 \le i \le \alpha$; $1 \le j \le \beta$), and there are no other genotypes. We label those zygotes with genotype A_iB_j by $e_{ij,k}$ with $k = 1, \ldots, \gamma_{ij}$, with coordinates $x_{ij,k}$.

Theorem 4.6.7 *If a nondegenerate population has s.g.s. but is not elementary then its coordinates can be labelled $\{x_{ij,k} : 1 \le i \le \alpha;\ 1 \le j \le \beta;\ 1 \le k \le \gamma_{ij}\}$ with $\alpha, \beta > 1$, so that the evolutionary operator is given by:*

$$x'_{ij,k} = c_{ij,k}p_iq_j \qquad (1 \le i \le \alpha;\ 1 \le j \le \beta;\ 1 \le k \le \gamma_{ij}) \qquad (4.6.11)$$

with

$$p_i = \sum_{j=1}^{\beta} \sum_{k=1}^{\gamma_{ij}} x_{ij,k}$$

$$q_j = \sum_{i=1}^{\alpha} \sum_{k=1}^{\gamma_{ij}} x_{ij,k} \qquad\qquad (4.6.12)$$

and the coefficients $c_{ij,k}$ are positive and satisfy

$$\sum_{k=1}^{\gamma_{ij}} c_{ij,k} = 1 \qquad (1 \le i \le \alpha;\ 1 \le j \le \beta). \qquad (4.6.13)$$

Conversely, if the evolutionary operator is of this form the population is nondegenerate with nonelementary s.g.s.

Proof We get the formulae (4.6.11), which describe the fertilization operator, from the Principle of Gene Inheritance. The coefficients $c_{ij,k}$ are positive because the population is nondegenerate.

The meiosis operator must look like

$$p_i = \sum_{j=1}^{\beta} \sum_{k=1}^{\gamma_{ij}} \pi_{ij,k} x_{ij,k} \qquad (1 \le i \le \alpha)$$

$$q_j = \sum_{i=1}^{\alpha} \sum_{k=1}^{\gamma_{ij}} \kappa_{ij,k} x_{ij,k} \qquad (1 \le j \le \beta).$$

$$(4.6.14)$$

with

$$0 \le \pi_{ij,k}, \kappa_{ij,k} \le 1 \qquad (4.6.15)$$

Invariance of the forms yields:

$$\sum_{k=1}^{\gamma_{ij}} \pi_{ij,k} c_{ij,k} = 1 = \sum_{k=1}^{\gamma_{ij}} \kappa_{ij,k} c_{ij,k}. \qquad (4.6.16)$$

But from (4.6.11) when we perform the external reduction to get a normal population we must have (4.6.5) and thus (4.6.13). It then follows from (4.6.13) and (4.6.5)-(4.6.16) that $\pi_{ij,k} = 1 = \kappa_{ij,k}$ for all ij, k. So (4.6.14) becomes (4.6.12).

Conversely, if the evolutionary operator has the pattern given in the theorem then the forms p_i and q_j are invariant and performing the external union of gametes $e_{ij,k}$ $(1 \le k \le \gamma_{ij})$ for each pair ij we obtain the population of Theorem 4.6.1 which is therefore a normal, nonelementary s.g.s. population. So the original population is nonelementary s.g.s. It is nondegenerate because all $c_{ij,k}$'s are positive. $\square$

Corollary 4.6.8 *If a nondegenerate population has nonelementary s.g.s. it is sufficient for normality that it be either internally or externally irreducible.*

This assertion is not correct for e.g.s.

Example The population with fertilization operator $x_1' = p_1^2$, $x_2' = p_2^2$, $x_3' = x_4' = p_1 p_2$ and the meiosis operator $p_1 = x_1 + \alpha x_3 + (1 - \alpha) x_4$, $p_2 = x_2 + (1 - \alpha) x_3 + \alpha x_4$ $(0 < \alpha < \frac{1}{2})$ has e.g.s. and is nondegenerate and externally irreducible, but is not internally irreducible.

Example The population with fertilization operator $x_1' = p_1^2 + p_1 p_3$, $x_2' = p_2^2 + p_1 p_2 + p_2 p_3$, $x_3' = p_3^2 + p_1 p_3 + p_2 p_3$, $x_4' = p_1 p_2$ and the meiosis operator $p_1 = x_1 + x_4$, $p_2 = x_2$, $p_3 = x_3$ has e.g.s. and is nondegenerate and internally irreducible but is not externally irreducible.

Observe that in contrast with an e.g.s. or a normal nonelementary s.g.s. the population of Theorem 4.6.7 can contain both splitting and nonsplitting heterozygotes. The zygotes $e_{ij,k}$ with $\gamma_{ij} > 1$, i.e. the nonisolated zygotes will be splitting, while the remaining $e_{ij,k}$'s (that is, the isolated ones) will be nonsplitting.

The assumption of nondegeneracy included in Theorem 4.6.7 made the formulation simpler but it is an easy condition to eliminate.

Let $\mathcal{A}$ be an evolutionary algebra with s.g.s. Suppose there are nondisappearing zygotes $e_1,\ldots,e_r$ and disappearing zygotes $e_{r+1},\ldots,e_n$. Denote by $\tilde{e}_l = e_l^2$ for $r + 1 \leq l \leq n$. Clearly, $\tilde{e}_l \equiv e_l \pmod{\mathrm{ann}\ \mathcal{A}}$ because $e_l^2 - e_l \in J^{\perp}$ and the algebra is conservative. So if for $x = \sum_{i=1}^{n} x_i e_i$ we denote by $Px = \sum_{i=1}^{r} x_i e_i$ then

$$x \equiv Px + \sum_{l=r+1}^{n} x_l \tilde{e}_l (\mathrm{mod}\ \mathrm{ann}\ \mathcal{A}).$$

Hence, we have

$$x^2 = (Px + \sum_{l=r+1}^{n} x_l \tilde{e}_l)^2 \tag{4.6.17}$$

But the vectors Px and $\tilde{e}_{r+1},\ldots\tilde{e}_n$ all lie in the subalgebra $[e_1,\ldots,e_r]$ generated by the nondisappearing zygotes. Formula (4.6.17) expresses the evolutionary operator of the whole population in terms of the evolutionary operator on a subpopulation consisting of only nondisappearing zygotes.

Chapter 5

The General Bernstein Problem

5.1   The Necessity of Mendel's First Law

We continue throughout this chapter to use the phrases *population* and *evolutionary algebra* to stand for a simplicial stochastic algebra. In particular, by regarding the set of vertices, the canonical basis, as an orthogonal basis, we obtain a natural inner product.

Mendel diallelic inheritance is equivalent to the fact that the evolutionary operator V is the Hardy-Weinberg operator. This operator is stationary (Bernstein), i.e. $V^2 = V$, non-degenerate and $p_{12,3} = 1$. Remarkably, these properties completely characterize Hardy-Weinberg Law, i.e. Mendel's First Law .

Theorem 5.1.1 *If a three-dimensional population is Bernstein, non-degenerate and*

$$p_{12,3} = 1, \tag{5.1.1}$$

then it is a Mendel two allele zygotic population.

The conditions of the theorem can be verified by an experiment. If an experiment shows that in the cross-breeding of non- vanishing types e_1, e_2 all descendants belong to the type e_3 and the population is stationary, then it obeys Mendel's law; all other heredity coefficients are defined uniquely. By the way, the situation is simplified if, in addition, it is known that the types e_1, e_2 are non-splitting, i.e. $e_1^2 = e_1$, $e_2^2 = e_2$, by Theorem 3.4.33. We will use this fact at the end of the proof of Theorem 5.1.1.

Proof Under the given conditions the evolutionary algebra $\mathcal{A}$ is Bernstein, stochastic, non-degenerate, $\dim \mathcal{A} = 3$, and the following equality holds for the canonical basis e_1, e_2, e_3:

$$e_1 e_2 = e_3 \tag{5.1.2}$$

This algebra can be neither unit nor constant. For, if it were a unit algebra, then $e_1 e_2 = 1/2(e_1 + e_2)$ contrary to (5.1.2). On the other hand, if it were constant, then it would follow from (5.1.2) that $\mathcal{A}^2 = [e_3]$, i.e. the types e_1, e_2 would be disappearing. Hence $\mathcal{A}$ cannot be of the type $(3,0)$ or $(1,2)$. Thus $\mathcal{A}$ is of the type $(2,1)$.

Let us prove that the algebra $\mathcal{A}$ is conservative. If not, then by the Corollary 3.4.24 it is exceptional. Then the manifold of its non-zero idempotents is a line l in unit plane $H = \{x | s(x) = 1\}$. Obviously, $l = \{x | x = e + \tau \bar{u}, -\infty < \tau < \infty\}$ where e is a non-zero idempotent, $\bar{u}$ is the basic vector of subspace U corresponding to this idempotent by the structure theory of Bernstein algebras (section 3.4) and $\dim U = 1$. Since the algebra was assumed to be exceptional, $H^2 \subset l$ by (3.4.27) and (3.4.13). So $H^2 = l$ because all points of line l are idempotents. Hence it follows from (5.1.2) that l passes through the vertex of the basis simplex (triangle) Δ. As is shown on Fig. 5.1, l intersects the edge $[e_1, e_2]$ at the interior point. Other configurations are impossible. Indeed, $l \cap \Delta = \operatorname{Im} V$. Fig. 5.2a would imply $\operatorname{Im} V = e_3$ and so both types e_1 and e_2 are disappearing. Analogously, Fig 5.2b and 5.2c would imply e_2 or e_1 are disappearing.

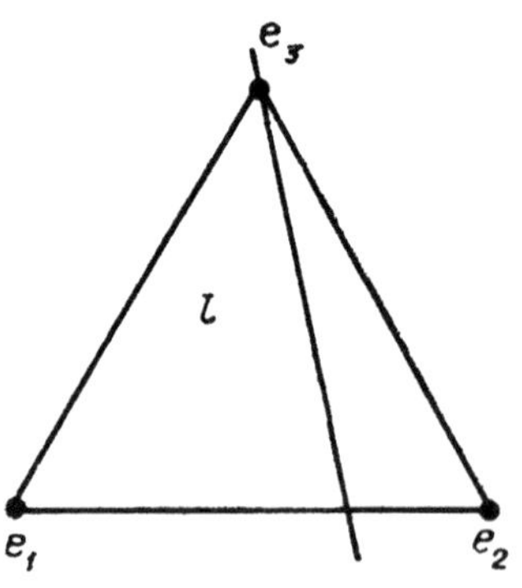

Fig. 5.1

Thus, the line l intersects the edge $[e_1, e_2]$ at an interior point $\xi e_1 + \eta e_2$, $\xi, \eta > 0$. Lying on l, this point is idempotent. Hence $\xi^2 e_1^2 + \eta^2 e_2^2 + 2\xi\eta e_3 =$

$\xi e_1 + \eta e_2$. The left hand side contains e_3 and the right hand sides does not, a contradiction. Therefore the algebra $\mathcal{A}$ is conservative.

Because algebra $\mathcal{A}$ is conservative, ann $\mathcal{A} = J^{\perp}$. The quotient algebra over this ideal is unique and has dimension 2.

Let E_1, E_2, E_3, as usual, be the images of basic vectors e_1, e_2, e_3 in the quotient algebra. Then (5.1.2) implies

$$E_3 = E_1 E_2 = \frac{E_1 + E_2}{2}$$

Hence $E_1 \neq E_2$, for otherwise the factor algebra has dimension one. Therefore E_1, E_2, E_3 are distinct and the algebra $\mathcal{A}$ is J-irreducible. The extremal points of convex hull D of the set $\{E_1, E_2, E_3\}$ are E_1 and E_2. By Corollary 3.8.11, e_1 and e_2 are idempotents. It remains to invoke Theorem 3.4.33. $\square$

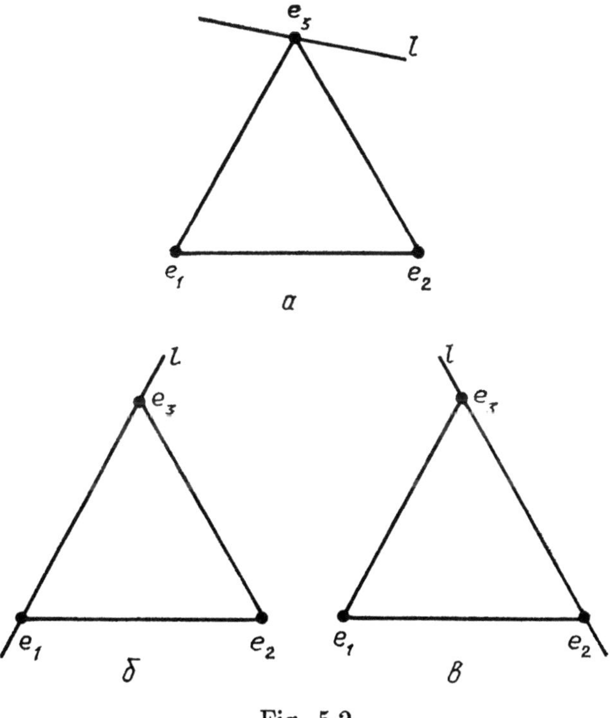

Fig. 5.2

Multiallele Mendel inheritance cannot be characterized as economically. However, in Theorem 3.4.34 it is possible to replace the condition of conservativity with the condition of non-negativity of the algebra structure constant. The proof remains essentially the same.

5.2   Theorem on Constant Inheritance

This theorem gives the solution of the Bernstein problem for the indecomposable case.

Theorem 5.2.1 *If a Bernstein evolutionary algebra is indecomposable, then it is constant.*

Hence in this case the probability distribution of the types among the first generation descendants does not depend on the distribution of the types in the parental generation. The existence of constant populations in nature is dubious (although such a population does have s.g.s., it is internally reducible). Probably the biological meaning of the Theorem 5.2.1 is that the "real" stationary evolutionary operators are decomposable.

To prove Theorem 5.2.1, we will use the following elementary lemma.

Lemma 5.2.2 *Let $\phi(x)$ be a polynomial of degree ≤ 2, K be a convex compact set. If $\phi|K$ achieves maximum and minimum at the interior, with respect to the affine hull aff K), then $\phi|K$ is constant.*

Proof. Let the extrema be achieved at x_0, x_1. For $x_0 = x_1$ the statement is obvious. For $x_0 \neq x_1$ draw the line through these points: $x_\tau = x_0 + \tau(x_1 - x_0)$ $(-\infty < \tau < \infty)$. Let it intersect the boundary ∂K (with respect to Aff K) at the points x_t, x_T (Fig. 5.3). Because $x_0, x_1 \in$ Int K, we have $t < 0$ and $T > 1$.

The function $\phi(x_\tau)$ in one variable τ is a polynomial of degree at most 2 and on the interval $[t, T]$ it achieves its maximum and minimum in the interior points $\tau = 0, \tau = 1$. Hence, $\phi(x_\tau)$ is constant, in particular, $\phi(x_0) = \phi(x_1)$, i.e. $\max(\phi|K) = \min(\phi|K)$, and so $\phi|K$ is constant. $\square$

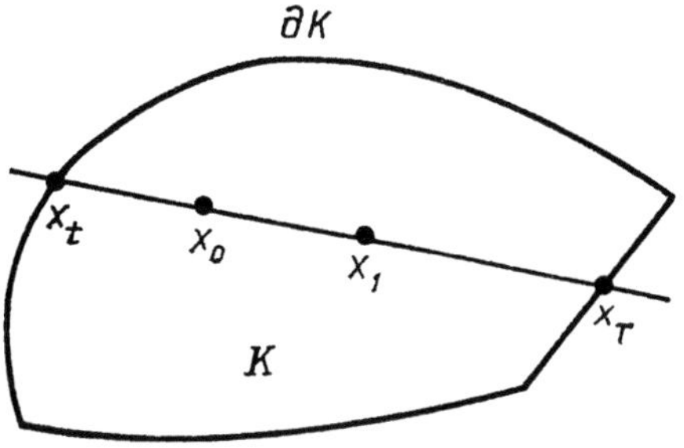

Fig. 5.3

Proof of Theorem 5.2.1 Now suppose that a Bernstein evolutionary algebra is indecomposable. By Theorem 3.8.4, all the idempotents lying in the basic simplex Δ are positive, i.e. lie in the interior of Δ (relative to the unit hyperplane). But because the algebra is Bernstein, all elements $x \in \Delta$ of the form $Vx = x^2$ are idempotents. Hence Im $V \subset$ Int Δ.

Let f be a linear form. Then $\phi(x) = f(x^2)$ is a quadratic form invariant for V, i.e. $\phi(Vx) = f((x^2)^2) = f(x^2) = \phi(x)$. Hence $\max(\phi|\Delta)$ and $\min(\phi|\Delta)$ are achieved on Im V, i.e. are achieved on some interior points of the simplex Δ. By Lemma 5.2.2, $\phi|\Delta$ is constant. Thus $f(Vx)$ is constant for any linear form f, therefore Vx is constant. $\square$

Corollary 5.2.3 *If in a Bernstein population the matrix $(p_{ii,j})_{i,j=1}^{n}$ of the hereditary coefficients of self-crossings is indecomposable (in particular, if all $p_{ii,j} > 0$), then the population is constant.*

Proof. In this case the evolutionary algebra is indecomposable by Corollary 3.8.5. $\square$

5.3 The Nonnegative Projection Associated with an Isolated Idempotent

Consider a Bernstein evolutionary operator V. Let $e \in \Delta$ be an equilibrium state: $Ve = e$, i.e $e^2 = e$, or $e \in \mathrm{Im}\ V$, which is equivalent because V is Bernstein. Consider a minimal face $\Gamma \subset \Delta$ containing e. By Theorem 3.8.3, the face Γ is invariant. Suppose that the operator $V|\Gamma$ is constant, i.e. $Vx = e$ for $x \in \Gamma$. We call the idempotent e *isolated*. Observe that this concept generalizes the idea of isolated vertex idempotent introduced for Theorem 3.8.13 and the results that follow it. Assume now that e is an isolated idempotent. By the structure theory of Bernstein algebras it corresponds to a projection $L \equiv L_e$. The projection L acts on the barideal $\mathcal{B}$ as $Ly = 2ey$. Its natural extension to the whole algebra $\mathcal{A}$ is the projection $\Lambda x = 2ex - s(x)e$. In this rank $\Lambda = m$, def $\Lambda = \delta$, where (m, δ) is the type of the algebra. But the projection Λ is generally not non-negative and we cannot use it in the stochastic situation.

Denote by P the coordinate projection to the linear hull of the face Γ, by Q the complementary coordinate projection, i.e. $Q = I - P$. Define the operator B *associated* with the isolated idempotent e by

$$Bx = 2ex - P(2ex) = Q(2ex).$$

Because $Q(e) = 0$,

$$B = Q\Lambda = \Lambda - P\Lambda. \tag{5.3.1}$$

Lemma 5.3.1 *The operator B associated with the isolated idempotent e is a non-negative projection. Moreover*

$$\ker B = \ker \Lambda \oplus [e]. \tag{5.3.2}$$

Proof It is clear from the definition that B is non-negative, since $e > 0$, $Q \geq 0$. Further, $B^2 = Q\Lambda Q\Lambda = Q\Lambda(I - P)\Lambda = Q\Lambda^2 - Q\Lambda P\Lambda = Q\Lambda - Q\Lambda P\Lambda = B - Q\Lambda P\Lambda$. Thus B is a projection if and only if $Q\Lambda P\Lambda = 0$. We have $\Lambda Px = 2ePx - s(Px)e$. But the subalgebra $[\Gamma]$ which contains both e and Px is constant. Hence $ePx = s(e)s(Px)e = s(Px)e$. Therefore

$$\Lambda Px = s(Px)e. \tag{5.3.3}$$

Hence $Q\Lambda Px = 0$, i.e. $Q\Lambda P = 0$ and $Q\Lambda P\Lambda = 0$.

To prove (5.3.2) we observe that (5.3.1) implies $\ker \Lambda \subset \ker B$. At the same time $e \in \ker B$ but $e \notin \ker \Lambda$ since $\Lambda e = e$. Hence $\ker \Lambda \oplus [e] \subset \ker B$. On the other hand, (5.3.1) implies

$$\Lambda B = \Lambda - \Lambda P \Lambda, \tag{5.3.4}$$

from which, by (5.3.3), $\Lambda B x = \Lambda x - s(P\Lambda x)e$ and $\ker B \subset \Lambda^{-1}([e]) = \ker \lambda \oplus [e]$. $\square$

Corollary 5.3.2 *rank $B = m - 1$, def $B = \delta + 1$.*

Corollary 5.3.3 *The operator B has the form*

$$B = \sum_{k=1}^{m-1} \langle \, , b_k^* \rangle b_k,$$

where $\{b_k\}_{k=1}^{m-1}$ is a non- negative Gaussian system of vectors, $\{b_k^*\}_{k=1}^{m-1}$ is a biorthogonal to it non-negative system of vectors, $\langle \, , \rangle$ is the usual scalar product.

This result follows from Lemma 5.3.1 and Theorem 3.7.8. Without loss of generality we will assume

$$s(b_k) = 1 \qquad (1 \leq k \leq m - 1). \tag{5.3.5}$$

We note that the dimension of the face Γ cannot exceed δ since for $x \in [\Gamma]$ we will have $\Lambda x = s(x)e$ whence $[\Gamma] \subset \Lambda^{-1}([e])$. The most interesting case is when $\dim \Gamma = 0$ i.e. e is a vertex idempotent.

Lemma 5.3.4 *If e is a vertex idempotent then the projections Λ and B commute.*

Proof Let us compute the commutator of Λ and B in the general case. By (5.3.1), (5.3.3) and (5.3.4), $[\Lambda, B] = \Lambda B - B\Lambda = P\Lambda - \Lambda P\Lambda = P\Lambda - s(P\Lambda \cdot)e$. In the vertex case, $Px = s(Px)e$, from which $P\Lambda x = s(P\Lambda x)e$, i.e. $[\Lambda, B] = 0$. $\square$

Corollary 5.3.5 *If e is a vertex idempotent then there exists an orthogonal decomposition $\operatorname{Im} \Lambda = \operatorname{Im} B \oplus [e]$.*

Proof Because $\ker \Lambda \subset \ker B$ and the operators commute, we have $\operatorname{Im} B \subset \operatorname{Im} \Lambda$. Also $e \perp \operatorname{Im} B$ by definition of operator B and $e \in \operatorname{Im} \Lambda$. Hence $\operatorname{Im} B \oplus [e] \subset \operatorname{Im} \Lambda$. Because $\operatorname{rank} \Lambda = m = \operatorname{rank} B + 1$, the assertion follows. $\square$

Suppose operator B is associated with vertex idempotent e_1. Consider the decomposition of the algebra $\mathcal{A} = [e_1] \oplus U \oplus W$ generated by this idempotent. In this case

$$[e_1] \oplus U = \operatorname{Im} \Lambda \tag{5.3.6}$$

and by Corollary 5.3.5 $[e_1] \oplus U = [e_1] \oplus \operatorname{Im} B$. But $\operatorname{Im} B = [b_1, \ldots, b_{m-1}]$ in terms of Corollary 5.3.3, and the vector system $\{b_k\}_{k=1}^{m-1}$ is linearly independent, i.e. is a basis in $\operatorname{Im} B$. Because the subspace U is selected in $\operatorname{Im} \Lambda$ by condition $s(x) = 0$, the vectors $u_k = b_k - e_1$ form a basis in U and augmenting it by e_1 we obtain a basis in $\operatorname{Im} A$ agreeing with decomposition (5.3.6). $\square$

Lemma 5.3.6 $e_1 b_k = (e_1 + b_k)/2$ *for* $1 \leq k \leq m - 1$.

Proof The equality $B b_k = b_k$ means, by definition of B, that $2 e_1 b_k = \beta_k e_1 + b_k$, where β_k are scalar coefficients. Applying s to both sides we get $\beta_k = 1$ by (5.3.5). $\square$

Lemma 5.3.7 $B(b_i b_k) = (b_i + b_k)/2$ *for* $1 \leq i, k \leq m - 1$.

Proof Because $U^2 \subset W$, we have $(b_i - e_1)(b_k - e_1) \in W$ which by Lemma 5.3.6 takes the form $b_i b_k - (b_i + b_k)/2 \in W$. But $W = \ker \Lambda \subset \ker B$. Hence $B(b_i b_k) = (B b_i + B b_k)/2$, with $B b_i = b_i$ and $B b_k = b_k$. $\square$

5.4 A Theorem on Two Non-Splitting Types

The theorem proved in this section describes a situation in which there are only two possibilities: Mendel's diallele Law or the Quadrille Law. Its proof makes a heavy use of the non-negative projection associated with a vertex idempotent developed in the previous section.

Theorem 5.4.1 *If a normal Bernstein population contains two nonsplitting types such that on crossing they reproduce all types except themselves then it is either a Mendel diallelic zygote population or a Quadrille population.*

Proof Denote the non-splitting types in the statement by e_1 and e_2. Then $e_1^2 = e_1$, $e_2^2 = e_2$ and

$$e_1 e_2 = \sum_{k=3}^{n} \pi_k e_k, \tag{5.4.1}$$

where all $\pi_k > 0$.

Consider the projection B associated with the vertex idempotent e_1. We know that $e_1 \perp \text{Im } B$. We will show that $e_2 \perp \text{Im } B$ as well. Multiplying (5.4.1) by e_1 and applying Theorem 3.4.33 we get $\frac{1}{2}(e_1 + \sum_{k=3}^{n} \pi_k e_k) = \sum_{k=3}^{n} \pi_k e_1 e_k$. Because all $\pi_k > 0$ and the left hand side is orthogonal to e_2, we have $e_1 e_k \perp e_2$ for $3 \le k \le n$, hence $e_2 \perp Be_k$ for $3 \le k \le n$. Because $Be_1 = 0$ and $Be_2 = 2\sum_{k=3}^{n} \pi_k e_k$, we have $e_2 \perp Be_k$ for all k and hence $e_2 \perp \text{im } B$. This means that in the decomposition of Corollary 5.3.3, all b_k are orthogonal to e_1 and e_2.

Now rewrite $e_1(e_1 e_2) = \frac{1}{2}(e_1 + e_1 e_2)$ as $B(e_1 e_2) = e_1 e_2$. On the other hand, by Corollary 5.3.3, $B(e_1 e_2) = \sum_{k=1}^{m-1} \lambda_k b_k$, where $\lambda_k = \langle e_1 e_2, b_k^* \rangle \ge 0$. Hence

$$e_1 e_2 = \sum_{k=1}^{m-1} \lambda_k b_k. \tag{5.4.2}$$

It is easy to see that all $\lambda_k > 0$. Indeed, if for example $\lambda_1 = 0$, then $b_1^* \perp e_1 e_2$, i.e. $b_1^* \in [e_1, e_2]$, from which $b_1^* \perp b_1$ while $\langle b_1, b_1^* \rangle = 1$.

Now clearly

$$\bigcup_{k=1}^{m-1} \text{supp } b_k = \{e_3, \ldots, e_n\}. \tag{5.4.3}$$

We can say that the sets $\text{supp } b_k, 1 \le k \le m-1$, form a cover of $\{e_3, \ldots, e_n\}$. This cover is minimal:

$$\bigcup_{k \ne i} \text{supp } b_k \ne \{e_3, \ldots, e_n\} \quad (1 \le i \le n-1), \tag{5.4.4}$$

because the system $b_1, \ldots, b_{m-1}$ is Gaussian and hence $\text{supp} b_i \not\subset \cup_{k \ne i} \text{supp } b_k$.

We square (5.4.2) using the appropriate formula from Theorem 3.4.33: $\sum_{i,k=1}^{m-1} \lambda_i \lambda_k b_i b_k = \frac{1}{4} e_1 + \frac{1}{4} e_2 + \frac{1}{2} \sum_{k=3}^{n} \pi_k e_k$.

Taking the inner product with e_2, we get $\sum_{i,k=1}^{m-1} \alpha_{ik} \lambda_i \lambda_k = \frac{1}{4}$, where $\alpha_{ik} = \langle b_i b_k, e_2 \rangle \ge 0$. Hence there exists a pair i_1, k_1 for which $\alpha_{i_1 k_1} > 0$. Applying Lemma 5.3.7 for $i = i_1$, $k = k_1$ we get $(b_{i_1} + b_{k_1})/2 = \alpha_{i_1 k_1} Be_2 + \cdots = 2\alpha_{i_1 k_1} \sum_{k=3}^{n} \pi_k e_k + \cdots$. But then $\text{supp} b_{i_1} \cup \text{supp } b_{k_1} = \{e_3, \ldots, e_n\}$.

Comparing this with (5.4.4) we see that either $i_1 = k_1$ and $m - 1 = 1$ or else $m - 1 = 2$, i.e. $m = 2$ or $m = 3$.

Case 1. $m = 2$. We will show that $\mathcal{A}^2 = [e_1, e_2, e_1 e_2]$. The inclusion $\mathcal{A}^2 \supset [e_1, e_2, e_1 e_2]$ is obvious because e_1 and e_2 are idempotents. On the other hand $\dim \mathcal{A}^2 \leq 3$ by Corollary 3.4.11.

The vectors $e_1, e_2, e_1 e_2$ form a basis in $\mathcal{A}^2$. Decomposing $x^2 = \phi_1(x)e_1 + \phi_2(x)e_2 + \phi_3(x)e_1 e_2$ we find in canonical coordinates $x'_1 = \phi_1(x)$, $x'_2 = \phi_2(x)$, $x'_k = \pi_k \phi_3(x)$, $3 \leq k \leq n$. If $n > 3$ the population is externally reducible. Therefore $n = 3$, $e_1 e_2 = e_3$ and by Theorem 3.4.33 we have the Mendel diallele zygotic population.

Case 2. $m = 3$. We will show that $\mathcal{A}^2 = [e_1, e_2, b_1, b_2]$. The inclusion $\mathcal{A}^2 \supset [e_1, e_2, b_1, b_2]$ is obvious because Im $B \subset \mathcal{A}^2$ (for $Bx = 2e_1 x - \langle 2e_1 x, e_1 \rangle e_1, e_1 = e_1^2$). It remains to show that $\dim \mathcal{A}^2 \leq 4$.

Suppose $\dim \mathcal{A}^2 > 4$. Then by Corollary 3.4.30 the algebra $\mathcal{A}$ is conservative. Therefore the population has s.g.s. and also is normal by assumption. We will use the explicit description of evolutionary operators of this class from Theorems 4.3.9 and 4.6.1. If the population has e.g.s. then the product of basis idempotents (homozygotes) e_1 and e_2 is a certain basis element (heterozygote). If, instead, the population obeys the Extended Quadrille Law, then either $e_1 e_2 = \frac{1}{2}(e_1 + e_2)$ which contradicts (5.4.1) or $e_1 e_2 = \frac{1}{2}(e_3 + e_4)$ which gives $n = 4$. Thus we always get $n \leq 4$ which contradicts our assumption.

The vectors e_1, e_2, b_1, b_2 are linearly independent by orthogonality of $[e_1, e_2]$ with $[b_1, b_2]$. Therefore they form a basis in $\mathcal{A}^2$. Decomposing $x^2 = \phi_1(x)e_1 + \phi_2(x)e_2 + \psi_1(x)b_1 + \psi_2(x)b_2$ we find in canonical coordinates $x'_1 = \phi_1(x)$, $x'_2 = \phi_2(x)$, and for $k \geq 3$,

$$x'_k = \begin{cases} \langle b_1, e_k \rangle \psi_1(x) & (e_k \in \operatorname{supp} b_1), \\ \langle b_2, e_2 \rangle \psi_2(x) & (e_k \in \operatorname{supp} b_2). \end{cases}$$

Because the population is externally irreducible, supp b_1 and supp b_2 have one element each. Therefore $b_1 = e_3$ and $b_2 = e_4$ (up to numeration). By (5.4.3), $n = 4$ and therefore $\mathcal{A}^2 = \mathcal{A}$. Hence the algebra $\mathcal{A}$ is of the type $(3,1)$ and is conservative (by Corollary 3.4.29). Because $n = 4$, the population is Quadrille. $\square$

5.5 The Solution of the Bernstein Problem for Exceptional Populations

The preceding chapter gives the solution of the Bernstein problem in the natural genetic setting of stationary gene structure (s.g.s.). Theorem 4.3.6 describes the evolutionary operators of populations having elementary gene structure (e.g.s.). Theorem 4.3.9 refines this description provided that the population is normal. Evolutionary operators of populations having nonelementary s.g.s. are described by Theorem 4.6.1 for the normal case and by Theorem 4.6.7 for the non-degenerate case. Furthermore, as explained there the non-degeneracy assumption can be easily eliminated. Thus, we have completely solved the Bernstein problem for the case of a *conservative* evolutionary algebra. This section presents the solution of the Bernstein problem for the complementary case of an *exceptional* evolutionary algebra. Thus, by Corollary 3.4.24, the Bernstein problem is solved for an algebra of the type (m, δ) for $m \leq 2$ or $\delta \leq 1$, in particular in dimensions $n \leq 4$. Biological interpretation of these results beyond the s.g.s. cases is unknown (and possibly does not exist at all). However the Bernstein problem is interesting from a purely mathematical point of view and in any case presents a good example of a mathematical problem motivated by applications but possibly extending beyond their boundaries.

Thus, let $\mathcal{A} = \mathcal{A}_V$ be an exceptional Bernstein evolutionary algebra. Consider the manifold of nonzero idempotents. Its parameterization (3.4.13), because of the algebra is exception $(U^2 = 0)$, takes the form $x = e + u$, $u \in U$, i.e. the manifold of nonzero idempotents in an exceptional algebra of the type (m, δ) is an affine plane P of dimension $m - 1$. The set of equilibrium states of the population (i.e. Im V because it is Bernstein) is the intersection of the plane P with the basic simplex Δ. Therefore Im V is a convex polyhedron. Consider its extremal points $e^1, \ldots, e^r$. Denote Γ^i the convex hull of the set supp e^i, i.e. the smallest face of the simplex Δ containing the point e^i. The face Γ^i is invariant by Theorem 3.8.3. The idempotent e^i is isolated, i.e. the face Γ^i contains no other idempotents $e \neq e^i$, since otherwise the line passing through the points e and e^i would lie in the plane P and in the face Γ^i and the point e^i would be an interior point of an interval contained in ImV, i.e. would not be extremal in Im V.

Because the operator V maps all points of the face Γ^i into the point e^i, it follows that $\Gamma^i \cap \Gamma^j = \emptyset$ for $i \neq j$, which implies that the system $e^1, \ldots, e^r$

is orthogonal:

$$\langle e^i, e^j \rangle = 0 \quad (i \neq j). \tag{5.5.1}$$

Write $V(x)$ as a linear combination of the extremal points of the polyhedron Im V:

$$V(x) = \sum_{i=1}^{r} \theta_i(x) e^i. \tag{5.5.2}$$

Because of (5.5.1), the coefficients $\theta_i(x)$ are defined uniquely:

$$\theta_i(x) = \frac{\langle V(x), e^i \rangle}{\langle e^i, e^i \rangle} \quad (1 \leq i \leq r). \tag{5.5.3}$$

The right hand sides of (5.5.3) are quadratic forms in x with non-negative coefficients. It follows from (5.5.2) that

$$x^2 = \sum_{i=1}^{r} \theta_i(x) e^i \tag{5.5.4}$$

for all $x \in \mathcal{A}$. Hence

$$\sum_{i=1}^{r} \theta_i(x) = s^2(x) \tag{5.5.5}$$

and

$$\theta_i(x^2) = s^2(x)\theta_i(x) \quad (1 \leq i \leq r) \tag{5.5.6}$$

(the quadratic forms $\theta_i(x)$ are invariant). Substituting (5.5.4) and (5.5.5) into (5.5.6) we obtain the identity

$$\sum_{j,k=1}^{r} \hat{\theta}_i(e^j, e^k)\theta_j(x)\theta_k(x) = \theta_i(x) \sum_{k=1}^{r} \theta_k(x), \tag{5.5.7}$$

where $\hat{\theta}_i$ is the symmetric bilinear form polar to θ_i. Notice further that by Corollary 3.4.10 the maximal functional rank of system of forms $\theta_i(x)$, $1 \leq i \leq r$, is m. Therefore $r \geq m$. On the other hand, $r \leq m$ because the points $e^1, \ldots, e^r$ lie in the $(m-1)$- dimensional affine plane P and are affinely independent by (5.5.1). Therefore $r = m$ and the forms $\theta_i(x)$ are functionally independent. Hence, (5.5.7) implies

$$\hat{\theta}_i(e^j, e^k) = (\delta_{ij} + \delta_{ik})/2 \quad (1 \leq i, j, k \leq m) \tag{5.5.8}$$

Let $E^i = [\Gamma^i]$, $1 \leq i \leq m$. The entire space decomposes into an orthogonal sum $E^0 \oplus E^1 \oplus \cdots \oplus E^m$ of these subspaces and the orthogonal complement E^0 of their sum. Let $x = x^0 \oplus x^1 \oplus \cdots \oplus x^m$ be the corresponding decomposition of a vector x.

It follows from (5.5.8) that $\theta_i(e^i) = 1$, $1 \le i \le m$ and from (5.5.5) that the elements of the matrix of the quadratic form θ_i do not exceed 1. Because vector e^i is stochastic, i.e. lies in Δ, in the above matrix all the elements of the diagonal block that corresponds to coordinate subspace E^i are equal to 1. Hence $\theta_i(x) = s^2(x^i)$, $1 \le i \le m$.

(5.5.8) further implies that $\hat{\theta}_i(e^j, e^k) = 0$ for $j, k \ne i$. The coefficients of the form $\hat{\theta}_i$ and the vectors e^j, e^k are non-negative, and therefore the coefficients that correspond to a pair of subspaces E^j, E^k, $j, k \ne i$ are zero. Hence, $\hat{\theta}_i(x^j, x^k) = 0$, $j, k \ne i, 0$.

Thus $\theta_i(x) = s^2(x^i) + 2\sum_{j \ne 0,i} \hat{\theta}_i(x^i, x^j) + \rho_i(x^0) + 2\bar{\rho}_i(x^0, \bar{x})$, where $\rho_i(x^0)$ is a quadratic form, $\bar{\rho}_i(x^0, \bar{x})$ is a bilinear form, $\bar{x} = x \ominus x^0 = x^1 \oplus \cdots \oplus x^m$.

Let us introduce linear operators (matrices) $T_{ji} \cdot E^i \to E^j$ ($j \ne 0, i \ne 0, j \ne i$) defined by the identities $\langle T_{ji}x^i, x^j \rangle = 2\hat{\theta}_i(x^i, x^j) - s(x^i)s(x^j)$ and linear operators $R_i^0 : e^0 \to E^0$, $\bar{R}_i : E^0 \to (E^0)^\perp$ ($i \ne 0$) defined by the identities $\langle R_i^0 x^0, x^0 \rangle = \rho_i(x^0)$ and $\langle \bar{R}_i x^0, \bar{x} \rangle = \bar{\rho}_i(x^0, \bar{x})$ (the operators R_i^0 are necessarily self-adjoint). The elements of the matrices T_{ji} lie in $[-1, 1]$ and the entries of matrices R_i^0, $\bar{R}_i$ lie in $[0, 1]$.

We have

$$\theta_i(x) = s(x^i)s(\bar{x}) + \sum_{j \ne 0,i} \langle T_{ji}x^i, x^j \rangle + \langle R_i^0 x^0, x^0 \rangle + 2\langle \bar{R}_i x^0, \bar{x} \rangle, \qquad (5.5.9)$$

from which by (5.5.5),

$$s^2(x) = s^2(\bar{x}) + \sum_{j,i=1}^{m} \langle T_{ji}x^i, x^j \rangle + \sum_{i=1}^{m} \langle R_i^0 x^0, x^0 \rangle + 2\sum_{i=1}^{m} \langle \bar{R}_i x^0, \bar{x} \rangle,$$

(if we set $T_{ii} = 0$). On the other hand, obviously $s^2(x) = s^2(\bar{x}) + s^2(x^0) + 2s(x^0)s(\bar{x})$. Therefore, $\sum_{j,i=1}^{m} \langle T_{ji}x^i, x^j \rangle = 0$, $\sum_{i=1}^{m} \langle R_i^0 x^0, x^0 \rangle = s^2(x^0)$ and $\sum_{i=1}^{m} \langle \bar{R}_i x^0, \bar{x} \rangle = s(x^0)s(\bar{x})$, from which $T_{ij} = -T_{ji}^*$, ($j \ne i$) and

$$\sum_{i=1}^{m} R_i^0 = \begin{pmatrix} 1 & \cdots & 1 \\ \cdot & \cdots & \cdot \\ 1 & \cdots & 1 \end{pmatrix}, \sum_{i=1}^{m} \bar{R}_i = \begin{pmatrix} 1 & \cdots & 1 \\ \cdot & \cdots & \cdot \\ 1 & \cdots & 1 \end{pmatrix}. \qquad (5.5.10)$$

Finally, (5.5.8) also implies

$$\langle T_{ji}e^i, e^j \rangle = 0 \qquad (j \ne i). \qquad (5.5.11)$$

Conversely, let an algebra be defined by formula (5.5.4), where $r = m$, $e^1, \ldots, e^m$ are pairwise orthogonal stochastic vectors, $\theta_i(x)$ are quadratic forms satisfying (5.5.9) and the conditions (5.5.10) and (5.5.11) are satisfied, the matrices R_i^0 are self-adjoint and the elements of the matrices T_{ji} do not exceed 1 in absolute value. Then the identity (5.5.5) is satisfied and therefore the algebra is stochastic; (5.5.8) is satisfied which implies (5.5.6), i.e. the algebra is Bernstein. Then (5.5.4) and (5.5.8) imply the multiplication table $e^j e^k = (e^j + e^k)/2$, $1 \leq j, k \leq m$. In particular, all the e^i are idempotents. A projection L which is associated, e.g., with idempotent e^1 acts on the differences $e^k - e^1$ as follows: $L(e^k - e^1) = 2e^1(e^k - e^1) = (e^k - e^1)$ $(2 \leq k \leq m)$. Thus the vectors $e^k - e^1$ are in $U = \operatorname{Im} L$. Therefore $\dim U \geq m - 1$. On the other hand, if a linear form f annihilates all of e^i, then by (5.5.4), $f(x^2) = 0$, i.e. $\dim N \geq n - m = \delta$. Hence $\dim U = m - 1$, $\dim N = \delta$, i.e. this is an exceptional algebra of the type (m, δ).

Thus we have obtained an *explicit description of all Bernstein exceptional stochastic algebras*, i.e. we have solved the Bernstein problem for this case.

An algebra generated by a stochastic projection belongs to the above class. Therefore the explicit type of stochastic projections mentioned in Theorem 3.7.11 must agree with the general result. We will omit the direct verification. Note that the algebras generated by stochastic projections are exactly those Bernstein stochastic algebras that are both conservative and exceptional (see Corollary 3.4.18).

The following theorem is an interesting consequence of the above description.

Theorem 5.5.1 *If a Bernstein population is exceptional and normal then it is unit, i.e.* $x'_j = x_j$, $1 \leq j \leq n$.

Proof The equation (5.5.4) is compatible with the definition of normality only for $r = n$, for the stochastic vectors $e^1, \ldots, e^r$ are pairwise orthogonal. Therefore the evolutionary algebra is of the type $(n, 0)$. The result follows from Theorem 3.4.3. $\square$

Recall that the normalization process leaves the algebra Bernstein and preserves its rank m. The latter follows from Corollary 3.4.9 in conjunction with Theorem 3.9.7. Because exceptional algebras are characterized by linearity of the manifold of nonzero idempotents, it follows also from Theorem 3.9.7 that the normalization process does not lead outside the class of exceptional algebras. By Theorem 5.5.1 this process always ends with a unit algebra.

5.6 Small Dimensions

As noted above, the general theorems related to the Bernstein problem include a complete solution for $n \leq 4$.

The cases $n = 1, 2$ are trivial. When $n = 2$ the only possible operators are the identity operator $x_1' = x_1$, $x_2' = x_2$ and the constant operators $x_1' = c_1$, $x_2' = c_2$ $(c_1, c_2 \geq 0, c_1 + c_2 = 1)$, for when $n = 2$ the algebra is either of the type (2,0) or (1,1) and in both cases Theorem 3.4.3 applies. The case $n = 1$ allows for only the identity operator, which is also a constant one.

Let us list all Bernstein evolutionary operators for the case $n = 3$. We will not present the case $n = 4$, for the corresponding list would be rather cumbersome and the only additional phenomenon (in comparison with $n = 3$) is the previously discussed Quadrille Law.

The case $n = 3$ allows for the types (3,0), (1,2) and (2,1). The operator of the type (3,0) is the identity and the operators of the type (1,2) are constant.

If the Bernstein evolutionary operator of the type (2,1) is conservative, i.e. has s.g.s., then by Corollary 4.3.3 it has e.g.s. Its type is described by Theorem 4.3.6 $x_1' = p_1 P_1$, $s_2' = p_2 P_2$ and $x_3' = p_{i_3} P_3$ where

$$p_1 = x_1 + \pi_1 x_3, \quad p_2 = x_2 + \bar{\pi}_1 x_3, \qquad (5.6.1)$$

$i_3 = 1$ or $i_3 = 2$ depending on whether $\pi_1 > 0$ or $\pi_1 = 0$, $P_j = t_{j1} p_1 + t_{j2} p_2$, $1 \leq j \leq 3$. Here all the coefficients are non-negative, $\pi_1 + \bar{\pi}_1 = 1$.

Suppose $\pi_1 > 0$. Then the last equation of (4.3.5) becomes $\bar{\pi}_1 t_{31} = 0$. Hence if also $\bar{\pi}_1 > 0$, then $t_{31} = 0$. The remaining equations of (4.3.5) simplify to $t_{11} = 1$, $t_{22} = 1$ and:

$$t_{12} + \pi_1 t_{32} = 1, \quad t_{21} + \bar{\pi}_1 t_{32} = 1. \qquad (5.6.2)$$

We obtain a family of evolutionary operators

$$x_1' = p_1^2 + t_{12} p_1 p_2, \quad x_2' = p_2^2 + t_{21} p_1 p_2, \quad x_3' = t_{32} p_1 p_2, \qquad (5.6.3)$$

where p_1, p_2 are linear forms determined by the formulas (5.6.1); the positive parameters $\pi_1, \bar{\pi}_1$ are determined by $\pi_1 + \bar{\pi}_1 = 1$; the non-negative parameters t_{12}, t_{21}, t_{32}, are determined by (5.6.2). When $t_{32} > 0$ the operator is normal and is described in Section 4.3 up to notation. If $t_{12} = t_{21} = 0$, $t_{32} = 2$, $\pi_1 = \bar{\pi}_1 = \frac{1}{2}$ it becomes the Hardy- Weinberg operator. In the general case it also has the genetic meaning indicated in Section 4.3.

If $\pi_1 > 0$, $\bar{\pi}_1 = 0$ (i.e. $\pi_1 = 1$) then t_{31} may be different from 0. In this case

$$p_1 = x_1 + x_3, \quad p_2 = x_2 \qquad (5.6.4)$$

and

$$x_1' = t_{11}p_1^2 + t_{12}p_1p_2, \quad x_2' = p_2^2 + p_1p_2, \quad x_3' = t_{31}p_1^2 + t_{32}p_1p_2, \qquad (5.6.5)$$

where t_{ik} are non-negative parameters given by $t_{11} + t_{31} = 1$, $t_{12} + t_{32} = 1$. Note that the middle equation in (5.6.5) simplifies: $x_2' = p_2$. This agrees with the invariance of the linear form p_2.

The population (5.6.4) (5.6.5) is internally reducible. It shows, by the way, that normal evolutionary operators do not form a dense subset in all stationary evolutionary operators. The normal case is not "generic" in this sense. On the other hand, the subset of the normal operators is open.

The last (in the context of s.g.s.) of the type (2,1) case is $\pi_1 = 0$, $\bar{\pi}_1 > 0$. It differs from the previous case only by the numeration and therefore need not be considered.

Let a stationary evolutionary operator of the type (2,1) be nonconservative. Hence it is exceptional (and by Theorem 5.5.1 cannot be normal if it differs from unity). Further we consider all exceptional operators including conservative (i.e. such that generated by stochastic projections). It follows from the are previous section that it has the form $x' = \theta_1(x)e^1 + \theta_2(x)e^2$, where e^1, e^2 are mutually orthogonal stochastic vectors, $\theta_1(x) = s(x^1)s(\bar{x}) + \langle T_{21}x^1, x^2 \rangle + \langle R_1^0 x^0, x^0 \rangle + 2\langle \bar{R}_1 x^0, \bar{x} \rangle$ and $\theta_2(x) = s(x^2)s(\bar{x}) + \langle T_{12}x^2, x^1 \rangle + \langle R_2^0 x^0, x^0 \rangle + 2\langle \bar{R}_2 x^0, \bar{x} \rangle$. Here $x = x^0 \oplus \bar{x}$, $\bar{x} = x^1 \oplus x^2$ according to the decomposition of the entire (3-dimensional) space into the orthogonal sum $E^0 \oplus E^1 \oplus E^2$, where $E^1 = [\text{supp } e^1]$, $E^2 = [\text{supp } e^2]$ and E^0 is the orthogonal complement of the sum $E^1 \oplus E^2$. Without loss of generality, we can assume $\dim E^1 \geq \dim E^2$. Moreover $\dim E^2 \geq 1$. Hence $\dim E^0 \leq 1$, i.e. $E^0 = 0$ or $\dim E^0 = 1$. Operators R_i^0, $\bar{R}_i$ appear only when $E^0 \neq 0$. In this case $\dim E^0 = 1$ and R_i^0, $\bar{R}_i$ can be defined by the stochastic matrices (1×1 and 1×2 respectively), i.e. $R_i^0 \geq 0$, $\bar{R}_{ij} \geq 0$ and

$$R_1^0 + R_2^0 = 1, \qquad \bar{R}_{1j} + \bar{R}_{2j} = 1 \ (j = 1, 2). \qquad (5.6.6)$$

The elements of the matrix T_{21} do not exceed 1 in absolute value, $T_{12} = -T_{21}^*$. Also

$$\langle T_{21}e^1, e^2 \rangle = 0. \qquad (5.6.7)$$

If $\dim E^1 = \dim E^2$ (hence equal to 1), then it follows from (5.6.7) that $T_{21} = 0$. If $\dim E^1 > \dim E^2$ (hence $\dim E^1 = 2$, $\dim E^2 = 1$), then $T_{21} = (\tau, \sigma)$ is a 1×2 matrix. Without loss of generality, we assume

$$e^1 = \alpha e_1 + \beta e_2, \qquad e^2 = e_3 \qquad (5.6.8)$$

(where e_1, e_2, e_3 is the canonical basis of the stochastic algebra in question), and obtain from (5.6.7)

$$\tau\alpha + \sigma\beta = 0. \tag{5.6.9}$$

Together with these possibilities for the dimensions of E^1 and E^2 we have the alternatives $\dim E^0 = 0$ or 1 with the corresponding consequences for $R_i^0, \bar{R}_i$ $(i = 1, 2)$.

Thus we have two cases:

Case 1. $\dim E^1 = \dim E^2 = \dim E^0 = 1$. Then we may assume $e^1 = e_1$, $e^2 = e_2$, and the evolutionary operators take the form

$$\begin{aligned}
x_1' &= x_1(x_1 + x_2) + R_1^0 x_3^2 + 2(\bar{R}_{11}x_1 + \bar{R}_{12}x_2)x_3, \\
x_2' &= x_2(x_1 + x_2) + R_2^0 x_3^2 + 2(\bar{R}_{21}x_1 + \bar{R}_{22}x_2)x_3, \\
x_3' &= 0
\end{aligned} \tag{5.6.10}$$

where R_i^0, $\bar{R}_{ij}$ $(i, j = 1, 2)$ are nonnegative parameters defined by conditions (5.6.6). The stochastic projections appear for $\bar{R}_{11} = \frac{1}{2}(R_1^0 + 1)$, $\bar{R}_{12} = \frac{1}{2}R_1^0$. They can be included in family (5.6.3) by putting $t_{32} = 0$ (hence $t_{12} = t_{21} = 1$) and $\pi_1 = R_1^0$.

Case 2. $\dim E^1 = 2$, $\dim E^2 = 1$, $E^0 = 0$. Then we can assume (5.6.8), and the evolutionary operator will be

$$\begin{aligned}
x_1' &= \alpha((x_1 + x_2)s + (\tau x_1 + \sigma x_1)x_3), \\
x_2' &= \beta((x_1 + x_2)s + (\tau x_1 + \sigma x_2)x_3), \\
x_3' &= x_3 s - (\tau x_1 + \sigma x_2)x_3,
\end{aligned}$$

where α, β are positive $\alpha + \beta = 1$, $|\tau| \leq 1$, $|\sigma| \leq 1$ and the condition (5.6.9) holds. Clearly, by (5.6.9), $\tau = k\beta$, $\sigma = -k\alpha$, where $|k| \leq \min(\alpha^{-1}, \beta^{-1})$. Finally we have

$$\begin{aligned}
x_1' &= \alpha((x_1 + x_2)s + k(\beta x_1 - \alpha x_2)x_3), \\
x_2' &= \beta((x_1 + x_2)s + k(\beta x_1 - \alpha x_2)x_3), \\
x_3' &= x_3 s - k(\beta x_1 - \alpha x_2)x_3,
\end{aligned} \tag{5.6.11}$$

The linear form $\beta x_1 - \alpha x_3$ in (5.6.11) is disappearing. All other disappearing forms are proportional to it, as should be the case for an exceptional algebra of the type $(2,1)$.

The example (4.1.7) above is obtained from (5.6.11) by putting $\alpha = \beta = \frac{1}{2}$, $k = 2$ (up to the numeration of the coordinates).

If (and only if) $k = 0$ we get a stochastic projection. It can be included in family (5.6.5) by putting $t_{11} = t_{12} = \alpha$ and $t_{31} = t_{32} = \beta$ up to substitution $x_2 \mapsto x_3 \ x_3 \mapsto x_2$.

Thus the following theorem has been proved.

Theorem 5.6.1 *The list of stationary evolutionary operators of the type (2,1) consists (up to the numeration of the coordinates) of (5.6.3), (5.6.5), (5.6.10), (5.6.11).*

We see the set of Bernstein evolutionary operators of the type (2,1) is connected. It consists of three 2-dimensional pieces (5.6.3), (5.6.5) and (5.6.11), and one 3-dimensional piece (5.6.10). In addition, when $n = 3$ there are also the identity operator and constant operators. Every type including (2,1) defines some connected component of the set of Bernstein evolutionary operators for $n = 3$. In contrast, the Quadrille Law is an isolated point of the set of Bernstein evolutionary operators of the type (3,1).

5.7 Estimate of the Number of Constant Subpopulations. Ultranormal Bernstein Populations

Theorem 5.7.1 *Every Bernstein population of rank m has at least m constant subpopulations.*

The cases of the Bernstein problem studied above support this theorem. The general proof will be obtained from combinatorial and topological considerations given below as a sequence of lemmas.

We shall call a non-empty subspace c of the affine space $\mathbf{R}^k$ a *affine cell* if it is bounded, open and its closure $\bar{c}$ is contractible (within itself to a point). The dimension k of the containing space is called *cell dimension*.

Let X be a Hausdorff topological space. A subspace $C \subset X$ is called a *k-dimensional cell* if there exists a k- dimensional affine cell c and a homeomorphism $f : \bar{c} \to \bar{C}$ such that $f(c) = C$ (and $f(\partial c) = \partial C$). An *elementary cell complex* on a space X is a finite collection of cells, with union X, such that 1) the boundary of each cell of the complex is a union of lower dimensional cells of the complex; 2) the intersection of the closures of any two cells is contractible. The maximal cell dimension is called the *dimension of the complex*.

Lemma 5.7.2 *A n-dimensional elementary cell complex contains at least $n + 1$ zero-dimensional cells.*

The proof of this purely topological lemma is given in the appendix to this chapter.

Now let V be a Bernstein evolutionary operator. Consider an arbitrary face Γ of the basic simplex Δ and let $C_\Gamma = \text{Int } \Gamma \cap \text{Im } V$ (as usual, Int Γ is the interior of the face Γ with respect to its affine hull). We will call the face Γ *essential* if $C_\Gamma \neq \emptyset$. For example, for the Hardy- Weinberg diallele operator, the essential faces consist of the 2- face Δ and the 0-faces $\{e_1, \}, \{e_2\}$ (see Fig. 2.1). Because all points $x \in \text{Im } V$ are fixed by $V = V^2$, each essential face Γ and its interior in Γ are invariant by Theorem 3.8.3 and the Remark after it. Conversely, if Int Γ is invariant, then the face Γ obviously is essential.

Lemma 5.7.3 *If Γ is an invariant face then*

$$\text{Im}(V|\Gamma) = \Gamma \cap \text{Im } V. \tag{5.7.1}$$

Proof If $x \in \Gamma \cap \text{Im } V$, then because the operator V is Bernstein, $x = Vx$, i.e. $x \in \text{Im}(V|\Gamma)$. Thus, $\Gamma \cap \text{Im } V \subset \text{Im}(V|\Gamma)$. The reverse inclusion is trivial. $\square$

Lemma 5.7.4 *An invariant face Γ is essential if and only if the operator $V|\Gamma$ is non-degenerate.*

Proof (5.7.1) implies $C_\Gamma = \text{Int } \Gamma \cap \text{Im}(V|\Gamma)$. Thus we may assume $\Gamma = \Delta$, and Δ obviously is essential if and only if V is non- degenerate, i.e. there are no disappearing forms among x_j. $\square$

Lemma 5.7.5 *If Γ is an essential face, then $\bar{C}_\Gamma = \Gamma \cap \text{Im } V = \text{Im}(V|\Gamma)$.*

For a non-essential face intersecting Im V this is obviously false.

Proof The set $\Gamma \cap \text{Im } V$ is closed and contains C_Γ, hence $\bar{C}_\Gamma \subset \Gamma \cap \text{Im } V$. To check the reverse inclusion, let $x \in \Gamma \cap \text{Im} V$ and let a sequence $\{x^i\}_{i=1}^{\infty} \subset \text{Int } \Gamma$ converge to x. Then the sequence $\{Vx^i\}$ is contained in C_Γ by invariance of Int Γ and obviously converges to $Vx = x$. Then $x \in \bar{C}_\Gamma$. $\square$

Corollary 5.7.6 *If Γ is an essential face then $\partial C_\Gamma = \partial \Gamma \cap \text{Im } V$.*

Proof $\partial C_\Gamma = \bar{C}_\Gamma \backslash C_\Gamma = (\Gamma \cap \text{Im } V) \backslash (\text{Int } \Gamma \cap \text{Im } V) = (\Gamma \backslash \text{Int } \Gamma) \cap \text{Im } V = \partial \Gamma \cap \text{Im } V$. $\square$

The role of essential faces in the topology of the set of equilibrium states is shown in the following lemma.

Lemma 5.7.7 *If Γ is an essential face, then C_Γ is a cell in Im V. Moreover if $V|\Gamma$ is an operator of the type $(m_\Gamma, \delta_\Gamma)$, then*

$$\dim C_\Gamma = m_\Gamma - 1. \tag{5.7.2}$$

Proof Letting $\Gamma = \Delta$ by Lemma 5.7.4 we will write C for C_Γ. By hypothesis, $C = \text{Int } \Delta \cap \text{Im } V \neq \emptyset$. By Lemma 5.7.5, $\bar{C} = \text{Im } V$. The set $\bar{C}$ is homeomorphic via $x = F(u) \equiv e + u + u^2 (u \in U)$ with the set $\bar{c} = \{u | e + u + u^2 \geq 0\} \subset U$ (by Corollary 3.4.9). Consider in U an open subset $c = \{u | F(u) \gg 0\}$. Obviously, $C = F(c)$, $c = F^{-1}(C)$. Because C is bounded, so is c. Moreover, $\bar{C}$ is contractible to any of its points x_0 by means of a homotopy $x_\tau = V((1 - \tau)x + \tau x_0))(0 \leq \tau \leq 1)$. Therefore c is contractible. Thus C is a cell in $\text{Im } V$ and $\dim C = \dim c = m - 1$. $\square$

Lemma 5.7.8 *If Γ_1 is an essential face and $\Gamma \subset \Gamma_1$, $\Gamma \neq \Gamma_1$, then $\dim C_\Gamma <$ $\dim C_{\Gamma_1}$.*

    Proof We have $C_\Gamma = \text{Int } \Gamma \cap \text{Im } V \subset \partial \Gamma_1 \cap \text{Im } V = \partial C_{\Gamma_1}$. $\square$
The following lemma results from the above considerations.

Lemma 5.7.9 *The system $\{C_\Gamma\}_{\Gamma \in E}$, where E is the set of essential faces, forms an $m - 1$-dimensional elementary cell complex on $\text{Im } V$.*

    Proof If Γ_1, Γ_2 are two distinct faces of a simplex, then their interiors do not intersect, and so $C_{\Gamma_1} \cap C_{\Gamma_2} = \emptyset$. Obviously $\cup_{\Gamma \in E} C_\Gamma \subset \text{Im } V$. Conversely, if $x \in \text{Im } V$ and Γ is the smallest face containing x, then $x \in C_\Gamma$ and hence Γ is essential. This proves that $\text{Im } V$ decomposes into the union over $\Gamma \in E$ of C_Γ. Obviously the boundary of Γ is the union of interiors of proper subfaces of Γ namely Γ_k. By Corollary 5.7.6 ∂C_Γ is the union of C_{Γ_k} where the non-essential subfaces Γ_k can be omitted, since of them $C_{\Gamma_k} = \emptyset$. By Lemma 5.7.8, the dimensions of the cells Γ_k are less than the dimension of cell C_Γ. Thus condition 1) from the definition of elementary cell complex is satisfied. To verify condition 2) we will use Lemma 5.7.5. $\bar{C}_{\Gamma_1} \cap \bar{C}_{\Gamma_2} = (\Gamma_1 \cap \Gamma_2) \cap \text{Im } V$ for any two essential faces Γ_1, Γ_2. The intersection $\Gamma_0 = \Gamma_1 \cap \Gamma_2$ is at least an invariant face. Let $\Gamma \subset \Gamma_0$ be the smallest face containing $\Gamma_0 \cap \text{Im } V$. This face is already essential and $\bar{C}_{\Gamma_1} \cap \bar{C}_{\Gamma_2} = \Gamma \cap \text{Im } V = C_\Gamma$. But $\bar{C}_\Gamma$ is contractible by Lemma 5.7.7. Thus 2) is satisfied.

Finally, if Γ is the smallest face containing $\text{Im } V$ then it is the largest essential face and by Lemma 5.7.8 the cell C_Γ has the highest dimension. Moreover if a linear form ϕ annihilates $\text{Im } V$, i.e. $\phi(x^2) = 0$ for $x > 0$ and $s(x) = 1$ then $\phi(x^2) = 0$ for all x, i.e. ϕ is disappearing. Therefore $[\Gamma] = \mathcal{A}^2$. Therefore $m_\Gamma = m$ and $\dim C_\Gamma = m - 1$ by (5.7.2). $\square$

Now to prove Theorem 5.7.1 it suffices to note that $\dim C_\Gamma = 0$ if and only if the operator V_Γ is of the type $(1, \delta_\Gamma)$, i.e. the operator is constant.

Thus by Lemma 5.7.2 the number of constant subpopulations cannot be less than $(m-1)+1 = m$.

We will call a population *ultranormal* if all of its subpopulations are normal. All constant subpopulations of an ultranormal populations are 1-dimensional.

Corollary 5.7.10 *An ultranormal population of rank m has at least m non-splitting types.*

The following conjecture seems plausible.

Conjecture 5.7.11 *An ultranormal Bernstein population has s.g.s.*

We will prove this conjecture for $m \leq 3$ or $\delta \leq 2$ (and hence of dimensions $n \leq 5$) using Corollary 5.7.10.

Lemma 5.7.12 *In an ultranormal Bernstein population of the type (m,δ) the inequality*

$$\dim N \leq \max(\delta - 2, 0) \qquad (5.7.3)$$

is satisfied for the dimension of the space N of disappearing linear forms.

Proof Let $e_1, \ldots e_m$ be basic idempotents. Then the coordinate notation of the evolutionary operator will include $x'_j = x_j^2 + \cdots (1 \leq j \leq m)$. Hence x_j^2 is not in x'_k for $k \neq j$. Hence quadratic forms $x'_1, \ldots, x'_m$ are linearly independent. Let us augment this system to a maximal linearly independent system

$$x'_1, \ldots, x'_m, x'_{m+1}, \ldots, x'_\nu \qquad (5.7.4)$$

i.e. to a basis in the subspace $Q' = [x'_1, \ldots, x'_n]$ of the space Q of all quadratic forms. The space Q' is the image of the homomorphism $\mathcal{A}^* \to Q$ mapping each linear form f to quadratic form $f(x^2)$. The kernel of this homomorphism is N. Hence $\dim N = n - \dim Q' = n - \nu$. Hence the basis in N consists of forms $f_j(x) = x_j - \sum_{k=1}^\nu a_{jk} x_k$, where $\nu + 1 \leq j \leq n$, defined by decomposition of quadratic forms $x'_{\nu+1}, \ldots, x'_n$ on basis (5.7.4): $x'_j = \sum_{k=1}^\nu \alpha_{jk} x'_k$. Moreover $\alpha_{jk} = 0$ for $1 \leq k \leq m$ because otherwise some x'_j $(j \geq \nu + 1)$ contains a square x_k^2 $(k \leq m)$. Therefore $f_j(x) = x_j - \sum_{k=m+1}^\nu \alpha_{jk} x_k$.

Suppose now that in violation of (5.7.3), $\dim N > \delta - 2$. On the other hand, $\dim N \leq \delta$ in any Bernstein algebra of the type (m,δ), so there are two possibilities.

Case 1. $\dim N = \delta - 1$. Then $\nu = n - \delta + 1 = m + 1$, $f_j(x) = x_j - \alpha_{j,m+1}x_{m+1}$, whence $x'_j = \alpha_{j,m+1}x'_{m+1}$ $(m + 2 \leq j \leq n)$ i.e. the population is externally reducible if $n \geq m + 2$. The case $n = m + 1$ is no exception, for then $\delta = 1$ and $\dim N = 0$. The case $n = m$, i.e. $\delta = 0$ is impossible.

Case 2. $\dim N = \delta$. Then $\nu = n - \delta = m$, $f_j(x) = x_j$, whence $x'_j = 0$ $(m+1 \leq j \leq n)$ i.e. the population contains disappearing types if $n \geq m+1$. The case $n = m$ is no exception for then $\delta = 0$ and $\dim N = 0$. $\square$

Theorem 5.7.13 *An ultranormal Bernstein population of the type (m, δ) where $m \leq 3$ or $\delta \leq 1$ has s.g.s.*

Proof Apply Theorem 3.4.27. For $m \leq 3$ the condition (3.4.37) is satisfied by the previous lemma, and therefore the algebra is conservative. On the other hand, if $\delta = 1$ then $N = 0$ and the conservatism follows from Corollary 3.4.29. For $\delta = 0$ the algebra is unit, i.e. again conservative. $\square$

Corollary 5.7.14 *An ultranormal Bernstein population of dimension $n \leq 5$ has s.g.s.*

For types (m, δ) where $m \leq 2$ or $\delta \leq 1$ (and thus $n \leq 4$) the presence of s.g.s. follows from normality by Corollary 3.4.24 and Theorem 5.5.1. It is possible that an even stronger conjecture than that formulated above holds.

Conjecture 5.7.15 *A normal Bernstein population has s.g.s.*

From the biological point of view the ultranormality conditions is not more burdensome than just normality, for if the latter is accepted as biologically necessary, then it applies to all populations. We note that if a population has s.g.s. and is normal, then it is ultranormal. This can be seen immediately from the explicit description of all such populations by Theorems 4.3.9 and 4.6.1.

The verification of either of these conjectures would strongly support our opinion that outside of the s.g.s. case the Bernstein problem loses any biological meaning, while counterexamples might lead to the discovery of biological meaningful stationary laws of heredity applicable outside of genetics. The latter possibility, although tempting, is considered improbable by the author.

Finally, we remark that if $N = 0$ (i.e. $A^2 = A$) and

$$\delta > \frac{(m - 1)(m - 2)}{2}, \tag{5.7.5}$$

then the population has s.g.s. by Corollary 3.4.28.

Conjecture 5.7.16 *If for a Bernstein population $N = 0$ then it has s.g.s.*

According to this conjecture, the inequality (5.7.5) is superfluous.

5.8 Appendix. The Proof of Topological Lemma 5.7.2

For $n = 1$ the assertion is trivial. Suppose $n > 0$ and the assertion is true for dimensions $< n$. Let the number of 0- dimensional cells not exceed n. Consider an n-dimensional cell τ^n. Its boundary has dimension $n - 1$ and contains a $(n - 1)$-dimensional cell τ^{n-1}. Consider the subcomplex $Y = \overline{\tau^{n-1}} = \tau^{n-1} \cup \partial\tau^{n-1}$, also of dimension $n - 1$. By the induction hypothesis it contains at least n 0-dimensional cells. Therefore all 0-dimensional cells of the original complex X are in Y. We will prove by induction on k that every k- dimensional cell $\tau^k \subset X$ is in Y. This will contradict $\tau^n \not\subset Y$.

Let $k > 0$, $\tau^k \not\subset Y$ but $\tau^j \subset Y$ for all $j < k$. Then $\tau^k \cap Y = \emptyset$ and $\partial\tau^k \subset Y$ since $\partial\tau^k$ consists of cells whose dimensions are less than k. Therefore $\overline{\tau^k} \cap Y = \partial\tau^k$, i.e. $\overline{\tau^k} \cap \overline{\tau^{n-1}} = \partial\tau^k$. This intersection is contractible by the definition of elementary cell complex. At the same time the boundary of a k-dimensional cell for $k > 0$ is not contractible. This contradiction implies $\tau^k \subset Y$.

Remark The lower bound for the count of zero-dimensional cells is sharp; it is achieved by n-dimensional simplex.

Remark The condition 2) that the intersection of closures of any two cells be contractible that we included in our definition of elementary cell complex is not standard, unlike condition 1) and, indeed need not hold. For example, decompose the sphere S^n into "northern" hemisphere τ^n_+, "southern" hemisphere τ^n_- and "equator" S^{n-1}, then analogously decompose S^{n-1} etc. We obtain a n-dimensional cell complex (in the sense of the standard definition) $X^n = \tau^n_+ \cup \tau^n_- \cup \cdots \cup \tau^0_+ \cup \tau^0_-$, which has only two 0- dimensional cells. It is not elementary as, for example, the intersection $\overline{\tau^n_+} \cap \overline{\tau^n_-}$ is not contractible.

Chapter 6

Recombination Processes

6.1 Linkage Distribution. Chromosome Structures

Consider a set $L = \{1, \ldots, l\}$ of autosomal loci. Recall from Section 1.1 that a zygote obtained from gametes of types g and h, will yield for the gene pool, by simple meiosis alone, gametes of types g and h again. However, when recombination occurs then gametes of mixed types are produced. Let $U|V$ be a binary partition of the set L, i.e. $U \cup V = L$, $U \cap V = \emptyset$. The crossing over pattern of type $U|V$ yields gamete types which we denote by $g_U h_V$ and $h_U g_V$, where $g_U h_V$ agrees with g at the loci in U and with h at the loci in V. Think of type $U|V$ as keeping together loci in U and loci in V by separating loci in U from those in V. In spite of the notation, the partition is unordered so that $U|V$ and $V|U$ describe the same partition with the same crossing over pattern. The number of partitions is 2^{l-1}. The special case of no crossing over corresponds to the *trivial partition* $L|\emptyset \equiv L$. The probability distribution $r = \{r(U|V)\}$, defined on the set of partitions and called the *linkage distribution*, is a fundamental object in the theory. In this section we will use it to describe crossings over and derive the chromosome structure.

For a subset K of L the partition $U|V$ of L induces the partition $U_1|V_1$ of K where $U_1 = U \cap K$ and $V_1 = V \cap K$ and we say that $U_1|V_1$ refines $U|V, U|V \succ U_1|V_1$. Observe that the subset K can be recovered from the symbol $U_1|V_1$ by $K = U_1 \cup V_1$ and, conversely, that, given K and U_1, the set V_1 is the complement $K\backslash V_1$. Beginning with a linkage distribution r, this mapping from partitions of L to those of K yields the *induced linkage distribution* r_K defined as

$$r(U_1|V_1) = \sum_{U|V \succ U_1|V_1} r(U|V), \qquad (6.1.1)$$

where the symbol $\succ$ denotes a partition refinement. Clearly $r_L = r$.

Lemma 6.1.1 *Let* $Q \subset K \subset L$. *Then*

$$(r_K)_Q = r_Q. \qquad (6.1.2)$$

Proof This is a special case of the functorial nature of induced distributions. Let $U_2|V_2$ be a partition of the set Q. Then $r(U_2|V_2) = \sum_{U|V \succ U_2|V_2} r(U|V) = \sum_{U_1|V_1 \succ U_2|V_2} (\sum_{U|V \succ U_1|V_1} r(U|V))$, where $U_1|V_1$ varies over partitions of K. $\square$

In general, for U_1, V_1 arbitrary subsets of L we introduce the convention $r(U_1|V_1) = 0$ when $U_1 \cap V_1 \neq \emptyset$. Intuitively, the symbol $U_1|V_1$ then corresponds to the impossible event that locus i is separated from itself when $i \in U_1 \cap V_1$. In particular, we have $r(i|i) = 0$ for all $i \in L$. Obviously, $r(i|j) = r(j|i)$ for any two loci i, j. In general, it turns out that the probability $r(i|j)$ may be considered as a distance (possibly degenerate) between loci i and j.

Lemma 6.1.2 *For any three loci* i, j, k, *we have*

$$r(i|j) = r(i|k) + r(k|j) - 2r(ij|k). \qquad (6.1.3)$$

Proof By Lemma 6.1.1 we may assume $L = \{i, j, k\}$. But then obviously

$$r(i|j) = r(ik|j) + r(i|kj),$$

$$r(i|k) = r(ij|k) + r(i|kj),$$

$$r(k|j) = r(ik|j) + r(k|ij),$$

and the lemma immediately follows. $\square$

Corollary 6.1.3 *For any three loci* i, j, k, *we have*

$$r(i|j) \leq r(i|k) + r(k|j).$$

Thus the functional $r(i|j)$ is a pseudometric on the loci. We will refer to it as the *Morgan metric*.

Two loci i, j are called *completely linked*, if their Morgan distance is zero. The following lemma is clearly implied by the triangle inequality.

Lemma 6.1.4 *Complete linkage is an equivalence relation.*

This lemma and its genetic meaning suggests identification of completely linked loci, but it is necessary to verify that the linkage distribution is not lost.

Observe first that if a partition $U|V$ separates two equivalent loci i, j then $r(U|V) = 0$ because $\sum_{U|V \succ i|j} r(U|V) = r(i|j) = 0$. On the other hand, if $\tilde{U}|\tilde{V}$ is a partition of the set of equivalence classes then it induces a unique partition $U|V$ of L by taking the union of the corresponding equivalence classes. This correspondence $\tilde{U}|\tilde{V} \mapsto U|V$ is a bijection onto partitions of L which are thus saturated by the equivalence relation. So we can define the distribution on equivalence classes by $r(\tilde{U}|\tilde{V}) = r(U|V)$.

It is easy to check that there is no complete linkage between distinct equivalence classes. In terms of Morgan metric, the identification of completely linked loci is the usual factorization procedure for removing the degeneracy from the pseudometric. We remark that the factor metric coincides with the Morgan metric generated by factor linkage distribution.

We will assume from now on that complete linkage has been removed. In particular, the Morgan metric will really be a metric rather than a pseudometric. No other restriction on linkage distribution will be assumed.

We have denoted the *trivial partition* $K|\emptyset$ by K. Then

$$r(K) = \sum_{U|V \succ K} r(U|V) = \sum_{U \supset K} r(U|V).$$

Obviously, $r(K) = 1$ if $|K| < 2$. Because there is no complete linkage,

$$r(K) < 1 \qquad (|K| \geq 2) \tag{6.1.4}$$

($|X|$ is the number of elements of a finite set X). Indeed, obviously, $r(K) \geq r(Q)$ $(K \subset Q)$. Therefore in (6.1.4) it suffices to consider the case $|K| = 2$. Let $K = \{i, j\}$, then $r(K) + r(i|j) = 1$ and $r(i|j) > 0$, so $r(K) < 1$.

Linkage distribution allows us to introduce the *chromosomal structure* formally.

Let $L_1|L_2$ be a fixed partition of the set L of loci. Every partition $U|V$ of L generates partitions $u_1|v_1$, $u_2|v_2$ of classes corresponding to L_1 and L_2 respectively. Conversely, if two partitions $u_1|v_1$, $u_2|v_2$ are given, then there exist exactly two such partitions $U|V$ that generate them; either $U = u_1 \cup u_2$, $V = v_1 \cup v_2$ or $U = u_1 \cup v_2$, $V = v_1|u_2$. We will call a partition $L_1|L_2$

chromosomal if for all partitions $U|V$,

$$r(U|V) = \frac{1}{2}r(u_1|v_1)r(u_2|v_2). \tag{6.1.5}$$

This says that a $U|V$ crossover pattern (with $U = u_1 \cup u_2$ and $V = v_1 \cup v_2$) is the outcome of three independent events: the two partition events $u_1|v_1$ and $u_2|v_2$ for L_1 and L_2, respectively, and the joining of u_1 and u_2 rather than the, equally likely, joining of u_1 with v_2.

Obviously, chromosomal partitions are non-trivial and hence can occur only when $l \geq 2$.

Two loci i, j are called *linked* if in every chromosome partition they belong to the same class. Obviously, being linked is an equivalence relationship. The equivalence classes under this relationship are called *linkage groups*, denoted $C_1, \ldots, C_s$. It may happen that all l loci belong to the same linkage group. This is equivalent to the absence of chromosomal partitions. We note that in presence of chromosomal partitions, (6.1.5) implies that $r(U|V) \leq \frac{1}{2}$ for all $U|V$.

Let us see how restricting the set of loci affects the chromosomal partitions.

Lemma 6.1.5 *If $K \subset L$, $|K| \geq 2$ then restrictions to K of chromosomal partitions of L are chromosomal partitions of K with respect to induced linkage distribution.*

Proof Let $L_1|L_2$ be a chromosome partition of set L and $K_1|K_2$ be its restriction to K ($K_1 \subset L_1, K_2 \subset L_2$). For any partition $U'|V'$ of the subset K we have by (6.1.1) and (6.1.5)

$$r(U'|V') = \sum_{U \supset U', V \supset V'} r(U|V) = \frac{1}{2} \sum_{u_i \supset u_i', v_i \supset v_i'} r(u_1|v_1)r(u_2|v_2),$$

where $u_i = U \cap L_i$, $u_i' = U' \cap K_i$, v_i, v_i' are analogously defined. But $\sum_{u_i \supset u_i', v_i \supset v_i'} r(u_i|v_i) = r(u_i'|v_i')$ for $i = 1, 2$.

Therefore $r(U'|V') = \frac{1}{2}r(u_1'|v_1')r(u_2'|v_2')$, i.e. the partition $K_1|K_2$ is chromosomal. $\square$

The converse of Lemma 6.1.5 is not true as the following example shows.

Example Suppose for three loci $\{1, 2, 3\}$ the probability of partitions $r(12|3), r(13|2), r(23|1), r(123)$ are distinct. Then the loci are linked. But if moreover $r(1|2) = r(13|2) + r(23|1) = \frac{1}{2}$ then the partition $1|2$ is chromosomal for the subset $\{1, 2\}$.

Therefore chromosome structure is not quite invariant; the linkage may not occur in an incomplete system of loci. But if it does occur, then by Lemma 6.1.5 it is also present in the complete system.

We will now introduce more general chromosomal partitions into an arbitrary number of classes. A partition $L_1|\cdots|L_t$ of set L is called *chromosomal* if it does not decompose any of the linkage groups $C_1,\ldots C_s$ (i.e. $L_1|\cdots|L_t \succ C_1|\cdots|C_s$) meaning each L is a union of $C's$; this implies $t \leq s$.

Let $L_1|\cdots|L_t$ be a chromosomal partition, $U|V$ be any partition of set L, and $u_i|v_i$ be the partition of the class L_i $(1 \leq i \leq t)$ induced from $U|V$. The following generalization of (6.1.5) holds.

Lemma 6.1.6 $r(U|V) = \frac{1}{2^{t-1}} \prod_{i=1}^{t} r(u_i|v_i).$

Proof For $t = 1, 2$, the assertion is trivial. Let us use induction on t. By Lemma 6.1.5 the partition $L_1|\cdots|L_{t-1}$ is chromosomal in the subset $\tilde{L} = \cup_{i=1}^{t-1} L_i$ and $\tilde{L}|L_t$ is obviously a chromosomal partition of L. Hence $r(U|V) = \frac{1}{2} r(\cup_{i=1}^{t-1} u_i| \cup_{i=1}^{t-1} v_i) r(u_t|v_t)$, and by the induction hypothesis the first multiplicand equals to $\frac{1}{2^{t-2}} \prod_{i=1}^{t-1} r(u_i|v_i)$. $\square$

Consider in particular the finest chromosomal partition $C_1|\cdots|C_s$ whose classes are the linkage groups. Every decomposition $U|V$ of the set L generates a system of partitions $u_i|v_i$ $(1 \leq i \leq s)$ in corresponding linkage groups and conversely is generated by this system, although not uniquely, but within a family of 2^{s-1} *conjugate* partitions. By Lemma 6.1.6 all conjugate partitions have the same probability

$$r(U|V) = \frac{1}{2^{s-1}} \prod_{i=1}^{s} r(u_i, v_i). \tag{6.1.6}$$

Corollary 6.1.7 *The number of chromosomes satisfies the inequality*

$$s \leq 1 + \min_{U|V} \log_2 \frac{1}{r(U|V)} \quad .$$

This means that $s \leq H + 1$, where H is the entropy of linkage distribution in bits. $\square$

Two loci i, j are called *independent* if they are not linked, i.e. belong to different linkage groups. By Lemma 6.1.5 independence is preserved by restricting the set of loci.

Lemma 6.1.8 *If in a subset $K \subset L$ every two loci are independent then the induced linkage distribution is uniform:* $r(K_1|K_2) = \frac{1}{2^{|K|-1}}$ *($K_1 \cup K_2 = K$).*

Proof By (6.1.6) applied to the case $L = K$,

$$r(K_1|K_2) = \frac{1}{2^{|K|-1}} \prod_{k \in K} r(k) = \frac{1}{2^{|K|-1}}.$$

$\square$

Corollary 6.1.9 *If loci i, j are independent then $r(i|j) = \frac{1}{2}$.*

The above theory does not assume any connection between linkage distribution and the numbering of loci. Once the linkage group partition is established, it is natural to number loci according to this partition. But the numbering of linked loci is still arbitrary. At the same time it is known that in real chromosomes the loci are linearly ordered. The usual argument is the observed additivity of the Morgan metric (although only approximate): $r(i|j) = r(i|k) + r(k|j)$ (with corresponding ordering of linked loci i, j, k). By Lemma 6.1.2 additivity holds if and only if for any three loci i, j, k

$$r(ij|k)r(jk|i)r(ki|j) = 0. \tag{6.1.7}$$

In general, the matrix ρ in a metric space is called *additive* if every three points can be ordered (and denoted x, y, z) so that

$$\rho(x, z) = \rho(x, y) + \rho(y, z). \tag{6.1.8}$$

Example Any subset of the real line $\mathbf{R}$ with the usual metric $|x - y|$.

We will call a metric space *rectilinear* if its metric is additive and there exists a linear order that agrees with the metric, i.e. $x \leq y \leq z$ if and only if (6.1.8) holds.

Lemma 6.1.10 *A metric space M is rectilinear if and only if it can be isometrically injected into $\mathbf{R}$.*

Proof Sufficiency is obvious. To show necessity define $\psi : M \to \mathbf{R}$ as

$$\psi(x) = \begin{cases} \rho(x, x_0) & x > x_0 \\ -\rho(x, x_0) & x < x_0 \\ 0 & x = x_0 \end{cases}$$

where x_0 is an arbitrary point. It is easy to check that ψ is an isometric injection. $\square$

The next example shows that a Morgan metric can be additive without being linear.

Example Consider the set $L = \{1,2,3,4\}$ with linkage distribution $r(12|34) = \alpha > \frac{1}{2}$, $r(13|24) = \beta > 0$ and $r(U|V) = 0$ for all other partitions. All loci are linked because $\alpha > \frac{1}{2}$ (see Corollary 6.1.7). Morgan metric is additive because $r(1|3) = r(2|4) = \alpha$, $r(1|2) = r(3|4) = \beta$, $r(1|4) = r(2|3) = \alpha + \beta$, but not rectilinear because there are two pairs of points with maximal distance. In this example there is no linear ordering of loci that agrees with distances even in the weak sense that $x < y < z$ implies $\rho(x,z) > \max(\rho(x,y),\rho(y,z))$.

Thus linear ordering of linked loci is an additional condition upon the theory. For the rest of this section we will impose this condition, assuming further that the numbering of loci agrees both with the linkage group partition and the order of the linked loci. This is important to note because the constructions which follow are not invariant under locus permutations.

For an arbitrary linkage group $C = (1,\dots,n)$, the partition $i|i+1$, $1 \leq i \leq n-1$, is called a *simple crossing over*. A system of simple crossings over $i_1|i_1+1;\dots;i_\nu|i_\nu+1$, $(1 \leq i_1 < \cdots < i_\nu \leq n-1)$ is called *crossing over of multiplicity ν (multiple if $\nu > 1$)*.

To every non-trivial partition $u|v$ of the group C corresponds a crossing over, generally speaking multiple, which is the system of those $i|i+1$ for which $u|v \succ i|i+1$. This correspondence between non-trivial partitions and crossings over is bijection: $u|v \succ j|k$ if and only if in the chain $(j,j+1),\dots,(k-1,k)$ an odd number of pairs is separated by simple crossings over. Therefore non-trivial partitions $u|v$ of the group C can be identified with crossings over and their probabilities called crossing over probabilities. We will denote the multiplicity of a crossing over $u|v$ by $\nu(u|v)$. The probability of the trivial partition is the complement of the sum of the crossing over probabilities. (6.1.6) shows that knowing the crossing over probabilities in each group one may recover the linkage distribution (but the chromosomal structure must be known a priori).

Given an order we will call a metric *additive* if (6.1.8) holds for $x < y < z$. The Morgan metric is usually not additive even within one linkage group.

By Lemma 6.1.1 again, the Morgan metric is additive within a linkage group when $i < k < j$ implies

$$r(ij|k) = 0. \tag{6.1.9}$$

Lemma 6.1.11 *The Morgan metric in a linkage group is additive if and only if the probabilities of all double crossings over are zero.*

Proof Suppose the metric is additive. Consider a double crossing over $u|v$. By definition of multiplicity there exist $i, j, 1 \leq i < j \leq n-1$, such that $u|v \succ i(j+1)|i+1$ or $u|v \succ ij|j+1$ and so $r(u|v) \leq r(i(j+1)|j)$ or $\leq r(ij|j+1)$. But the latter are zero by (6.1.9).

Conversely, suppose the probabilities of all double crossing over are zero. Then the probabilities of all multiple crossings over are zero, because each one of them contains a double one. Then for $i < k < j$, $r(ij|k) = \sum_{u|v \succ ij|k} r(u|v) = 0$, i.e. (6.1.9) is satisfied, from which additivity of the metric follows. $\square$

One might also use the inequality

$$r(ij|k) \leq \sum_{i \leq \alpha < k \leq \beta < j} r(\alpha(\beta+1)|\beta(\alpha+1)),$$

which implies that if the probabilities of double crossings over do not exceed some δ, then

$$r(ij|k) \leq (k-i)(j-k)\delta, \qquad (6.1.10)$$

and the Morgan metric is additive with accuracy up to the order of δ (with limited number of loci). Exactly because of Lemma 6.1.2 and (6.1.10) $|r(i|j) - r(i|k) - r(k|j)| \leq 2(k-j)(j-k)\delta$.

In contrast with the case when the Morgan metric is additive, if simple crossings over are independent and their probabilities are p_i then for any partition $u|v$

$$r(u|v) = \prod_{u|v \succ i|i+1} p_i \prod_{u|v \nsucc j|j+1} q_j \qquad (q_j = 1 - p_j). \qquad (6.1.11)$$

Denote $\max p_i = \epsilon$. Then the probabilities of double crossing over do not exceed ϵ^2 and Morgan metric is additive up to the order of ϵ^2.

We will call the linkage distribution (6.1.11) *linear*.

Lemma 6.1.12 (Trow's Formula.) *If the linkage distribution in a linkage group is linear then for any two loci i, k $(i < k)$ in this group*

$$r(i|k) = \frac{1}{2}\left(1 - \prod_{j=i}^{k-1}(1 - 2p_j)\right). \qquad (6.1.12)$$

Proof If we omit the extreme loci $1, n$, the linkage distribution remains linear. For example, if we omit n then for any crossing over $u|v$ in the system $\{1, \ldots, n-1\}$ we have (assuming for concreteness $n-1 \in u$)

$$r(u|v) = r(u|vn) + r(un|v)$$

$$= \prod_{u|v \succ i|i+1} p_i \prod_{u|v \not\succ j|j+1} q_j p_n + \prod_{u|v \succ i|i+1} p_i \prod_{u|v \not\succ j|j+1} q_j q_n = \prod_{u|v \succ i|i+1} p_i \prod_{u|v \not\succ j|j+1} q_j.$$

By induction it suffices to prove the assertion for the extreme $i = 1$, $k = n$:

$$r(1|n) = \sum_{\nu(u|v) \equiv 1 \,(\mathrm{mod}\,2)} \prod_{u|v \succ i|i+1} p_i \prod_{u|v \not\succ j|j+1} q_j$$

$$= \frac{1}{2}\Big(\prod_{i=1}^{n-1}(q_i + p_i) - \prod_{i=1}^{n-1}(q_i - p_i)\Big) = \frac{1}{2}\Big(1 - \prod_{i=1}^{n-1}(1 - 2p_i)\Big).$$

$\square$

Corollary 6.1.13 *For linear linkage distribution, the function* $d(i,k) = -\ln|1 - 2r(i|k)|$ *is an additive metric agreeing with locus numeration.*

Proof From Trow's formula, $d(i,k) = -\sum_{j=i}^{k-1} \ln|1 - 2p_j|$, $i < k$. $\square$

Corollary 6.1.14 *For linear linkage distribution, for three or more loci,* $\min_{i,k} r(i|k) \geq \frac{1}{2}$.

$\square$

We will call the function $d(i,k)$ *Trow's metric*.

A special case of linear linkage distribution in a group is the *binomial distribution* in which all simple crossings over are independent and have the same probability p. Then $r(u|v) = p^\nu q^{n-\nu-1}$, where $q = 1-p$ and $\nu = \nu(u|v)$. Trow's formula for binomial distribution takes the form $r(i|k) = \frac{1-(1-2p)^{k-i}}{2}$, $i < k$, and Trow's metric is proportional to the difference $k - i$, i.e. the number of loci in the half-interval $(i, k]$.

Corollary 6.1.14 can be strengthened for binomial linkage distribution for three loci

$$\min_{i,k} r(i|k) = \left\{ \begin{array}{ll} p & p \leq \frac{1}{2}, \\ 2pq & p \geq \frac{1}{2}. \end{array} \right\}. \tag{6.1.13}$$

We can motivate the formula relating $d(i,k)$ with $r(i|k)$ in Corollary 6.1.13 by beginning with d instead. Suppose that the probability of a single crossing over between i and k is given by $d(i,k)$ and that the total number of crossovers in Poisson distributed so that the probability of ν crossovers is given by $e^{-d}d^\nu/\nu!$. Then, as in the proof of Lemma 6.1.12, i and k are separated if an odd number of crossovers occurs between them:

$$r(i|k) = \sum_{\nu \equiv 1\,(\mathrm{mod}\,2)} e^{-d}d^\nu/\nu! = e^{-d}\sinh d = \frac{1}{2}(1 - e^{-2d}). \tag{6.1.14}$$

Solving for d we get the formula of Corollary 6.1.13 (times $\frac{1}{2}$). (6.1.14) is called *Kosambi's Formula*.

6.2   Evolutionary Equations. Reiersöl Algebra

At locus i of $L = \{1,\ldots,l\}$ we denote by Γ_i the set of alternative alleles of which we assume there are m_i. The set of gametes is the symmetric Cartesian product

$$\Gamma = \prod_{i=1}^{l} \Gamma_i. \tag{6.2.1}$$

So a typical gamete $g \in \Gamma$ is a choice function with allele $a_i(g) \in \Gamma_i$ at locus i for $i = 1,\ldots,l$.

We denote by $\mathbf{R}(\Gamma)$ the vector space of formal linear combinations $p = \sum_{g \in \Gamma} p(g)g$ with real coefficients. As usual we write $p \geq 0$ if $p(g) \geq 0$ for all $g \in \Gamma$ and we define as usual the linear functional $s(p) = \sum_{g \in \Gamma} p(g)$. We regard the elements of the simplex $\Delta(\Gamma) = \{p|p \geq 0, s(p) = 1\}$ as probability distributions on the set of gametes and call them the *gamete states* of the population. $(\mathbf{R}(\Gamma), \Delta(\Gamma))$ is a simplicial space with weight s and canonical basis Γ.

On the product set Γ we denote by F^i the projection to factor Γ_i which associates to gamete g its allele $a_i(g) \equiv F^i(g)$ at locus i. Extending linearly we get $F^i : \mathbf{R}(\Gamma) \to \mathbf{R}(\Gamma_i)$ which is clearly a stochastic homomorphism. F^i associates to the gamete distribution p in $\Delta(\Gamma)$ the induced marginal distribution on the alleles at locus i (compare (2.2.11)).

In general, for a non-empty subset $K \subset L$ we denote by Γ_K the product of the factors for loci in K:

$$\Gamma_K = \prod_{i \in K} \Gamma_i. \tag{6.2.2}$$

To cover the case $K = \emptyset$ we define the *empty gamete* to be the scalar 1 and the space $\mathbf{R}(\Gamma_\emptyset)$ to be $\mathbf{R}$. The set projection $F^K : \Gamma \to \Gamma_K$ associates to gamete g the K- locus sub-gamete $g_K = F^K(g)$ which agrees with g for all loci $i \in K$. Again we extend linearly to define the surjective stochastic homomorphism of simplicial spaces, $F^K : (\mathbf{R}(\Gamma), \Delta(\Gamma)) \to (\mathbf{R}(\Gamma_K), \Delta(\Gamma_K))$ by $F^K(p)(h_K) = \sum\{p(g)|g \in \Gamma, F^K(g) = h_K\}$. When $K = L$ the operator is the identity and when $K = \emptyset$ it coincides with s.

If K is a linkage group then the sub-gamete g_K is called a *chromosome*. Chromosomes belonging to the same linkage group are called *homologous*.

Let K_1 and K_2 be two non-empty subsets of the set L and K is their union. If the intersection of these subsets is empty we can define the product of a K_1-gamete g_1 and a K_2- gamete g_2 as the K-gamete $g_1 g_2$ which agrees with g_1 on K_1 and with g_2 on K_2. In other words, we use the multiplication in the commutative semi-group of words. In particular, for any partition $U|V$ of L and any gamete g the product of sub-gametes $g_U g_V$ is equal to g. Extending linearly from the two canonical bases we get an isomorphism

$$\mathbf{R}(\Gamma_U) * \mathbf{R}(\Gamma_V) \to \mathbf{R}(\Gamma). \tag{6.2.3}$$

where $*$ denotes the symmetric tensor product (see Section 3.6) We can similarly identify

$$\mathbf{R}(\Gamma_1) * \cdots * \mathbf{R}(\Gamma_l) \cong \mathbf{R}(\Gamma) \tag{6.2.4}$$

by regarding the gamete g as the commutative product of the alleles $a_i(g)$.

Now we consider the symmetric tensor square, $\mathbf{R}(\Gamma) * \mathbf{R}(\Gamma)$, generated by the set of *gamete pairs* $g * h$, subject to the identification

$$g * h = h * g. \tag{6.2.5}$$

It is a step toward a formal description of zygotes. A further identification is based on the the chromosome structure. In the fertilized egg it is our inability to distinguish which sets of chromosomes came from each parent (we are considering only autosomal loci) which led to (6.2.5). But also for two pairs of homologous chromosomes the absence of any linkage between them (by definition) means we cannot observe the original association in the separate parental gametes. We will call two gamete pairs *equivalent* if one can be obtained from the other by transpositions of homologous chromosomes. A class of equivalent gamete pairs is called a *zygote*. We denote by Z the set of zygotes and consider the corresponding space $\mathbf{R}(Z)$. It is a quotient space of $\mathbf{R}(\Gamma) * \mathbf{R}(\Gamma)$. A zygote ζ obtained from a gamete pair $g * h$ under the quotient map is denoted $\zeta = g \circ h$. Thus, for any chromosomal partition $L_1|L_2$ of L we have, in addition to the symmetric condition (6.2.5)

$$g \circ h = g_{L_1} h_{L_2} \circ g_{L_2} h_{L_1}. \tag{6.2.6}$$

The quotient map defines a stochastic homomorphism of simplicial spaces $(\mathbf{R}(\Gamma) * \mathbf{R}(\Gamma), S_\pi) \to (\mathbf{R}(Z), S_\zeta)$ where S_π and S_ζ are the simplices on the canonical bases.

We obtain the evolutionary operator of this two-level population by defining the fertilization and meiosis operators.

The fertilization operator $\phi : \Delta(\Gamma) \to S_\zeta$ is defined via the formula

$$\phi(p) = p \circ p = \sum_{g,h \in \Gamma} p(g)p(h)(g \circ h). \tag{6.2.7}$$

Of course, many gamete pairs determine the same zygote by (6.2.6) so we can also write

$$\phi(p) = \sum_{\zeta \in Z} \sum_{g \circ h = \zeta} p(g)p(h)\zeta. \tag{6.2.8}$$

For the meiosis operator we first define on the space of gamete pairs the partial meiosis operator associated with a $U|V$-crossing over:

$$\mu_{U|V} : \begin{array}{ccc} \mathbf{R}(\Gamma) * \mathbf{R}(\Gamma) & \longrightarrow & \mathbf{R}(\Gamma) \\ g * h & \longmapsto & \frac{1}{2}(g_U h_V + g_V h_U) \end{array} . \tag{6.2.9}$$

Formally, $\mu_{U|V}$ is defined via the homomorphism

$$\mathbf{R}(\Gamma) * \mathbf{R}(\Gamma) \stackrel{F^U * F^V}{\longrightarrow} \mathbf{R}(\Gamma_U) * \mathbf{R}(\Gamma_V) \tag{6.2.10}$$

composed by (6.2.3).

The meiosis operator itself is the weighted average of the $U|V$-operators with weights given by the linkage distribution r:

$$\mu = \sum_{U|V} r(U|V)\mu_{U|V} : \begin{array}{c} \mathbf{R}(\Gamma) * \mathbf{R}(\Gamma) \longrightarrow \mathbf{R}(\Gamma) \\ g * h \mapsto \frac{1}{2} \sum_{U|V} r(U|V)(g_U h_V + g_V h_U) \end{array} . \tag{6.2.11}$$

While each $U|V$ operator is defined only on gamete pairs, the total meiosis operator μ factors to define a linear one on the space $\mathbf{R}(Z)$. This is a consequence of Lemma 6.1.6. Alternatively, suppose that $L_1|L_2$ is a chromosomal partition with $U|V$ and $U'|V'$ both inducing $u_1|v_1$ and $u_2|v_2$ on L_1 and L_2 but with $U = u_1 \cup u_2$ and $U' = u_1 \cup v_2$. Then $\mu_{U|V}$ and $\mu_{U'|V'}$ have the same weight in μ by (6.1.5) and their sum maps $g * h$ to the same element.Thus, μ factors by(6.2.6).

Composing, as usual, we get the gametic evolutionary operator $V_\gamma = \mu\phi$ with

$$\begin{aligned} V_\gamma(p) &= \sum_{g,h} p(g)p(h) \sum_{U|V} \frac{1}{2} r(U|V)(g_U h_V + g_V h_U) \\ &= \sum_{U|V} r(U|V) \sum_{g,h} p(g)p(h)g_U h_V. \end{aligned} \tag{6.2.12}$$

One can also write

$$V_\gamma = \sum_{U|V} r(U|V)(F^U * F^V).$$
(6.2.13)

This dynamical system we will call the *recombination process* on the gamete level.

Alternatively, we can parameterize the ordered pairs g, h such that $g_U h_V$ is a given gamete γ. These are the pairs $g = \gamma_U \chi_V$, $h = \chi_U \gamma_V$ for $\chi \in \Gamma$:

$$p'(\gamma) = \sum_{U|V} r(U|V) \sum_\chi p(\gamma_U \chi_V) p(\chi_U \gamma_V).$$
(6.2.14)

The evolutionary operator comes from a stochastic algebra structure on $(\mathbf{R}(\Gamma), \Delta(\Gamma))$ with the table of multiplication

$$g \times h = \frac{1}{2} \sum_{U|V} r(U|V)(g_U h_V + h_U g_V).$$
(6.2.15)

Therefore,

$$xy = \frac{1}{2} \sum_{U|V} r(U|V)(F^U(x)F^V(y) + F^U(y)F^V(x))$$
(6.2.16)

for $x, y \in \mathbf{R}(\Gamma)$. It is the bilinearization of the formula (6.2.13).

The obtained stochastic algebra is called the *Reiersöl algebra* and is denoted $\mathcal{A}(L; r)$, where L is the set of loci, r is the linkage distribution. The multiplication map R is related to μ by the canonical fertilization operator. Similarly, we define the $U|V$- *recombination operator* $R_{U|V}$ by

$$R_{U|V}(x, y) = \frac{1}{2}(F^U(x)F^V(y) + F^U(y)F^V(x))$$
(6.2.17)

so that R is the weighted average of the partial recombination operators. It is called the *recombination operator* as well.

The main lemma about Reiersöl algebras concerns naturality of the construction with respect to projection to subsystems.

Lemma 6.2.1 *With $K \subset L$ the map $F^K : \mathcal{A}(L; r) \to \mathcal{A}(K; r_K)$ is a stochastic algebra homomorphism, where r_K is the induced linkage distribution (see (6.1.1).*

Proof If $U|V$ is a partition of L and $u|v$ is the induced partition of K then the recombination operators satisfy

$$F^K R_{U|V} = R_{u|v}(F^K * F^K). \tag{6.2.18}$$

By bilinearity it suffices to note that both sides map (g, h) to $\frac{1}{2}(g_u h_v + h_u g_v)$ for any pair of gametes g, h. Now by (6.1.1) multiplying the left by $r(U|V)$ and summing on partitions $U|V$ of L is the same as multiplying the right by $r(u|v)$ and summing on partitions $u|v$ of K. The result is $F^K R = R_K(F^K * F^K)$ where R and R_K are the multiplications of $\mathcal{A}(L; r)$ and $\mathcal{A}(K; r_K)$ respectively. This says that F^K is an algebra homomorphism. $\square$

Since $V_\gamma(p) = p^2$ we now have

Corollary 6.2.2 F^K *maps an evolutionary trajectory in* $\Delta(\Gamma)$ *to the corresponding evolutionary trajectory in* $\Delta(\Gamma_K)$.

It means that the recombination processes in subsystems of loci are induced by the process in the whole system.

For inductive proofs on the number of loci which we will systematically use it will be useful to rewrite (6.2.13) isolating the term which corresponds to the trivial partition:

$$p' = V_\gamma(p) = r(L)p + \sum_{U|V \neq L} r(U|V)F^U(p)F^V(p). \tag{6.2.19}$$

By the results of section 1.3 the evolutionary behavior of the gamete pair and zygote populations are obtained automatically from the gamete level dynamics. Indeed, by the above constructions we have

Lemma 6.2.3 *The gamete pair algebra is the duplication of the gamete algebra.*

It remains to remark that the natural map $g*h \mapsto g \circ h$ defines a stochastic algebra homomorphism of the gamete pair algebra onto the zygote algebra.

6.3   Structure of the Set of Equilibrium States. Convergence to Equilibrium

The set of equilibrium states of a recombination process is described by the following theorem.

Theorem 6.3.1 *A state p is an equilibrium if and only if*

$$p = \prod_{i \in L} p^i, \tag{6.3.1}$$

where p^i is the state of ith locus, i.e. the element of the basic simplex corresponding to the single-locus algebra, $\Delta(\Gamma_i)$).

In other words, equilibrium states are exactly those in which the loci are statistically independent, i.e. the probability of each gamete is equal to the product of probabilities of its genes. Recall that we are excluding complete linkage but not linkage in general.

Proof Assuming (6.3.1) $F^K(p) = \prod_{i \in K} p^i$ and so $R_{U|V}(p,p) = p$ (cf. (6.2.17)). As $V(p) = R(p,p)$ is the weighted average of the $R_{U|V}(p,p)$'s it follows that p is an equilibrium.

For the converse we use induction. Assume p is an equilibrium and $U|V$ is a nontrivial partition. By Corollary 6.2.2, the sub-gametes $F^U(p)$ and $F^V(p)$ are equilibria for the corresponding subsystems. Obviously, if $i \in U$ then $F^i(F^U(p)) = F^i(p) \equiv p^i$. By inductive hypothesis $F^U(p) = \prod_{i \in U} p^i$ and similarly for $F^V(p)$. Now apply (6.2.19), observing that $p' = p$ (p is an equilibrium) and $1 - r(L) = \sum_{U|V \neq L} r(U|V)$, so $p = r(L)p + (1 - r(L)) \prod_{i \in L} p^i$. As we have assumed no complete linkage, $r(L) < 1$, and (6.3.1) follows. $\square$

From (6.3.1) we see that the set of equilibrium states is homeomorphic to the product $\Delta(\Gamma_1) \times \cdots \times \Delta(\Gamma_l)$, but because the homeomorphism is the restriction of the l-linear map $\mathbf{R}(\Gamma_1) \times \cdots \times \mathbf{R}(\Gamma_l) \to \mathbf{R}(\Gamma)$ the image is not a convex subset of $\Delta(\Gamma)$. Instead we have

Corollary 6.3.2 *The set of equilibrium states of a recombination process does not depend on the linkage distribution. It is a connected semi- algebraic sub-manifold of the simplex $\Delta(\Gamma)$ whose dimension is $\sum_{i=1}^{l} m_i - l$.*

Recall that the dimension of $\Delta(\Gamma)$ is $\prod_{i=1}^{l} m_i - 1$. Notice that $m = \sum_{i=1}^{l} m_i$ is the total number of genes and l is the number of loci.

Corollary 6.3.3 *Every substate $F^K(p)$, $K \subset L$, of an equilibrium state p is an equilibrium state.*

The converse (for $K \neq L$) is false (see Section 6.5).

From (6.3.1) it is easy to obtain the description of equilibrium states on zygote level, i.e. generalize the Hardy-Weinberg Law to multilocus autosomal systems. By Theorem 1.3.3 the equilibrium at the gamete pair or systems. By Theorem 1.3.3 the equilibrium at the gamete pair or zygote level is just the image of a gamete equilibrium under the appropriate identification map ϕ.

Let $d(\zeta)$ be the number of pairs of different homologous chromosomes in zygote ζ. Then the number of gamete pairs generating ζ is $2^{d(\zeta)}$ and the equilibrium states on the zygote level are

$$x(\zeta) = 2^{d(\zeta)} \prod_{i=1}^{l} p(A^i)p(a^i)$$

where A^i, a^i are allele pairs comprising ζ, $p(\cdot)$ - the probabilities.

Let us now prove that a recombination process always converges to an equilibrium. For the case $l = 1$ this is well known and is also obvious in the present context since in the absence of nontrivial partitions the evolutionary equation (6.2.19) takes the form $p' = p$.

Theorem 6.3.4 *Every trajectory of the recombination process converges to an equilibrium state defined by the initial gene probabilities. The gene probabilities of each locus are preserved in the evolutionary process.*

Proof Let $\{p_t\}_{t=0}^{\infty}$ be a trajectory, i.e. by (6.2.19)

$$p_{t+1} = r(L)p_t + \sum_{U|V \neq L} r(U|V)F^U(p_t)F^V(p_t). \qquad (6.3.2)$$

For the induction hypothesis suppose that the assertion holds for fewer than l loci. By Corollary 6.2.2, $\{F^U(p_t)\}_{t=0}^{\infty}$ and $\{F^V(p_t)\}_{t=0}^{\infty}$ are trajectories for the subsystems corresponding to U and V. If the partition $U|V$ is non-trivial then by the induction hypothesis these trajectories converge. Therefore

$$p_{t+1} = r(L)p_t + R_t \qquad (6.3.3)$$

where $\{R_t\}_{t=0}^{\infty}$ converges. Because $r(L) < 1$, $\{p_t\}_{t=0}^{\infty}$ also converges (see the appendix to chapter 2).

Denote the state of $i$the locus in $t$the generation by $p_t^i = F^i(p_t)$. This trajectory is in the one-locus algebra and does not depend on t. $\square$

As $p_\infty^i = p_0^i$ the equilibrium state p_∞ can be computed via (6.3.1):

$$p_\infty = \prod_{i=1}^{l} p_0^i = \prod_{i=1}^{l} F^i(p_0). \qquad (6.3.4)$$

Let us consider another approach to convergence to equilibrium,
Statistical mechanics has Bolzman's "H-theorem" about the rise of entropy. A similar theorem holds in "population mechanics".

The entropy of the state $p = \{p(g)\}$ is defined to be $H(p) = -\sum_g p(g) \ln p(g)$.

Theorem 6.3.5 *(H-Theorem) If $p' \neq p$, i.e. p is not an equilibrium state, then $H(p') > H(p)$.*

Proof By (6.2.13) p' is the weighted average of the products $F^U(p)F^V(p)$. By strict convexity of the entropy function $H(p') \geq \sum_{U|V} r(U|V)H(F^U(p)F^V(p))$ with strict inequalities unless all the weight is concentrated at points with the same entropy. Furthermore, by elementary information theory $H(F^U(p)F^V(p)) = H(F^U(p)) + H(F^V(p)) \geq H(p)$ with strict inequality unless p equals the tensor product, i.e. the two complementary systems are independent.

Thus, we have that $H(p') \geq H(p)$ and that equality implies $F^U(p)F^V(p) = p$ whenever $r(U|V) > 0$, i.e. $p' = \sum_{U|V} r(U|V)F^U(p)F^V(p) = p$. $\square$

From the H-theorem it is easy to prove the convergence to equilibrium. Indeed on every trajectory these is a limit of entropy, $\bar{H}$. On the set Ω of limit points of the trajectory the entropy is constant and equals $\bar{H}$. But Ω is invariant for evolutionary operator. By the H- theorem its points are fixed, i.e. they are equilibrium states. On a trajectory and therefore in Ω gene probabilities do not change, so by Theorem 6.3.1, Ω has a single point.

6.4 Exact Linearization. Evolutionary Spectrum

The transition from non-linear equation (6.3.2) to equation (6.3.3), linear if R_t is considered known, allows for an exact linearization of the evolutionary equation. However, the resulting linear equation is of high order, i.e. describes the evolution not from the beginning, but from a certain remote generation. The prehistory of this generation can be described by retrospective extrapolation of the linear process, but this requires special analysis described in Section 6.5. Finally we will obtain the "explicit evolutionary formula", i.e. an explicit solution of the evolutionary equation with arbitrary initial condition. Thus the evolutionary equation of the recombination process is "fully integrable". Exact linearization of evolutionary process is easily established.

Theorem 6.4.1 *To every Reiersöl algebra $A(L; r)$ corresponds a polynomial $\Phi(\lambda)$ such that all trajectories $p = \{p_t\}_{t=0}^{\infty}$ satisfy*

$$[\Phi(T)p]_t = 0, \tag{6.4.1}$$

where T is the shift operator, i.e. $(Tp)_t = p_{t+1}$.

    Proof For a single locus, $p_{t+1} - p_t = 0$, i.e. $\Phi(\Lambda) = \lambda - 1$. Suppose that for algebras with fewer than l loci such polynomials exist and are monic.

    Let Φ_1, Φ_2 be two polynomials, W_1, W_2 be the sets of their roots, $\nu_1 : W_1 \to \mathbf{N}$ and $\nu_2 : W_2 \to \mathbf{N}$ be root multiplicities. We define the *quasitensor product* $\Phi = \Phi_1 \diamond \Phi_2$ as the polynomial whose roots are $W = W_1 W_2 = \{\lambda | \lambda = \lambda_1 \lambda_2, \lambda_1 \in W_1, \lambda_2 \in W_2\}$ with multiplicity $\nu : W \to \mathbf{N}$ defined as $\nu(\lambda) = \max_{\lambda = \lambda_1 \lambda_2}(\nu_1(\lambda_1) + \nu_2(\lambda_2) - 1)$. Obviously this operation is commutative. We will also introduce the *least common multiple*, denoted by $\Phi_1 \vee \Phi_2$, to be the polynomial whose roots are $W_1 \cup W_2$ and the multiplicity

$$\nu(\lambda) = \begin{cases} \nu_1(\lambda) & \lambda \in W_1 \backslash W_2 \\ \nu_2(\lambda) & \lambda \in W_2 \backslash W_1 \\ \max(\nu_1(\lambda), \nu_2(\lambda)) & \lambda \in W_1 \cap W_2 \end{cases}$$

The operation $\vee$ is commutative and associative.

    If a scalar sequence $\{a_t\}_{t=0}^{\infty}$ satisfies $[\Phi(T)a]_t = 0$ then

$$a_t = \sum_{\lambda \in W} \theta_\lambda(t)\lambda^t$$

where W are the roots of $\Phi(\lambda)$, $\theta_\lambda(t)$ are polynomials in t of degree $< \nu(\lambda)$, $\nu(\lambda)$ is the multiplicity of root λ, and conversely any sequence of this form is a solution. This is a classical theorem due to Euler from the theory of linear recursive (difference) equations with constant coefficients. This theorem implies if $\{a_t\}_{t=0}^\infty$ and $\{b_t\}_{t=0}^\infty$ satisfy $[\Phi(T)a]_t = 0$ and $[\Psi(T)b]_t = 0$ then $c_t = a_t b_t$ satisfies $[(\Phi \diamond \Psi)(T)c]_t = 0$ and $d_t = a_t + b_t$ satisfies $[(\Phi \vee \Psi)(T)d]_t = 0$.

Let $K \subset L$, $K \neq L$. Denote by $\Phi_K(\lambda)$ the polynomial corresponding by induction hypothesis to the algebra $\mathcal{A}(K; r_K)$. If $p = \{p_t\}_{t=0}^\infty$ is a trajectory in the algebra $\mathcal{A}(L; r)$ then for any nontrivial partition $U|V$ of the system L, $[\Phi_U(T)F^U(p)]_t = 0$ and $[\Phi_V(T)F^V(p)]_t = 0$. By (6.3.2) and the above constructions we can let

$$\Phi_L(\lambda) = (\cdots \vee [\Phi_U(\lambda) \diamond \Phi_V(\lambda)] \vee \cdots)(\lambda - r(L)). \qquad (6.4.2)$$

$\square$

This inductive construction uniquely defines the polynomial $\Phi(\lambda)$ for each Reiersöl algebra $\mathcal{A}(L; r)$. We will call it the *evolutionary polynomial*, its roots *evolutionary roots*, and the set of roots *evolutionary spectrum* of the algebra (or the recombination process). Denoting the evolutionary spectrum by $W(L; r)$ we have

$$W(L; r) = \{r(L)\} \bigcup_{U|V \neq L} W(U; r_U)W(V; r_V) \qquad (6.4.3)$$

for $l > 1$ and $W(L; r) = \{1\}$ for $l = 1$. This immediately implies the following theorem.

Theorem 6.4.2 *The evolutionary spectrum of any Reiersöl algebra $\mathcal{A}(L, r)$ is such that*

$$1 \in W(L; r) \subset [0, 1]. \qquad (6.4.4)$$

Example $\Phi_L(\lambda) = \lambda - 1$ if $l = 1$ and $\Phi_L(\lambda) = (\lambda - 1)(\lambda - r(12))$ if $l = 2$. In the case $l = 3$ $\Phi_L(\lambda) = (\lambda - 1)(\lambda - r(12))(\lambda - r(13))(\lambda - r(23))(\lambda - r(123))$ if all the $r(ij)$ are distinct. If, for instance, $r(12) = r(13) \neq r(23)$ then $\Phi_L(\lambda) = (\lambda - 1)(\lambda - r(12))(\lambda - r(23))(\lambda - r(123))$ and if $r(12) = r(13) = r(23)$ then $\Phi_L(\lambda) = (\lambda - 1)(\lambda - r(12))(\lambda - r(123))$. We see that $\Phi_L(\lambda)$ has no multiple roots if and only if $r(123)$ is different from all the $r(ij)$.

It follows from formula (6.4.2) that $\Phi_L(\lambda)$ is divisible by $\Phi_K(\lambda)$ if $K \subset L$.

Formula (6.4.4) provides a recursive description of the evolutionary spectrum. The first step toward the explicit evolutionary formula is to compute the evolutionary spectrum. The result is quite explicit.

We consider all possible partitions $K_1|\cdots|K_m$ of all subsystems $K \subset L$ satisfying $|K_i| \geq 2$. We will call such partition *admissible* and the number m the *degree of partition*. Obviously $m \leq [l/2]$ ([.] is the integer part). We also introduce the *empty* (i.e. having no classes) partition and assign to it $m = 0$. For $m = 1$ there is one admissible partition, namely the trivial partition, for each subsystem $K \subset L$, $|K| \geq 2$.

Theorem 6.4.3 *The evolutionary spectrum consists of products*

$$\lambda_{K_1|\cdots|K_m} = \prod_{i=1}^{m} r(K_i), \tag{6.4.5}$$

corresponding to all admissible partitions.

The empty partition corresponds to the root $\lambda_\emptyset = 1$.

Proof As usual we will use induction on l. We recall that the probability $r(K)$ is invariant under projection.

By (6.4.3) every evolutionary root λ (except $\lambda_L = r(L)$ accounted for in (6.4.5)) is generated by some non-trivial partition $U|V$ and roots chosen one from each $W(U; r_U)$ and $W(V; r_V)$. If the chosen roots are $\lambda_{Q'_1|\cdots|Q'_i}(\cup Q'_j \subset U)$ and $\lambda_{Q''_1|\cdots|Q''_k}(\cup Q''_j \subset V)$ then their product is $\lambda_{Q'_1|\cdots|Q'_i|Q''_1|\cdots|Q''_k}$.

Conversely, let $\lambda_{K_1|\cdots|K_m}$ be given and assume $m \geq 1$, $K_1 \neq L$. Consider the partition $U|V$ in which $U = K_1$, $V = L\backslash K_1$. Because $\cup_{i>1}K_i \subset V$, we have $\lambda_{K_2|\cdots|K_m} \in W(V; r_V)$. Also $\lambda_{K_1} \in W(U; r_U)$. Hence $\lambda_{K_1|\cdots|K_m} = \lambda_{K_1}\lambda_{K_2|\cdots|K_m} \in W(L; r)$. $\square$

Recall that $r(K_i)$ is $\sum r(U|V)$ summed over all partitions of L with $K_i|\emptyset \prec U|V$. Regard the probabilities of the linkage distribution for L as variables subject only to the relation that the total sum is 1. Then by Theorem 6.4.3 the root $\lambda_{K_1|\cdots|K_m}$ is a homogeneous polynomial of degree m in these variables. A decrease in degree due to the total sum identity is impossible, since otherwise the root would be divisible by the sum of all $r(U|V)$ which is impossible since it does not include the probability $U|V$ partitioning every K_i (recall $|K_i| \geq 2$). This consideration allows us to call m the *degree of the evolutionary root* $\lambda_{K_1|\cdots|K_m}$; the roots of degree $m > 1$ are products (but not arbitrary) of first degree roots.

Theorem 6.4.4 *To distinct admissible partitions correspond roots distinct as polynomials of probabilities.*

Proof Let polynomials $\lambda_{K_1|\cdots|K_m}$ and $\lambda_{Q_1|\cdots|Q_n}$ be equal. $m = n$ from the above. We will show that some subsystem $Q_j \subset K_1$. Suppose that no $Q_j \subset K_1$. Choose $q_j \in Q_j \backslash K_1$ and define partition $U|V$ in which $U = \{q_1, \ldots, q_n\}$. Obviously $K_1 \subset V$ and hence $r(U|V)$ is in $r(K_1)$ and hence in the whole $\lambda_{K_1|\cdots|K_m}$. But $Q_j \not\subset U$, (since $|Q_j| \geq 2$ and $Q_j \cap Q_i = \emptyset$ for $i \neq j$) and $Q_j \not\subset V$ (since $q_j \in U$). Hence $r(U|V)$ is not in any $r(Q_j)$ and not in the whole $\lambda_{Q_1|\cdots|Q_n}$, a contradiction. Analogously some subsystem $K_i \subset Q_j$. Because $K_1 \cap K_i = \emptyset$ for $i \neq 1$, we must have $i = 1$ and $Q_j = K_1$. Therefore $K_1|\cdots|K_m = Q_1|\cdots|Q_n$. $\square$

Theorem 6.4.4 does not mean that the evolutionary polynomial never has multiple roots. It only means that generically (in the space of linkage distributions) the enumeration of evolutionary roots by means of Theorem 6.4.3 is bijective.

We consider an important example which is not generic.

Example Suppose all l loci are independent. Then it follows from Lemma 6.1.8 and Theorem 6.4.3 that

$$\lambda_{K_1|\cdots|K_m} = 2^{m-\sum_{i=1}^{m}|K_i|}. \tag{6.4.6}$$

Thus the description of evolutionary roots by means of admissible partitions here is not bijective.

The criterion for the absence of multiple roots is as follows.

Theorem 6.4.5 *An evolutionary polynomial has no multiple roots if and only if for all admissible partitions $K_1|\cdots|K_m$ and for each $K \supset \cup_{i=1}^{m}K_i$ (except $m = 1$, $K = K_1$)*

$$r(K) \neq \lambda_{K_1|\cdots|K_m}. \tag{6.4.7}$$

Proof Quasi-tensor product and the least common multiple of polynomials have no multiple roots if and only if these polynomial have no multiple roots. Inductive use of (6.4.2) does not introduce multiple roots if and only if (6.4.7) holds. $\square$

Corollary 6.4.6 *When the loci are independent, the evolutionary polynomial has no multiple roots.*

Proof In this case the right hand side of (6.4.7) is given by (6.4.6) and the left hand side is $2^{1-|K|}$. $\square$

6.5 Explicit Evolutionary Formula

If the linkage distribution is generic then, by Theorems 6.4.4 and 6.4.5 evolutionary roots $\lambda_{K_1|\ldots|K_m}$ are distinct and simple. By Theorems 6.4.1 and 6.4.3 each trajectory has the form

$$p_t = \sum_{K_1|\cdots|K_m} C_{K_1|\ldots|K_m}(p_0)(\prod_{i=1}^{m} r(K_i))^t, \qquad (6.5.1)$$

where the constants $C_{K_1|\cdots|K_m}(p_0)$ depend only on the initial state p_0 and the linkage distribution. It suffices to compute these constants to turn (6.5.1) into the explicit evolutionary formula, but this is non-trivial.

The linear equation (6.4.1), whose general solution is (6.5.1), is not equivalent to the original non-linear equation (6.3.2), for the solution of the latter is uniquely defined by the initial value p_0, while the linear equation admits arbitrary values of p_t for t less than the degree of the evolutionary polynomial. Hence (6.3.2) establishes non-trivial relations among the desired constants:

$$C_{K_1|\ldots|K_m}(p_0) =$$

$$\frac{\sum_{K_1|\cdots|K_m \prec U|V \neq L} r(U|V)C_{\ldots|K_i^U|\ldots}(F^U(p_0))C_{\ldots|K_i^V|\ldots}(F^V(p_0))}{\prod_{i=1}^{m} r(K_i) - r(L)}, \qquad (6.5.2)$$

where K_i^U and K_i^V are those K_i that are respectively in U and V.

The equation (6.5.2) inductively defines all constants $C_{K_1|\cdots|K_m}$, except C_L. In that case it is meaningless because the denominator is zero. However the initial condition ((6.5.1) when $t = 0$) gives one more constraint

$$p_0 = \sum_{K_1|\cdots|K_m} C_{K_1|\cdots|K_m}(p_0), \qquad (6.5.3)$$

from which we can find C_L if the other constants are known.

The presence of the extra equation (6.5.3) causes a significant difficulty, since (6.5.3), unlike (6.5.2), does not lend itself to induction on degree m, which is the natural measure of complexity of $C_{K_1|\cdots|K_m}$. There is an interconnection here, for example, computing the constants C_K, i.e. $m = 1$, cannot be separated from computing the other constants. Only the case $m = 0$ can be considered alone.

Theorem 6.5.1 $C_\emptyset(p_0) = \prod_{i=1}^{1} F^i(p_0)$.

Proof When $m = 0$, (6.5.2) takes the form

$$C_\emptyset(p_0) = \frac{\sum_{U|V \neq L} r(U|V) C_\emptyset(F^U(p_0)) C_\emptyset(F^V p_0)}{1 - r(L)}.$$

If we observe that $1 - r(L) = \sum_{U|V \neq L} r(U|V)$, the assertion is proved by induction on l. $\square$

Remark Theorem 6.5.1 can also be proved by considering the convergence of p_t as $t \to \infty$. Because all evolutionary roots of degree $m > 0$ are less than 1, it follows from (6.5.1) that $p_\infty \equiv \lim_{t \to \infty} p_t = C_\emptyset(p_0)$ and it remains to apply (6.3.4).

An important step in computing the constants $C_{K_1|\cdots|K_m}$ is the construction of adequate measures of disequilibrium. The simplest of them, the *absolute measure of disequilibrium* of state p, is defined as

$$E_L(p) \equiv p - \prod_{i=1}^{l} F^i(p) = p - C_\emptyset(p). \tag{6.5.4}$$

Obviously a state p is an equilibrium if and only if $E_L(p) = 0$. However besides $E_L(p)$ we will need relative measures of disequilibrium. We define the *measure of disequilibrium of state p relative to subsystem* $K \subset L, |K| \geq 2$ to be

$$E_K(p) \equiv E_K(F^K(p)) = F^K(p) - \prod_{i \in K} F^i(p).$$

The state $F^K(p)$ (the restriction of state p to subsystem K) is an equilibrium if and only if $E_K = 0$. Obviously $E_K(p) \in \mathcal{A}(K; r_K)$.

The measures of disequilibrium are *natural* in the sense that $F^Q(E_K(p)) = E_Q(F^K(p)) = E_Q(p)$ (when $Q \subset K, |Q| \geq 2$), $F^j(E_K(p)) = 0$ (when $j \in K$). Also $s(E_K(p)) = 0$.

We define the *measure of disequilibrium of state p relative to an admissible partition* $K_1|\cdots|K_m, m > 0$, to be $E_{K_1|\cdots|K_m}(p) = \prod_{i=1}^{m} E_{K_i}(p)$, where the product belongs to $\mathcal{A}(K, r_K)$, $K = \cup_{i=1}^{m} K_i$. Here we use the identification generalizing (6.2.3):

$$\mathbf{R}(\Gamma_{K_1}) * \cdots * \mathbf{R}(\Gamma_{K_m}) \simeq \mathbf{R}(\Gamma_K), \tag{6.5.5}$$

The number m we will, as usual, call the *degree (of the measure of disequilibrium)*.

The desired constants will be expressed in terms of *normalized measures of disequilibrium*, defined as

$$\overline{E_{K_1|\cdots|K_m}}(p) = E_{K_1|\cdots|K_m}(p) \prod_{j\notin K} F^j(p),$$

belonging to the original algebra $\mathcal{A}(L;r)$. For any admissible partition $U|V$ of system L we have the *recovery equation*

$$\overline{E_{K_1|\cdots|K_m}}(F^U(p))\overline{E_{J_1|\cdots|J_n}}(F^V(p)) = \overline{E_{K_1|\cdots|K_m|J_1|\cdots|J_n}}(p), \qquad (6.5.6)$$

where $\cup K_j \subset U$, $\cup J_i \subset V$. The normalizations on left hand side are in subsystems of U and V, respectively.

Theorem 6.5.2 *When $m > 0$,*

$$C_{K_1|\cdots|K_m}(p_0) = \sum_{Q_1|\cdots|Q_n} A^{Q_1|\cdots|Q_n}_{K_1|\cdots|K_m} \overline{E_{Q_1|\cdots|Q_n}}(p_0), \qquad (6.5.7)$$

where

1) $Q_1|\cdots|Q_n$ runs over all admissible partitions satisfying a) for every $1 \le j \le n$, $Q_j \subset K_i$ for some i, b) for every $1 \le i \le m$, $Q_j \subset K_i$ for some j.

2) The coefficients $A^{Q_1|\cdots|Q_n}_{K_1|\cdots|K_m}$ depend only on the probabilities in distribution r_K, $K = \bigcup_{i=1}^{m} K_i$.

Condition b) implies that in all terms of (6.5.7) $n \ge m$.

The induction on the number of loci l quickly proves all assertions of the theorem except the last one, instead of which we get a weaker dependence on the complete distribution r instead of the restriction r_K. The latter refinement is non- trivial, yet essential.

The minimum number of loci in the statement of the theorem is 2. In this case the theorem becomes

$$C_{12}(p_0) = E_{12}(p_0). \qquad (6.5.8)$$

To see this, we apply the condition (6.5.3) which for $l = 2$ takes the form $p_0 = C_\emptyset(p_0) + C_{12}(p_0)$ from which, by (6.5.4), the assertion reduces to $C_{12}(p_0) = p_0 - C_\emptyset(p_0) = E_{12}(p_0)$.

For the induction step we first apply (6.5.2). From there, using the recovery equation (6.5.6), we obtain the required expansion of all the constants except C_L and the induction formula for the coefficients

$$A^{Q_1|\cdots|Q_n}_{K_1|\cdots|K_m}(r) =$$

$$\frac{\sum_{K_1|\cdots|K_m \prec U|V \neq L} r(U|V) A_{K_i^V|\cdots}^{\cdots|Q_j^V|\cdots}(r_V) A_{K_i^U|\cdots}^{\cdots|Q_j^U|\cdots}(r_U)}{\prod_{i=1}^m r(K_i) - r(L)}. \tag{6.5.9}$$

On the right hand side on (6.5.9) there may be summands for which all K_i are in the same class of partition $U|V$. Here we set $A_{\emptyset}^{\emptyset} = 1$. In particular for $m = 1$ this is the case for all summands and (6.5.9) reduces to

$$A_K^{Q_1|\cdots Q_n}(r) = \frac{\sum_{K \subset U \neq L} r(U|V) A_K^{Q_1|\cdots|Q_N}(r_U)}{r(K) - r(L)}$$

which agrees, inductively, with the equality of all $A_K^{Q_1|\cdots|Q_n}(r_U)$ for $\sum_{K \subset U \neq L} r(U|V) = r(K) - r(L)$. Thus after the theorem is proved, the equation (6.5.9) will retain content only for $m > 1$.

To use (6.5.3) we rewrite it using Theorem 6.5.1 as

$$C_L(p_0) = E_L(p_0) - \sum_{K_1|\cdots|K_m \neq \emptyset, L} C_{K_1|\cdots|K_m}(p_0). \tag{6.5.10}$$

Now induction becomes possible for C_L because it has been done for all the constants on right hand side of (6.5.10). The latter also gives us

$$A_L^L = 1 \tag{6.5.11}$$

and

$$A_L^{Q_1|\cdots|Q_n}(r) = - \sum_{K_1|\cdots|K_m \neq L} A_{K_1|\cdots|K_m}^{Q_1|\cdots|Q_n}, \quad (Q_1|\cdots|Q_n \neq L). \tag{6.5.12}$$

To prove the last assertion of Theorem 6.5.2 we will need a few lemmas.

Lemma 6.5.3 *If $l > 1$ then there exists a non-equilibrium state p such that all of its non-trivial substates $F^K(p)$ ($K \neq L$) are equilibria.*

Proof First we remark that the system of linear equations

$$F^K(x) = 0 \quad (K \neq L), \tag{6.5.13}$$

has a non-trivial solution, for example at locus i define q_i to be $+1$ and -1 at the first and second alleles and to be 0 at the remaining alleles of Γ_i. Then define x by $\sum_g \prod_{i=1}^l q(i)(g_i)$ ($g \in \Gamma$). Let p^i be a state of ith locus with positive probabilities of all genes. Then $p(\epsilon) = \epsilon x + \prod_{i=1}^l p^i \in \Delta(\Gamma)$ for ϵ

small enough, since $s(x) = 0$ by (6.5.13). Obviously, $F^K(p(\epsilon)) = \prod_{i\in K} p^i$, i.e. the sub-states $F^K(p(\epsilon))$ are equilibria. However $E_L(p(\epsilon)) = \epsilon x \neq 0$. $\square$

Lemma 6.5.3 is interesting from the biological point of view because it demonstrates the evolutionary role of the number of loci; the absence of equilibrium may not be detected until the number of loci under observation is large enough.

Lemma 6.5.4 *For every admissible partition* $Q_1|\cdots|Q_n$, $n > 0$, *there exists a state* p *such that* $E_{K_1|\cdots|K_m}(p) \neq 0$ *if every* K_i *contains some* Q_j *and* $E_{K_1|\cdots|K_m} = 0$ *otherwise.*

Proof In subsystem Q_j select a disequilibrium state p_j all of whose non-trivial sub-states are equilibria. Let $Q = \cup_{i=1}^n Q_j$ and in the complementary subsystem select an equilibrium state p_0. Let $p = \prod_{i=0}^n p_j$. Let $K_1|\cdots|K_m$ be an admissible partition such that every K_i contains some Q_j. For example, let $K_1 \supset Q_1$. Then $F^{K_1}(p)$ is not an equilibrium state since its sub-state p_1 is not an equilibrium. Similarly for $F^{K_2}(p)$ etc. Hence $E_{K_1|\cdots|K_m}(p) \neq 0$.

Suppose some K_i, e.g. K_1, contains no Q_j. Then $K_1 \cap Q_j \neq Q_j$, $1 \leq j \leq n$, and hence the state $F^{K_1}(p) = \prod_{j=0}^n F^{K_1\cap Q_j}(p_j)$, $Q_0 = L\backslash Q$, is an equilibrium. Hence $E_{K_1}(p) = 0$ and $E_{K_1|\cdots|K_m}(p) = 0$. $\square$

Corollary 6.5.5 *The normalized measures of disequilibrium, considered as functions from* $\Delta(\Gamma) \to \mathbf{R}(\Gamma)$, *are linearly independent.*

Proof Suppose

$$\sum_{K_1|\cdots|K_m \neq \emptyset} \alpha_{K_1|\cdots K_m} \overline{E_{K_1|\cdots|K_m}}(p) \equiv 0. \tag{6.5.14}$$

Consider the family of admissible partitions $K_1|\cdots|K_m$ for which $\alpha_{K_1|\cdots|K_m} \neq 0$ and order them according to Lemma 6.5.4. Let $Q_1|\cdots|Q_n$ be a maximal element of this family and p be the state satisfying the conditions of Lemma 6.5.4. Substituting p in (6.5.14) we get $\alpha_{Q_1|\cdots|Q_n} = 0$. $\square$

We can now complete the

Proof of Theorem 6.5.2 We first note that (6.5.1) implies $F^U(p_t) = \sum_{K_1|\cdots|K_m} F^U\{C_{K_1|\cdots|K_m}(p_0)\}(\prod_{i=1}^m r(K_i))^t$. But the only evolutionary roots in the subsystem U are $\prod_i r(K_i)$, where $\cup K_i \subset U$. Hence

$$F^U\{C_{K_1|\cdots|K_m}(p_0)\} = \begin{cases} 0 & \cup K_i \not\subset U \\ C_{K_1|\cdots|K_m}(F^U(p_0)) & \cup K_i \subset U \end{cases}.$$

This indicates that the constants $C_{K_1|\cdots|K_m}(p_0)$ are *natural*. Moreover the coefficients $A^{Q_1|\cdots|Q_n}_{K_1|\cdots|K_m}$ depend only on natural arguments. To prove it, apply F^K to the expansion (6.5.7). Because the constants $C_{K_1|\cdots|K_m}(p_0)$ and the measures of disequilibrium are natural we get in subsystem K the equation $C_{K_1|\cdots|K_m}(p) = \sum_{K_1|\cdots|K_m} A^{Q_1|\cdots|Q_n}_{K_1|\cdots|K_m}(r)\overline{E_{Q_1|\cdots|Q_n}}(p)$, which holds for all the states p of the subsystem because F^K is surjective. By linear independence of the normalized measures of disequilibrium we obtain $A^{Q_1|\cdots|Q_n}_{K_1|\cdots|K_m}(r) = A^{Q_1|\cdots|Q_n}_{K_1|\cdots|K_m}(r_K)$, which is the desired result. $\square$

Now we can to compute the coefficients $A^{Q_1|\cdots|Q_n}_{K_1|\cdots|K_m}$ using equations (6.5.9), (6.5.11) and (6.5.12). By the last assertion of Theorem 6.5.2, in (6.5.9) we can let $L = K$. The same can be done in (6.5.11), (6.5.12) by relabeling.

We will use (6.5.9) for induction on m. Let $w = \{1,\ldots,m\}$. Because $L = K$, the partition $U|V \succ K_1|\cdots|K_m$ of system L can be described as partitions $u|v$ of the set w.

Consider the following dichotomic process. Split the set w (class of rank 0) into two non-empty classes u, v (classes of rank 1), repeat the process with each of them obtaining classes of rank 2 if possible, etc. until we reach a stage where every class has a single element. Because the choice of the dichotomy is arbitrary, the process has a family $\mathcal{T}_m$ of realizations. Each realization $T \in \mathcal{T}_m$ can be viewed as a binary tree whose root is class w of rank 0 and the end vertices are the singleton classes, possibly of different ranks.

Denote the set of non-end vertices of tree $T \in \mathcal{T}_m$ by T'. To every vertex $z \in T'$, splitting into classes x and y, assign a function of linkage distribution

$$\phi_{T,z}(r) = \frac{r(\cup_{i\in x}K_i|\,\cup_{i\in y}K_i)}{\prod_{i\in z}r(K_i) - r(\cup_{i\in z}K_i)}. \tag{6.5.15}$$

Induction on m easily gives the following theorem from (6.5.9).

Theorem 6.5.6 *If the numeration Q_j is such that $Q_1,\ldots,Q_{n_1} \subset K_1,\ldots,Q_{n_{m-1}+1},\ldots,Q_n \subset K_m$ then*

$$A^{Q_1|\cdots|Q_n}_{K_1|\cdots|K_m} = \Big(\sum_{T\in\mathcal{T}_m} \prod_{z\in T'} \phi_{T,z}\Big) A^{Q_1|\cdots|Q_{n_l}}_{K_1} \cdots A^{Q_{n_{m-1}+1}|\cdots|Q_n}_{K_m}. \tag{6.5.16}$$

Letting $Q_1 = K_1,\ldots,Q_m = K_m$ and using (6.5.11) we get the following corollary.

Corollary 6.5.7 $A_{K_1|\cdots|K_m}^{K_1|\cdots|K_m} = \sum_{T \in \mathcal{T}_m} \prod_{z \in T'} \phi_{T,z}.$

Therefore the equation (6.5.16) can be rewritten as

$$A_{K_1|\cdots|K_m}^{Q_1|\cdots|Q_n} = A_{K_1|\cdots|K_m}^{K_1|\cdots|K_m} A_{K_1}^{Q_1|\cdots|Q_{n_1}} \cdots A_{K_m}^{Q_{n_{m-1}+1}|\cdots|Q_n}. \tag{6.5.17}$$

Now it is clear that all coefficients can be represented via "diagonal" coefficients of Corollary 6.5.7. But we must use (6.5.12) written as

$$A_K^{Q_1|\cdots|Q_n} =$$

$$- \sum_{J \subset K, J \neq K} A_J^{Q_1|\cdots|Q_n} - \sum_{\cup K_i = K; m \geq 2} A_{K_1|\cdots|K_m}^{Q_1|\cdots|Q_n} \quad (Q_1|\cdots|Q_m \neq K).$$

When $n = 1$ this equation takes the form $A_K^Q = \sum_{Q \subset J \not\subset K} A_J^Q, (K \neq Q)$. Also, by (6.5.11), $A_Q^Q = 1$. For $n > 1$ the condition $Q_1|\cdots|Q_n \neq K$ is automatic.

In both cases we have an equation of the form

$$\alpha_K = - \sum_{Q \subset J \not\subset K} \alpha_J + \omega_K \quad (K \supset Q) \tag{6.5.18}$$

relative to $\alpha_K = A_K^{Q_1|\cdots|Q_n}$, with $Q = \cup_{j=1}^n Q_j, Q_j$ are fixed. Its solution is

$$\alpha_K = \sum_{Q \subset I \subset K} (-1)^{|K \setminus I|} \omega_I, \tag{6.5.19}$$

which is directly verified by induction on $|K| \geq |Q|$ (when $K = Q$, (6.5.18) implies $\alpha_Q = \omega_Q$). Thus for $n = 1$ we have the following theorem.

Theorem 6.5.8 $A_K^Q = (-1)^{|K \setminus Q|}.$

The following corollary follows from this theorem and (6.5.17).

Corollary 6.5.9 $A_{K_1|\cdots|K_m}^{Q_1|\cdots|Q_m} = (-1)^{|K \setminus Q|} A_{K_1|\cdots|K_m}^{K_1|\cdots|K_m}$ where $K = \cup K_i, Q = \cup Q_i.$

Thus for $m = n$ the problem is solved.

For $n > 1$ by (6.5.19) we have

$$A_K^{Q_1|\cdots|Q_m} = - \sum_{\cup K_i = K; m \geq 2} A_{K_1|\cdots|K_m}^{K_1|\cdots|K_m} A_{K_1}^{\cdots|Q_{1j}|\cdots} \cdots A_{K_m}^{\cdots|Q_{mj}|\cdots}, \tag{6.5.20}$$

where Q_{ij} are the Q_j contained in K_i.

The equation (6.5.20) can be iterated similarly to (6.5.9). We will describe the corresponding combinatorial structure.

For fixed K, $Q_1|\cdots|Q_n$ we take an arbitrary admissible partition $K_1|\cdots|K_m$, $m \geq 2$, of the set of loci K, containing $Q_1|\cdots|Q_n$. Each K_i containing more than one Q_j we subject to similar partition etc. while possible. Each realization of this process is a tree with root K. We will denote the set of these trees by $T_K^{Q_1|\cdots|Q_n}$.

To each nonend vertex J of tree $T \in T_K^{Q_1|\cdots|Q_n}$ assign the value $\alpha_{T,j} = A_{J_1|\cdots|J_q}^{J_1|\cdots|J_q}$, where $J_1|\cdots|J_q$ is the partition of J corresponding to this vertex.

The following theorem follows from (6.5.20) by induction on n.

Theorem 6.5.10 *When $n > 1$*

$$A_K^{Q_1|\cdots|Q_n} = (-1)^{|K\setminus Q|} \sum_{T\in T_K^{Q_1|\cdots|Q_n}} (-1)^{|T'|} \prod_{J\in T'} a_{T,J} \quad (Q = \cup_{j=1}^n Q_j).$$

The multiplicand $(-1)^{|K\setminus|}$ arises by Theorem 6.5.8 from the end vertices.

The problem of "integrating" the equations of recombination process in explicit form is completely solved. The assumption of no multiple evolutionary roots in (6.5.1) can be removed by taking the limit. Then we may or may not get secular terms of the form $t\lambda^t, t^2\lambda^t,\ldots$. They appear only at the poles of the function $\phi_{T,z}$ defined by (6.5.15). This means that a root of higher degree coincides with a corresponding root of first degree. It is called a *resonance*. In particular, there is no resonance on roots of the form $r(K)$ when $|K| \leq 3$ except the case $r(K) = r(Q)$ where Q is a resonance root, $|Q| > 3$. Corresponding expressions for constants C_K do not depend on the linkage distribution at all. Theorems 6.5.2 and 6.5.8 easily imply

$$C_K(p_0) = \overline{E_K}(p_0) \tag{6.5.21}$$

when $|K| = 2$ and

$$C_K(p_0) = \overline{E_K}(p_0) - \sum_{Q\subset K;|Q|=2} \overline{E_Q}(p_0) \tag{6.5.22}$$

when $|K| = 3$. When $|K| = 4$ we first get resonance, precisely when $|K_1| = |K_2| = 2$

$$C_{K_1|K_2}(p_0) = \frac{r(K_1|K_2)}{r(K_1)r(K_2) - r(K_1 \cup K_2)}\overline{E_{K_1|K_2}}(p_0)$$

and

$$C_K(p_0) = \sum_{Q \subset K; |Q|=2} (-1)^{|K \setminus Q|} \overline{E_Q}(p_0)$$

$$+ \sum_{K_1 \cup K_2 = K} \frac{r(K_1|K_2)}{r(K_1)r(K_2) - r(K_1 \cup K_2)} \overline{E_{K_1|K_2}}(p_0).$$

In the important case of independent loci there is no resonance by Corollary 6.4.6. Here $\phi_{T,Z} = \frac{2}{2^{|z|-1}-1}$ and the denominator is not zero, for $|z| \geq 2$ when $z \in T'$.

We will now write the explicit evolutionary formulas for $l \leq 4$. The initial distribution is denoted by p, the distribution of genes in ith locus by $p^i = F^i(p)$.

For $l = 1$ we have $p_t = p^1$.

For $l = 2$ by (6.5.21) (or (6.5.8)) $p_t = p^1 p^2 + E_{12}(p)(r(12))^t$, i.e.

$$p_t = p^1 p^2 + (p - p^1 p^2)(r(12))^t \quad .$$

For $l = 3$ by (6.5.21) and (6.5.22) we have

$$p_t = p^1 p^2 p^3 + E_{12}(p)p^3\{r(12)\}^t + E_{13}(p)p^2\{r(13)\}^t + E_{23}(p)p^1\{r(23)\}^t +$$

$$(E_{123}(p) - E_{12}(p)p^3 - E_{13}(p)p^2 - E_{23}(p)p^1)\{r(123)\}^t.$$

It can be written in another form:

$$p_t = p^1 p^2 p^3 + (F^{12}p - p^1 p^2)p^3\{r(12)\}^t + (F^{13}p - p^1 p^3)p^2\{r(13)\}^t +$$

$$(F^{23}p - p^2 p^3)p^1\{r(23)\}^t + (p - p^1 F^{23}p - p^2 F^{13}p - p^3 F^{12}p + 2p^1 p^2 p^3)\{r(123)\}^t.$$

For $l = 4$ we will omit the terms that differ from the "typical" ones only by numeration:

$$p_t = p^1 p^2 p^3 p^4 + (F^{12}(p) - p^1 p^2)p^3 p^4\{r(12)\}^t + \cdots$$

$$+(F^{123}(p) - p^1 F^{23}(p) - p^2 F^{13}(p) - p^3 F^{12}p + 2p^1 p^2 p^3)p^4\{r(123)\}^t + \cdots$$

$$+\frac{r(12|34)}{r(12)r(34) - r(1234)}(F^{12}(p) - p^1 p^2)(F^{34}(p) - p^3 p^4)\{r(12)\}^t\{r(34)\}^t + \cdots$$

$$+(p - p^1 F^{234}(p) - \cdots + p^1 p^2 F^{34}(p) + \cdots - 3p^1 p^2 p^3 p^4 +$$

$$\frac{r(12|34)}{r(12)r(34) - r(1234)}(F^{12}(p) - p^1 p^2)(F^{34}(p) - p^3 p^4) + \cdots)\{r(1234)\}^t.$$

The explicit evolutionary formula immediately implies the stability of recombination process.

Theorem 6.5.11 *A trajectory depends continuously on the initial state p_0.*

In particular, if a population is shaken slightly out of equilibrium, it again tends to an equilibrium state close to the original one. Therefore we consider evolution near the manifold of equilibrium states. The role of small parameters is played by first degree measures of disequilibrium. Measures of disequilibrium of higher degrees have higher orders of smallness. Disregarding those we get the following result.

Theorem 6.5.12 *Near the manifold of equilibrium states, a trajectory of the recombination process has the following form, with accuracy up to measures of disequilibrium of higher degrees.*

$$p_t \approx \prod_{i=1}^{l} F^i(p_0) + \sum_{K \supset L; |K| \geq 2} \{r(K)\}^t \sum_{Q \subset K; |Q| \geq 2} (-1)^{|K \setminus Q|} E_Q(p_0) \prod_{i \notin Q} F^i(p_0).$$

6.6 The Rate of Convergence to Equilibrium

Consider the greatest probability that a certain subsystem K, $|K| \geq 2$, of locus system will not spit on crossing over:

$$\kappa = \max r(K) < 1. \tag{6.6.1}$$

This is the greatest evolutionary root of first degree. Since $r(K) \geq r(Q)$ for $K \subset Q$, the maximum in (6.6.1) is achieved when $|K| = 2$. All evolutionary roots of degree higher than one are less than κ because of (6.4.5). Therefore resonance roots are also less than κ. We have proved the following theorem without the assumption of a genericity.

Theorem 6.6.1 *For any trajectory we have the asymptotic expansion*

$$p_t = p_\infty + \kappa^t \sum_{r(K)=\kappa} C_K(p_0) + o(\kappa^t).$$

Also the following lemma holds.

Lemma 6.6.2 *There exists a state p_0 for which $\sum_{r(K)=\kappa} C_K(p_0) \neq 0$.*

Proof Otherwise, (6.5.7) and the linear independence of the normalized measures of disequilibrium imply

$$\sum_{K \succ Q_1|\cdots|Q_n; r(K)=\kappa} A_K^{Q_1|\cdots|Q_n} = 0 \tag{6.6.2}$$

for any admissible partition $Q_1|\cdots|Q_n$. Choose $Q \subset L$ such that $r(Q) = \kappa$. $|Q| = \max$. Then by (6.6.2), $A_Q^Q = 0$ which contradicts Theorem 6.5.8. $\square$

Theorem 6.6.1 and Lemma 6.6.1 imply:

Theorem 6.6.3 *The global (over all trajectories) estimate of the rate of convergence to equilibrium is* $\max_{p_0}(\lim_{t\to\infty} ||p_t - p_\infty||^{1/t}) = \kappa$.

Proof The limit equals the maximal root $\lambda = \lambda(p_0)$ such that p_t contains a term $t^\nu \lambda^t$. Obviously, $\lambda(p_0) \leq \kappa$ and by Theorem 6.6.1 $\lambda(p_0) = \kappa$ on the complement of the manifold where

$$\sum_{r(K)=\kappa} C_K(p_0) = 0. \tag{6.6.3}$$

This complement is non-empty by Lemma 6.6.2. $\square$

Theorem 6.6.3 allows up to call κ the *stabilization coefficient* for the whole population. For "typical" trajectories satisfying Lemma 6.6.2, the distance to a limit equilibrium asymptotically equals to const$\cdot\kappa^t$. For "exceptional" trajectories, not satisfying the lemma, originating on the manifold (6.6.3) the distance to the limit equilibrium is $o(\kappa^t)$. By the way this shows that the variety (6.6.3) and its complement are invariant for the evolutionary operator.

The stabilization coefficient can be arbitrarily close to 1 and the rate of convergence to equilibrium arbitrarily slow (on exponential scale). For two loci ($l = 2$) the stabilization coefficient $\kappa = r(12)$ can be anything smaller than 1 and the convergence can be arbitrarily fast. For $l > 2$ this is impossible as the following theorem shows.

Theorem 6.6.4 *The minimal (over all linkage distributions) possible stabilization coefficient of l loci is*

$$\kappa_l = \frac{[\frac{l-1}{2}]}{2[\frac{l-1}{2}] + 1}.$$

Proof To get the inequality $\kappa \geq \kappa_l$ for any linkage distribution use the obvious equality $\sum_{|K|=2} r(K) = \sum_{U|V} \nu(U|V) r(U|V)$, where $\nu(U|V)$ is the number of subsets K, $|K| = 2$ in one class of the partition $U|V$. Then $(l(l-1)/2)\kappa \geq \min \nu(U|V)$. But $\nu(U|V) = |U|(|U|-1)/2 + |V|(|V|-1)/2$ and since $|U| + |V| = l$, $\nu(U|V) = l(l-1)/2 - |U||V|$. If l is even then $\max |U||V| = l^2/4$; this maximum is achieved when $|U| = |V| = l/2$ and hence $\min \nu(U|V) = l(l-2)/4$, $\kappa \geq (l-2)/2(l-1)$. If l is odd, then $\max |U||V| = (l^2-1)/4$; this maximum is achieved when $|U| = (l-1)/2$, $|V| = (l+1)/2$ and hence $\min \nu(U|V) = (l-1)^2/4$, $\kappa \geq (l-1)/2l$. In both cases the assertion holds.

We will now construct for each l a linkage distribution such that $\kappa = \kappa_l$. With $\binom{n}{m}$ denoting the binomial coefficient, for even l let

$$r(U|V) = \begin{cases} 2\binom{l}{l/2}^{-1} & |U| = |V| = l/2 \\ 0 & \text{otherwise} \end{cases}. \qquad (6.6.4)$$

Then for $|K| = 2$, $r(K) = 2\binom{l}{l/2}^{-1}\binom{l-2}{l/2-2} = (l-2)/2(l-1)$.
For odd l let

$$r(U|V) = \begin{cases} 2\binom{l}{(l-1)/2}^{-1} & |U| = (l-1)/2, |V| = (l+1)/2 \\ 0 & \text{otherwise} \end{cases}. \qquad (6.6.5)$$

Then for $|K| = 2$, $r(K) = \binom{l}{(l-1)/2}^{-1}\left(\binom{l-2}{(l-3)/2} + \binom{l-2}{(l-5)/2}\right) = (l-1)/2l$. $\square$

Corollary 6.6.5 *If $l > 2$ then $\kappa \geq 1/3$ and this estimate is achieved for some linkage distribution when $l = 3$.*

Corollary 6.6.6 *As $l \to \infty$ the smallest possible stabilization coefficient increases monotonically and approaches $\frac{1}{2}$.*

An alternative to Theorem 6.6.1 is the trivial estimate $\kappa \geq r(L)$, which is stronger if $r(L) \geq \frac{1}{2}$. The boundary $\kappa = \frac{1}{2}$ from the biological point of view separates the area of "positive interference", $\kappa > \frac{1}{2}$ from the area of "negative interference", $\kappa < \frac{1}{2}$. With positive interference of two loci 1, 2, a "linkage phase" (12) is more probably that "rejection phrase" (1|2), with negative interference the opposite is the case. Positive interference apparently occurs much more often that negative in real populations. We will give some theoretical justification for that. For example, Corollary 6.1.14 implies the estimate

$$\kappa \geq \frac{1}{2} \qquad (6.6.6)$$

for linear linkage distributions and $l \geq 3$. (6.1.12) implies an exact formula, which implies (6.6.6), for binomial distribution and $l \geq 3$:

$$\kappa = \begin{cases} q & p \leq \frac{1}{2}, \\ 1 - 2pq & p \geq \frac{1}{2}. \end{cases} \tag{6.6.7}$$

When at least two linkage groups at present, (6.6.6) holds no matter what linkage distribution. Finally if all l loci are independent then $\kappa = \frac{1}{2}$ by Corollary 6.1.9.

6.7 Combining Recombination and Mutation

Suppose that mutation occurs in each locus independently. We denote the mutation rate matrix at ith locus by $T^{(i)} = (\tau_{kj}^{(i)})_{kj=1}^{m^i}$, $1 \leq i \leq l$. The element $\tau_{kj}^{(i)}$ is defined as the probability of the mutation $j \to k$ in the set of alleles Γ_i. Then with probability $\tau_{k_1 j_1}^{(1)} \ldots \tau_{k_l j_l}^{(l)}$ a gamete with alleles $j_1, \ldots, j_l$ mutates into a gamete with alleles $k_1, \ldots, k_l$. The mutation matrix is thus defined at the gamete level (the linear *mutation operator* in the space $\mathbf{R}(\Gamma)$) as the Kronecker product $T = T^{(1)} \otimes \cdots \otimes T^{(l)}$. The matrix T is stochastic (by columns) and all the diagonal entries are positive because the $T^{(i)}$s all have these properties (see Section 2.5).

Lemma 6.7.1 *The mutation operator T is a stochastic endomorphism of the Reiersöl algebra $\mathcal{A}(L; r)$ for any linkage distribution.*

Proof Consider a pair of gametes g, h and a partition $U|V$. Independence of the mutation process at each locus separately easily implies that for any gamete $\tilde{g}$ the probability that $g_U h_V$ mutates to $\tilde{g}$ is the product of the probability that the subgamete g_U mutates to $\tilde{g}_U$ and that h_V mutates to $\tilde{h}_V$. Taking expected value this becomes

$$T(F^U(g)F^V(h)) = F^U(Tg)F^V(Th). \tag{6.7.1}$$

So from (6.2.16) we see that $T(xy) = (Tx)(Ty)$ in the Reiersöl algebra $\mathcal{A}(L; r)$. $\square$

The genetical meaning of this lemma is that the recombination and mutation processes (independent at each locus) make independent contributions to evolution. The evolutionary equation with mutation taken into consideration has the form $p' = (Tp)^2$ with the assumption that first mutation, then fertilization occurs, and finally by meiosis the new gamete state is produced. But by Lemma 6.7.1 we can write $p' = T(p^2)$ placing the mutation process after fertilization and meiosis, i.e. the two processes commute. Thus in ith generation

$$p_t = T^t p^{[t]}, \qquad (6.7.2)$$

where $p^{[t]}$ is the state in the tth generation with purely recombination process $(p^{[t+1]} = (p^{[t]}))^2$ defined by the same initial state p_0 that the combined process. As in the one locus case (see Section 2.5) there exists $T^\infty = \lim_{t\to\infty} T^t$, We have proved the following theorem.

Theorem 6.7.2 *During the combination of recombination and mutation processes, each trajectory converges to some equilibrium state.*

By (6.7.2) and (6.3.4) this limit state is $p_\infty = T^\infty(\prod_{i=1}^l F^i(p_0)) = \prod_{i=1}^l F^i(T^\infty p_0)$. Only the states of this form are equilibria. The set of equilibrium states still does not depend on the linkage distribution and is a connected semi-algebraic submanifold of the simplex. Its dimension does not exceed rank $T^\infty - 1$. In particular, if rank $T^\infty = 1$ then it contains a single point. The latter condition is satisfied if all the matrices $T^{(i)}$ are indecomposable. The genetical meaning of this is that in every locus for every pair of alleles A and a the mutation $A \to a$ occurs with positive probability either directly or through some chain of mutations $A \to \cdots \to a$.

As can be seen from (6.7.2) the stabilization coefficient in the combined process is $\max(\kappa, \rho)$, where κ is the stabilization coefficient of recombination process, $\rho = \rho_T$ is the largest absolute value of non-1 eigenvalues of matrix T. Obviously $\rho_T = \max(\rho_{T^{(1)}} \ldots \rho_{T^{(l)}})$.

6.8 Appendix. Recombination Among Sex Chromosomes

Let $L = \{1, \ldots, l\}$ be a set of loci of the X-chromosome. We assume that a part of them, the subset $L_{XY} = \{k+1, \ldots, l\}$ for concreteness, belongs also to the Y-chromosome and $L_X = \{1, \ldots, k\}$ is the set of purely X loci. While a typical X gamete, denoted g or h, lies in the product $\Gamma = \prod_{i=1}^l \Gamma_i$ as before, a typical Y gamete, denoted y, lies in the subproduct $\Gamma^Y = \prod_{i=k+1}^l \Gamma_i$. We denote the unit simplices in $\mathbf{R}(\Gamma) \equiv \mathbf{R}(\Gamma_X)$ and $\mathbf{R}(\Gamma_Y)$ by Δ (or Δ_X) and

Δ_Y respectively. The state of the gene pool is now a triple $P = (p, q_X, q_Y)$ in $\Delta \times \Delta_X \times \Delta_Y$ with p and q_X the distribution of X gametes produced by the females and the males respectively, and q_Y the distribution of Y gametes (produced only by males).

The statistics of recombination for females is described by female linkage distribution $r(U|V)$ defined, as in the autosomal situation, for all partitions $U|V$ defined on the set L of all loci. For males we define male linkage distribution $\rho(u, v)$ on the set of ordered pairs (u, v), where u is a subset of L_{XY} and v is its complement. For an XY gamete pair in the male, gy, a (u, v) recombination event produces X gametes of the form $g_{L_X \cup v} y_u$ and Y gametes of the form $g_u y_v$. Contrast this with a (v, u) recombination event. In general case $\rho(u, v) \neq \rho(v, u)$. As before we assume that the probability $\rho(u, v)$ does not depend on genes linked among themselves or with X or Y chromosome. It is the case $\rho_1 = \rho_2$ in terms of Section 2.3.

We give the evolutionary equations at gamete level

$$p' = \sum_{U|V} r(U|V)\frac{1}{2}(F^U(p)F^V(q_X) + F^U(q_X)F^V(p)),$$

$$q'_X = \sum_u \rho(u, v)F^{L_X \cup v}(p)F^u(q_Y), \qquad (6.8.1)$$

$$q'_Y = \sum_u \rho(u, v)F^u(p)F^v(q_Y).$$

The equations (6.8.1) can be written as a single expression $\gamma' = \gamma^2$ in terms of an algebra on the space $\mathbf{R}(\Gamma) \times \mathbf{R}(\Gamma_X) \times \mathbf{R}(\Gamma_Y)$ with multiplication table

$$(\gamma \times \tilde{\gamma})_\Gamma = \sum_{U|V} r(U|V)\frac{1}{4}(g_U \tilde{h}_V + \tilde{g}_U h_V + h_U \tilde{g}_V + \tilde{h}_U g_V),$$

$$(\gamma \times \tilde{\gamma})_{\Gamma_X} = \sum_u \rho(u,v)\frac{1}{2}(g_{L_X \cup v}\tilde{y}_u + \tilde{g}_{L_X \cup v} y_u), \qquad (6.8.2)$$

$$(\gamma \times \tilde{\gamma})_{\Gamma_Y} = \sum_u \rho(u,v)\frac{1}{2}(g_u \tilde{y}_v + \tilde{g}_u y_v),$$

where $\gamma = (g,h,y)$ and $\tilde{\gamma} = (\tilde{g},\tilde{h},\tilde{y})$ are gamete triples. Observe that we have symmetrized the three formulae to get a commutative multiplication. We will denote the algebra (6.8.2) by $\mathcal{A}(L; r, \rho)$.

If $K \subset L$ then the *induced linkage distribution* r_K is defined and if $K_{XY} = K \cap L_{XY} \neq \emptyset$ then ρ_K is also defined. The naturally defined operator $F^K : \mathcal{A}(L; r, \rho) \to \mathcal{A}(K; r_K, \rho_K)$ ($\to \mathcal{A}(K; r_K)$ if $K_{XY} = \emptyset$) is an algebra homomorphism.

Just as in (6.2.19) we isolate in equations (6.8.1) the summands corresponding to the trivial partition $\emptyset | L$ or trivial pairs $(\emptyset, L_{XY})$, $(L_{XY}, \emptyset)$. We have

$$p' = (r/2)(p + q_X) + \sum_{U|V \neq \emptyset|L} \cdots,$$

$$q'_X = \rho p + \tilde{\rho}F^{L_X}(p)q_Y + \sum_{(u,v)\neq\emptyset,L_{YX},u\neq\emptyset} \cdots, \qquad (6.8.3)$$

$$q'_Y = \tilde{\rho}F^{L_{XY}}(p) + \rho q_Y + \sum_{u \neq L_{XY}} \cdots,$$

where $r = r(\emptyset|L)$, $\rho = \rho(\emptyset, L_{XY})$, $\tilde{\rho} = \rho(L_{XY},\emptyset)$, and $\cdots$ are the terms of the same form as in the original equations corresponding to nontrivial pairs (partitions). In these terms all projection operators act non-trivially, mapping into algebras with fewer loci. These operators, being algebra homomorphisms, map trajectories to trajectories. For one locus trajectories converge, as shown in Section 2.3. Therefore we can assume inductively that trajectories converge when the number of loci is less than l. But the isolated part in evolutionary equations contains non-linearity $F^{L_X}(p)q_Y$. To overcome this we apply F^{L_X} to get $(F^{L_X}(p), F^{L_X}(q^X), 1)$ in the algebra without XY-loci. Thus we consider the evolution in the system defined only by X-loci

$$p' = (r/2)(p + q_X) + \sum_{U|V \neq \emptyset|L} \cdots,$$

$$(q_X)' = p. \qquad (6.8.4)$$

In this case the isolated part is obviously linear and is obtained by action on the state $P = (p, q^X)$ by the matrix

$$S = \begin{pmatrix} \frac{r}{2} & \frac{r}{2} \\ 1 & 0 \end{pmatrix} \otimes I_l$$

where I_l is the unit matrix of the order l. A trajectory $\{P_t\}_{t=0}^{\infty}$ satisfies $P_{t+1} = SP_t + H_t$. Here $\{H_t\}_{t=0}^{\infty}$ converges by the induction hypothesis. This implies that the trajectory converges, for the eigenvalues of S,

$$\lambda_{1,2} = \frac{r \pm \sqrt{r^2 + 8r}}{4},$$

lie in the disk $|\lambda| < 1$ (provided that $r < 1$, i.e. the loci are not completely linked; when $r = 1$, (6.8.4) contains no non-linear terms and the situation simplifies to single locus, i.e. the trajectories still converge).

Returning to the general case, when XY-loci are present, we can, as shown above, write $F^{Lx}(p_t) = (F^{Lx}(p_0))_t$ in the form $\bar{p} + o(1)$, where $\bar{p} = \lim_{t\to\infty}(F^{Lx}(p_0))_t$, and then by (6.8.3) for trajectory $\{P_t\}_{t=0}^{\infty}$ we have the equation

$$P_{t+1} = TP_t + H_t \tag{6.8.5}$$

where

$$T = \begin{pmatrix} \frac{r}{2}I_l & \frac{r}{2}I_l & 0 \\ \rho I_l & 0 & \tilde{\rho}\bar{p} \\ \tilde{\rho}F^{Lxy} & 0 & \rho \end{pmatrix},$$

and $\{H_t\}_{t=0}^{\infty}$ converges. We investigate the spectrum of matrix T to prove by (6.8.5) that $\{P_t\}_{t=0}^{\infty}$ also converges.

Let (p, q_X, q_Y) be an eigenvector (not necessarily a state) corresponding to eigenvalue λ:

$$\begin{aligned} (r/2)p + (r/2)q_X &= \lambda p, \\ \rho p + \tilde{\rho}\bar{p}q_Y &= \lambda q_X, \\ \tilde{\rho}F^{Lxy}(p) + \rho q_Y &= \lambda q_Y. \end{aligned}$$

Applying operator F^{Lxy} to the first two equations we see that the vector $(F^{Lxy}(p), F^{Lxy}(q_X), q_Y)$ is an eigenvector (with the same eigenvalue λ) for the matrix

$$T_Y = \begin{pmatrix} r/2 & r/2 & 0 \\ \rho & 0 & \tilde{\rho} \\ \tilde{\rho} & 0 & \rho \end{pmatrix} \otimes I_{l-k} \equiv \tau \otimes I_{l-k}.$$

We denote the spectral radius of matrix τ (and hence T_Y) by R. The row norm of this matrix is $||\tau|| = \max(r, \rho + \tilde{\rho}) \leq 1$, and $R \leq ||\tau||$. We show that the eigenvalues of matrix τ always lie in the disk $|\lambda| < 1$. This will be the case first if $||\tau|| < 1$. Further, if $||\tau|| = 1$ then if $r > 0$, $\tilde{\rho} > 0$ then the matrix is indecomposable, and then by Taussky's theorem $R < 1$ if $\min(r, \rho + \tilde{\rho}) < 1$. It remains to consider the cases when $||\tau|| = 1$ and 1) $r = 0$, 2) $\tilde{\rho} = 0$, 3) $r > 0$, $\tilde{\rho} > 0$, $\min(r, \rho + \tilde{\rho}) = 1$, i.e. $r = 1$, $\tilde{\rho} > 0$, $\rho = 1 - \tilde{\rho}$. Correspondingly 1) $R = \rho$, 2) $R = \max(\rho, (r \pm \sqrt{r^2 + 8r\rho})/4)$, 3) $R = 1$. Hence, if $R = 1$ then $\tilde{\rho} = 0$, $\rho = 1$ or $r = 1$, $\tilde{\rho} > 0$, $\rho = 1 - \tilde{\rho}$.

(6.8.5) implies $P_t = \sum_{i=0}^{t-1} T^{t-i-1} H_i + T^t P_0$. If $R < 1$ then there exists $P_\infty \equiv \lim_{t \to \infty} P_t = (1 - T)^{-1} H_\infty$. The case $R = 1$ requires special consideration.

If $r = 1$, $\tilde{\rho} > 0$, $\rho = 1 - \tilde{\rho}$ then the equations (6.8.3) contains no nonlinear terms other than the isolated ones. (X-loci are completely linked, XY-loci are also completely linked). Therefore we have

$$
\begin{aligned}
p' &= \frac{1}{2}(p + q_X), \\
q'_X &= (1 - \tilde{\rho})p + \tilde{\rho} F^{L_X}(p) q_Y, \\
q'_Y &= \tilde{\rho} F^{L_{XY}}(p) + (1 - \tilde{\rho}) q_Y,
\end{aligned}
$$

from which

$$
\begin{aligned}
(F^{L_{XY}}(p))' &= \frac{1}{2}(F^{L_{XY}}(p) + F^{L_{XY}}(q_X)), \\
(F^{L_{XY}}(q_X))' &= (1 - \tilde{\rho})F^{L_{XY}}(p) + \tilde{\rho} q_Y \\
q'_Y &= \tilde{\rho} F^{L_{XY}}(p) + (1 - \tilde{\rho}) q_Y.
\end{aligned}
$$

This is a linear system with the matrix $T_0 \otimes I_{l-k}$ where

$$
T_0 = \begin{pmatrix} \frac{1}{2} & \frac{1}{2} & 0 \\ 1 - \tilde{\rho} & 0 & \tilde{\rho} \\ \tilde{\rho} & 0 & 1 - \tilde{\rho} \end{pmatrix}
$$

is the same as in section 2.3, where it is shown that T_0^t has a limit as $t \to \infty$. So the sequence of vectors $\{F^{L_{XY}}(P_t)\}_{t=0}^{\infty}$ converges. On the other hand $\{F^{L_X}(P_t)\}_{t=0}^{\infty}$ converges. Therefore $\{P_t\}_{t=0}^{\infty}$ also converges.

If $\tilde{\rho} = 0$, $\tilde{\rho} = 1$, then $(q_X)' = p$, $(q_Y)' = q_Y$. Thus $((q_Y)_t)$ is constant and $(p_t, (q_X)_t)$ satisfies (6.8.4) and converges. We have proved the following theorem.

Theorem 6.8.1 *All trajectories of the recombination process for the sex chromosomes converge.*

By the same technique as in autosomal case one can estimate the rate of convergence to equilibrium. Moreover, one can obtain an explicit evolutionary formula, but it will be even more complicated than in the autosomal case. We will not consider these questions.

Chapter 7

Evolution in Genetic Algebras

7.1 Reiersöl Algebras are Genetic

To prove the assertion in the title, we will use induction on l, the number of loci. When $l = 1$, the assertion is trivial, for the algebra is unit. Let $l > 1$, and suppose that all the Reiersöl algebras with fewer than l loci are genetic. Lemma 6.2.1 implies that the operator F^K is a baric algebra homomorphism $\mathcal{A}(L;r)$ onto $\mathcal{A}(K;r_K)$. If $K \neq L$ then the algebra $\mathcal{A}(K;r_K)$ is genetic by induction. Thus it satisfies Holgate's condition, i.e its barideal is a principal nilalgebra and its Lie hull is solvable. Therefore, if $s(p) = 0$ then $(F^K(p))^{N_K} = 0$ for some exponent N_K. Put $N = \max_K N_K$. Then $(F^K(p))^N = 0$ for all $K \neq L$, i.e. by Lemma 6.2.1 $F^K(p^N) = 0$ $(K \neq L)$, but then, by equation (6.2.16) (and $F^\emptyset = s$), $p^{N+1} = 0$.

Note further that for any algebra homomorphism $h : \mathcal{A}_1 \to \mathcal{A}_2$ there is the following connection between the operators of multiplication in $\mathcal{A}_1$ and in $\mathcal{A}_2$:

$$hM_x = M_{hx}h. \tag{7.1.1}$$

Thus if the Lie $\mathcal{A}_2$ is solvable then some derived subalgebra of the Lie $\mathcal{A}_1$ lies in the annihilator of the homomorphism f, i.e. all of all the operators in it map $\mathcal{A}_1$ into $\ker f$. For the Reiersöl algebra $\mathcal{A}(L;r)$ this means by induction hypothesis that the images of the operators in a derived subalgebra of the Lie $\mathcal{A}(L;r)$ lie in the subspace $Q = \cap_{K \neq L} \ker F^K$. But if $q \in Q$ then it follows from (6.2.16) that $pq = \frac{1}{2}r(L)s(p)q$ for all p, i.e. the operators $M_p|Q$ are scalar. Hence the next derived of Lie $\mathcal{A}(L;r)$ is equal to zero.

Corollary 7.1.1 *The algebra of gamete pairs is genetic.*

Proof By Lemma 6.2.3 it is the duplication of the Reiersöl algebra, and by Theorem 3.6.5, the duplication of a genetic algebra is genetic. $\square$

Corollary 7.1.2 *An algebra of zygotes is genetic.*

Proof It is a baric quotient algebra of the algebra of gamete pairs. $\square$
Now we will compute the train roots of the Reiersöl algebra.

Theorem 7.1.3 *The set of the train roots of the Reiersöl algebra $\mathcal{A}(L;r)$ is*

$$\{1,\frac{1}{2},\frac{1}{2}r(K), K \subset L, |K| \geq 2\}. \tag{7.1.2}$$

Hence the train roots different from 1 and $\frac{1}{2}$ are the evolutionary roots of the first degree reduced by half (and $\frac{1}{2}$ is half of the evolutionary root of degree zero).

Proof When $l = 1$ the assertion is trivial. We will use induction on l. Suppose $\mu \neq 1$ is a train root of algebra $\mathcal{A}(L;r)$. Then for any p, $s(p) = 1$, there exists a non-trivial solution q of the equation $pq = \mu q$ in the barideal $s(q) = 0$. Consider the alternatives:

1. $F^W(q) = 0$ for all $W \neq L$. Then $pq = \frac{1}{2}r(L)q$, hence $\mu = \frac{1}{2}r(L)$.

2. There exists $W \neq L$ for which $F^W(q) \neq 0$. Moreover, $W \neq \emptyset$, for $s(q) = 0$. By Lemma 6.2.1, $F^W(p)F^W(q) = \mu F^W(q)$, and moreover $s(F^W(p)) = 1$, $s(F^W(q)) = 0$. Thus μ is a train root in the barideal of the algebra $\mathcal{A}(L_W;r_W)$ with fewer loci. But then $\mu = \frac{1}{2}$ or $\mu = \frac{1}{2}r_W(K)$ for some $K \subset W$, $|K| \geq 2$, i.e. $\mu = \frac{1}{2}r(K)$ by Lemma 6.1.1.

Thus the set of the train roots is contained in the set (7.1.2). We will now show that all the elements of the set (7.1.2) are train roots.

As we have seen in the proof of Lemma 6.5.3, the system of equations $F^W(q) = 0$, $(W \neq L)$, has a nontrivial solution q. For this solution, alternative 1 above gives the train root $\mu = \frac{1}{2}r(L)$. Let $K \subset L$, $K \neq L$. Then the algebra $\mathcal{A}(K;r_K)$ contains an element $q_K \neq 0$ such that $F^W(q_K) = 0$ $(W \subset K, W \neq K)$. Take any product element $p \in \mathcal{A}(L;r)$, $s(p) = 1$ (i.e. an equilibrium by Theorem 6.3.1), and put $q = q_K F^{L\setminus K}(p)$. Then if both classes of the partition $U|V$ of L intersect K then $F^U(q) = 0$ and $F^V(q) = 0$. Therefore the only contributions to pq come from those partitions in which one class does not intersect K and hence the second class contains $K : pq = \sum_{K \subset V} r(U|V)\frac{1}{2}(F^U(p)F^V(q) + F^U(q)F^V(p))$. Moreover, $F^U(q) = 0$, so in fact

$$pq = \frac{1}{2} \sum_{K \subset V} r(U|V)F^U(p)F^V(q). \qquad (7.1.3)$$

But $F^V(q) = q_K F^{V \setminus K}(p)$. Thus in all the addends of (7.1.3) $F^U(p)F^V(q) = q_K F^{L \setminus K}(p) = q$ and $pq = (\frac{1}{2}\sum_{K \subset V} r(U|V))q = \frac{1}{2}r(K)q$. Since obviously $q \neq 0$, $\mu = \frac{1}{2}r(K)$ is a train root. Finally, to obtain the train root $\mu = \frac{1}{2}$, consider arbitrary $q^1,\ldots,q^l$ $(s(q^i) = 0, 1 \leq i \leq l)$ in the algebras corresponding to separate loci and let $q = \sum_{i=1}^{l} F^{L \setminus i}(p)q^i$ for any p, $s(p) = 1$. Then for any partition $U|V$, $F^V(q) = \sum_{i \in V} F^V(F^{L \setminus i}(p))q^i$ and $F^U(p)F^V(q) = \sum_{i \in V} F^{L \setminus i}(p)q^i$. Transposing U and V we get $F^V(p)F^U(q) = \sum_{i \in U} F^{L \setminus i}(p)q^i$. Hence $F^U(p)F^V(q) + F^V(p)F^U(q) = q$ and therefore $pq = \frac{1}{2}q$. $\square$

7.2 Idempotents. Convergence of Trajectories

The problems raised in the title are simplified for a genetic algebra because of the triangular form of the evolutionary operator in its eigenbasis. We will call the self-map $\tilde{V}x = x^2$ of the unit hyperplane the *evolutionary operator* in a baric algebra $(\mathcal{A}, \sigma)$. If the algebra is stochastic then the evolutionary operator V in the usual sense is a restriction of the operator $\tilde{V}$ to the basic body. Focusing upon the unit hyperplane makes sense because of the following lemma.

Lemma 7.2.1 *In any train algebra $(\mathcal{A}, \sigma)$ the non-zero idempotents lie in the unit hyperplane.*

Proof If $x^2 = x$ and $\sigma(x) \neq 1$ then $\sigma(x) = 0$, but then the operator M_x is nilpotent. Therefore $x^{n+1} = 0$ which implies that $x = 0$. $\square$

Below $(\mathcal{A}, \sigma)$ will denote a genetic algebra, $v_0, v_1, \ldots, v_{n-1}$ will denote its eigenbasis (generally speaking, it lies in the complexification $(\mathcal{A}^{(c)}, \sigma^{(c)})$). The evolutionary operator expressed in this basis has the form

$$\xi_k' = \lambda_{00,k} + 2\sum_{i \leq k} \lambda_{0i,k}\xi_i + \sum_{1 \leq j, i < k} \lambda_{ji,k}\xi_i\xi_j \quad (1 \leq k \leq n-1, \ \xi_0' = \xi_0 = 1),$$

i.e.

$$\xi_k' = 2\tau_k\xi_k + f_k(\xi_1,\ldots,\xi_{k-1}) \qquad (1 \leq k < n), \qquad (7.2.1)$$

where $\tau_1, \ldots, \tau_{n-1}$ are the train roots in the barideal and

$$f_k(\xi_1, \ldots, \xi_{k-1}) = \lambda_{00,k} + 2 \sum_{i<k} \lambda_{0i,k} \xi_i + \sum_{1 \leq j,i < k} \lambda_{ji,k} \xi_i \xi_j \quad (1 \leq k < n) \quad (7.2.2)$$

are polynomials of degree not higher than two. These polynomials can be assumed to be real if all the train roots are real, since in this case the eigenbasis can be chosen to be real. In the future, we will assume that either all the train roots are real, or all the objects of interest (idempotents, trajectories, and so on) are complex, otherwise the formulations of the theorems would not be convenient, though the techniques still would work.

The system of equations of the manifold of non-zero idempotents has the triangular form

$$\xi_k = 2\tau_k \xi_k + f_k(\xi_1, \ldots, \xi_{k-1}) \qquad (1 \leq k \leq n - 1). \qquad (7.2.3)$$

It can be solved recursively, if all the train roots are different from $\frac{1}{2}$: $\xi_k = f_k(\xi_1, \ldots, \xi_{k-1})/(1 - 2\tau_k)$. Thus the following theorem holds.

Theorem 7.2.2 *If all the train roots of a genetic algebra are different from $\frac{1}{2}$, then it contains a unique non-zero idempotent (real, if all the train roots are real, complex in the general case).*

Theorem 7.2.3 *The dimension of the manifold of non-zero idempotents in a genetic algebra does not exceed the number of train roots equal to $\frac{1}{2}$.*

Proof Let some train roots be $\frac{1}{2}$, say $\tau_{k_1}, \ldots, \tau_{k_r}$. Then the system (7.2.3) uniquely defines the coordinates $\xi_1, \ldots \xi_{k_1-1}$ (if $k_1 > 1$). If moreover $f_{k_1}(\xi_1, \ldots, \xi_{k_1-1}) \neq 0$ (when $k_1 = 1$ this inequality becomes $\lambda_{00,1} \neq 0$ since $f_1 = \lambda_{00,1}$ is constant) then the system is incompatible, i.e., there are no non-zero idempotents. On the other hand if $f_{k_1}(\xi_1, \ldots, \xi_{k_1-1}) = 0$ then we choose an arbitrary ξ_{k_1} and thus uniquely define $(\xi_{k_1+1}, \ldots, \xi_{k_2-1})$. These are obviously polynomials in ξ_{k_1}. If a polynomial $\phi(\xi_{k_1}) = f_{k_2}(\xi_1, \ldots, \xi_{k_1-1}, \xi_{k_1}, \ldots, \xi_{k_2-1})$ is not identically zero, then by setting it to zero we get a finite (possible empty) set of complex (real) values of ξ_{k_1} for which the equation can be solved for ξ_{k_2}, and moreover for each of them the value of ξ_{k_2} is arbitrary at this stage. But if $\phi(\xi_{k_1}) \equiv 0$ then ξ_{k_1} and ξ_{k_2} are arbitrary at this stage. Continuing this argument, we complete the proof. $\square$

The conditions sufficient to achieve this estimate are given by he following theorem.

Theorem 7.2.4 *Suppose that in a genetic algebra $(\mathcal{A}, \sigma)$ the train roots τ_k, $1 \leq k \leq n-1$, are real with $\tau_{k_1} = \cdots = \tau_{k_r} = \frac{1}{2}$ and the other $\tau_k \neq \frac{1}{2}$. Let $v_0, v_1, \ldots, v_{k-1}$ be the (real) eigenbasis and $E = [v_{k_1}, \ldots, v_{k_r}]$. If all the vectors $v_j v_i$ ($0 \leq j, i \leq k_p - 1; p = 1, \ldots r$), are orthogonal to E with respect to the inner product making the given basis orthonormal, then $\dim(\mathrm{Id}_1 \, \mathcal{A}) = r$ and the manifold of non-zero idempotents is the graph of a polynomial map.*

Proof The condition that all $v_j v_i$ are orthogonal to E means in terms of structure constants that $\lambda_{ji,k_p} = 0$ for $0 \leq j, i < k_p; p = 1, \ldots, r$, i.e. all the polynomials $f_{k_p} = 0$, $1 \leq p \leq r$. But then the system (7.2.3) is solvable. The unknowns $\xi_{k_1}, \ldots, \xi_{k_r}$ are free and the rest are polynomials in them. $\square$

We note that the bound in Theorem 7.2.3 is achieved for Bernstein genetic algebras by Theorem 3.5.19 and Corollary 3.4.9.

Now we consider the question of convergence of trajectories in genetic algebras. The evolutionary equation here has the form $x_{t+1} = \tilde{V}_{x_{t+1}} = x_t^2$ and $\sigma(x_0) = 1$ and so $\sigma(x_t) = 1$ for all t. If the trajectory converges then its limit is a non-zero idempotent. Hence if an algebra has no non-zero idempotents then none of the trajectories converge.

Example Consider a 2-dimensional genetic algebra with eigenbasis v_0, v_1 and multiplication table $v_0^2 = v_0 + \lambda v_1$, $v_0 v_1 = \frac{1}{2} v_1$, $v_1^2 = 0$. If $\sigma(x) = 1$ then $x = v_0 + \xi v_1$ and $x^2 = v_0 + (\xi + \lambda) v_1$. When $\lambda \neq 0$ the algebra has no non-zero idempotents, and a trajectory starting at x has the form $x_t = v_0 + (\xi + t\lambda) v_1$ and obviously diverges.

The sufficiency condition for convergence of all trajectories is given by the following theorem.

Theorem 7.2.5 *If all the train roots of a genetic algebra for the barideal lie in the circle $|\tau| < \frac{1}{2}$ then all the trajectories converge.*

Proof Let $\{x_t\}_{t=0}^{\infty}$ be a trajectory. Then in eigencoordinates by (7.2.1)

$$\xi_{t+1,k} = 2\tau_k \xi_{t,k} + f_k(\xi_{t,1}, \ldots, \xi_{t,k-1}) \quad (t = 0, 1, \ldots; 1 \leq k < n). \qquad (7.2.4)$$

The sequence $\{\xi_{t,k}\}_{t=0}^{\infty}$ converges if all sequence $\{\xi_{t,1}\}_{t=0}^{\infty}, \ldots, \{\xi_{t,k-1}\}_{t=0}^{\infty}$ converge (see appendix to chapter 2). Furthermore, $\{\xi_{t,1}\}_{t=0}^{\infty}$ converges because f_1 is constant. So we can use induction on k. $\square$

There is even a necessary and sufficient condition for convergence of all trajectories. Consider the real case for concreteness.

Theorem 7.2.6 *Suppose all the train roots of a genetic algebra are real. All trajectories converge if and only if all train roots different from $\frac{1}{2}$ lie in the* *circle $|\tau| < \frac{1}{2}$ and the dimension of the manifold of non-zero idempotents is the same as the number of train roots equal to $\frac{1}{2}$.*

Proof Suppose all the trajectories converge. Consider the family of trajectories arising by fixing the initial values $\xi_{0,1}, \ldots, \xi_{0,k-1}$. These determine the limiting value $\xi_{\infty,1}, \ldots, \xi_{\infty,k-1}$. The convergence for all $\xi_{0,k}$ implies that $|\tau_k| < \frac{1}{2}$ or $\tau_k = \frac{1}{2}$. In the former case $\xi_{\infty,k}$ is determined as well, while in thelatter $f_k(\xi_{\infty,1}, \ldots, \xi_{\infty,k-1}) = 0$, i.e. f_k becomes zero on all solutions of the system

$$\xi_j = 2\tau_j\xi_j + f_j(\xi_1, \ldots, \xi_{j-1}) \quad (1 \le j < k) \tag{7.2.5}$$

($f_1 = 0$ for $k = 0$). This means that in the process of solving the complete system of equations (7.2.3) the unknowns $\xi_{k_1}, \ldots, \xi_{k_r}$, corresponding to train roots equal to $\frac{1}{2}$, remain free. Hence $\dim(\mathrm{Id}_1 \, \mathcal{A}) = r$.

Conversely, suppose these necessary conditions hold. Then for any k_p the polynomial f_{k_p} becomes zero on all solutions of system (7.2.5). If $k_1 = 1$, this implies convergence of the sequence $\{\xi_{t,1}\}_{t=0}^{\infty}$ which corresponds to the case $\tau_1 = \frac{1}{2}$. If $k_1 > 1$ then $|\tau_1| < \frac{1}{2}$ and $\{\xi_{t,1}\}_{t=0}^{\infty}$ still converges. In both cases the rate of convergence is exponential.

We will use induction on k supposing that $\{\xi_{t,1}\}_{t=0}^{\infty}, \ldots, \{\xi_{t,k-1}\}_{t=0}^{\infty}$ converge exponentially fast. Then $\{f_k(\xi_{t,1}, \ldots, \xi_{t,k-1})\}_{t=0}^{\infty}$ converges and its limit is equal to zero if $\tau_k = \frac{1}{2}$, i.e $k = k_p$ for some p. Moreover in the latter case $\sum_{t=0}^{\infty} f_k(\xi_{t,1}, \ldots, \xi_{t,k-1})$ converges. But then $\{\xi_{t,k}\}_{t=0}^{\infty}$ also converges at an exponential rate. $\square$

Corollary 7.2.7 *A Reiersöl algebra $\mathcal{A}(L,r)$ with l loci and m total number of genes has $m - l$ train roots equal to $\frac{1}{2}$.*

Proof In this algebra all train roots are real by Theorem 7.1.3, all trajectories converge by Theorem 6.3.4, and the dimension of manifold of non-zero real idempotents is $m - l$ by Corollary 6.3.2. $\square$

7.3 Exact Linearization. Evolutionary Spectrum

Theorem 6.4.1 about exact linearization of the evolutionary equation in Reiersöl algebra is easily generalized for any genetic algebra.

Theorem 7.3.1 *To every genetic algebra $(\mathcal{A}, \sigma)$ corresponds a polynomial $\Phi(\lambda)$ such that all trajectories satisfy the linear equations $[\Phi(T)x]_t = 0$.*

Proof We use induction on dimension. In a 2-dimensional algebra we have in eigencoordinates one scalar equation $\xi_{t_1,1} = 2\tau_1\xi_{t,1} + f_1$, where f_1 is constant. The desired polynomial has the form $\Phi_1(\lambda) = (\lambda - 1)(\lambda - 2\tau_1)$. (When $f_1 = 0$ the multiplicand $\lambda - 1$ is superfluous but it will be useful later.)

Stepping from $n - 1$ to n we use the fact that if $v_0, v_1, \ldots, v_n$ is an eigenbasis then $I_{n-1} = [v_{n-1}]$ is a baric ideal (because $v_0 v_{n-1} = \tau_{n-1} v_{n-1}$, $v_i v_{n-1} = 0$ $(i > 0)$) and in baric quotient algebra $\mathcal{U}(\sigma)/I_{n-1}$ the eigenbasis is formed by images of vectors $v_0, v_1, \ldots, v_{n-2}$. If $\xi_0, \ldots, \xi_{n-1}$ are the coordinates of a vector $x \in \mathcal{A}$ then $\xi_0, \ldots, \xi_{n-2}$ are the coordinates of its image $\tilde{x}$ in the quotient algebra. The trajectories in the algebra $(\mathcal{A}, \sigma)$ are mapped into trajectories in the quotient algebra by the natural homomorphism.

Let $\{x_t\}$ be a trajectory in $(\mathcal{A}, \sigma)$, i.e. equations (7.2.4) hold. By the induction hypothesis, the factored trajectory $\{\tilde{x}_t\}$ satisfies a linear equation $[\Phi_{n-1}(T)\tilde{x}]_t = 0$, and we can assume that $\lambda - 1$ divides the polynomial $\Phi_{n-1}(\lambda)$. Therefore each sequence of coordinates $\{\xi_{t,i}\}_{t=0}^\infty$, $1 \leq i \leq n - 2$, satisfies this equation. Also notice that $\xi_{t+1,n-1} = 2\tau_{n-1}\xi_{\tau,n-1} + f_{n-1}(\xi_{t,1}, \ldots, \xi_{t,n-2})$. Because of (7.2.2), the sequence

$$\eta_t = f_{n-1}(\xi_{t,1,\ldots,\xi_{t,n-2}}), \tag{7.3.1}$$

is annihilated by the quasitensor square $\Phi_{n-1} \diamond \Phi_{n-1}$. It is essential that $\lambda = 1$ is a root of Φ_{n-1}, and hence of $\Phi_{n-1} \diamond \Phi_{n-1}$, and also Φ_{n-1} divides $\Phi_{n-1} \diamond \Phi_{n-1}$. Therefore $\Phi_{n-1} \diamond \Phi_{n-1}$ annihilates not only quadratic, but also linear terms in (7.3.1). It remains to let $\Phi_n(\lambda) = (\lambda - 2\tau_{n-1})\Phi_{n-1}(\lambda) \diamond \Phi_{n-1}(\lambda)$, for the smallest possible h. $\square$

The polynomial Φ_n constructed in the proof is determined uniquely by the genetic algebra $(\mathcal{A}, \sigma)$, and depends only on its train roots. It may happen that in a particular case there exists a polynomial of a lesser degree annihilating all the trajectories. We will call the polynomial Φ_n the *maximal evolutionary polynomial* of a genetic algebra; the set of its roots will be called the *maximal evolutionary spectrum*. A minimal (with respect to the degree) polynomial annihilating all the trajectories will be called the *minimal evolutionary polynomial* and the set of its roots the *minimal evolutionary spectrum*. The polynomials annihilating all the trajectories form an ideal in the ring of polynomials. Hence they are multiples of the minimal evolutionary polynomial, and the latter is unique up to the constant factor and therefore can be taken to be monic. The maximal evolutionary polynomial is divisible by the minimal one, and as the result, the minimal evolutionary spectrum is contained in the maximal one. Let us describe the maximal evolutionary spectrum and then the minimal one.

Consider the decreasing sequence of baric ideals $I_k = [v_k, \ldots, v_{n-1}]$, $1 \leq k \leq n-1$, $I_n = 0$ and the corresponding increasing sequence of quotient algebras $(\mathcal{A}_k, \sigma_k) = (\mathcal{A}, \sigma)/I_k$, $1 \leq k \leq n$. We will denote the maximal evolutionary polynomial of the quotient algebra $(\mathcal{A}_k, \sigma_k)$ by $\Phi_k(\lambda)$, as in the proof of Theorem 7.3.1. The maximal evolutionary spectrum will be denoted by W_k. Because of the recursive relation $\Phi_k(\lambda) = (\lambda - 2\tau_k)\Phi_{k-1}(\lambda) \diamond \Phi_{k-1}\lambda)$, $2 \leq k \leq n-1$, $\Phi_1(\lambda) = \lambda - 1$, we have

$$W_k = \{2\tau_k\} \cup W_{k-1}^2 \ (2 \leq k \leq n-1), \ W_1 = \{1\}.$$

Theorem 7.3.2 *The maximal evolutionary spectrum W_k of genetic algebra $(\mathcal{A}, \sigma)$ with the train roots $1, \tau_1, \ldots, \tau_{n-1}$, $n \geq 2$ coincides with the set of numbers of the form $\lambda_{\epsilon_1, \epsilon_2, \ldots, \epsilon_{n-1}} = (2\tau_1)^{\epsilon_1}(2\tau_2)^{\epsilon_2} \cdots (2\tau_{n-1})^{\epsilon_{n-1}}$, where $\epsilon_1, \epsilon_2, \ldots, \epsilon_{n-1}$ are non-negative integers satisfying the inequality $\|\epsilon\| \equiv \epsilon_1 + 2\epsilon_2 + \cdots + 2^{n-2}\epsilon_{n-1} \leq 2^{n-2}$.*

For the quotient algebra $(\mathcal{A}_k, \sigma_k)$, train roots are $1, \tau_1, \ldots, \tau_{k-1}$. Because for $(\mathcal{A}_2, \sigma_2)$ the assertion clearly holds and $(\mathcal{A}_n, \sigma_n) = (\mathcal{A}, \sigma)$, we can use induction on k (replacing n by k in the statement of the theorem). But $\lambda_{\epsilon_1, \ldots, \epsilon_{k-1}} \lambda_{\eta_1, \ldots, \eta_{k-1}} = \lambda_{\epsilon_1 + \eta_1, \ldots, \epsilon_{k-1} + \eta_{k-1}, 0}$ for $\|\epsilon + \eta\| \leq 2^{k-2} + 2^{k-2} = 2^{k-1}$. Now to see $\subset$ we note that $2\tau_k = \lambda_{0, \ldots, 0, 1}$.

To prove $\supset$ we note that if $\omega_1 + 2\omega_2 \cdots + 2^{k-1}\omega_k \leq 2^{k-1}$ and $\omega_k \geq 1$ then $\omega_k = 1$ and $\omega_i = 0$ for $1 \leq i \leq k-1$, i.e. in this case $\lambda_{\omega_1, \ldots \omega_k} = 2\tau_k$. Hence it remains to consider the case $\omega_k = 0$ and for this the following purely combinatorial lemma suffices.

Lemma 7.3.3 *If an integer non-negative vector $\omega = (\omega_1, \ldots, \omega_{k-1})$ satisfies $\|\omega\| \leq 2^{k-1}$, then it is a sum $\omega = \epsilon + \eta$ of two integer non-negative vectors satisfying $\|\epsilon\|, \|\eta\| \leq 2^{k-2}$.*

Proof Obviously $\omega_{k-1} \leq 2$. Suppose $\omega_i < 2$, $1 \leq i \leq k-2$. If $\omega_{k-1} = 0$ then $\|\omega\| \leq 2^{k-2}$ and we can let $\epsilon = \omega$, $\eta = 0$. If $\omega_{k-1} = 1$ then we can let $\epsilon = (\omega_1, \ldots, \omega_{k-2}, 0)$, $\eta = (0, \ldots, 0, \omega_{k-1})$. If $\omega_{k-1} = 2$ then all other $\omega_i s$ are zero and we can let $\epsilon = \eta = \frac{1}{2}\omega$.

Now let us reduce the general case to this case. Suppose for some index i, $1 \leq i \leq k-2$, the coordinate $\omega_i \geq 2$. Consider the vector $t_i(\omega) = (\omega_1, \ldots, \omega_{i-1}, \omega_i - 2, \omega_{i+1} + 1, \ldots, \omega_{k-1})$. Obviously $\|t_i(\omega)\| = \|\omega\|$. We can make all coordinates $\omega_i < 2$, $1 \leq i \leq k - 2$, by applying a sequence of $t_i s$. It remains to show that if a decomposition exists for $t_i(\omega)$ then it exists for ω.

Suppose $t_i(\omega) = \epsilon + \eta$, $||\epsilon||, ||\eta|| \leq 2^{k-2}$. Obviously $\epsilon_{i+1} \geq 1$ or $\eta_{i+1} \geq 1$. Suppose first is the case. Then $\omega = t_i^{-1}(\epsilon) + \eta = (\epsilon_1 + \eta_1, \ldots, \epsilon_i + 2 + \eta_i, \epsilon_{i+1} - 1 + \eta_{i+1}, \ldots, \epsilon_{k-1} + \eta_{k-1})$. $\square$

When describing the minimal evolutionary spectrum it is necessary to consider the possibility that some of the structure constants $\lambda_{ji,k}$ may become zero. Therefore it is convenient to introduce a family of binary trees. A binary tree is said to *belong* to eigenbasis $v_0, \ldots, v_{n-1}$ of a genetic algebra if every vertex corresponds to a basis element in such a way that if a certain vertex corresponds to v_k and its successors to v_i and v_j then $i, j < k$ and $\lambda_{ij,k} \neq 0$. We distinguish the left and right successor so the construction applies to non-commutative algebras. It will be useful in the next section. Fig 7.1 shows trees belonging to eigenbasis v_0, v_1, v_3 of 3-dimensional genetic algebra with multiplication table $v_0^2 = v_0$, $v_0 v_1 = \lambda_{01,1} v_1 + \lambda_{01,2} v_2$, $v_0 v_2 = \lambda_{02,2} v_2$, $v_1^2 = v_1 v_2 = v_2^2 = 0$, where all $\lambda_{ij,k} \neq 0$.

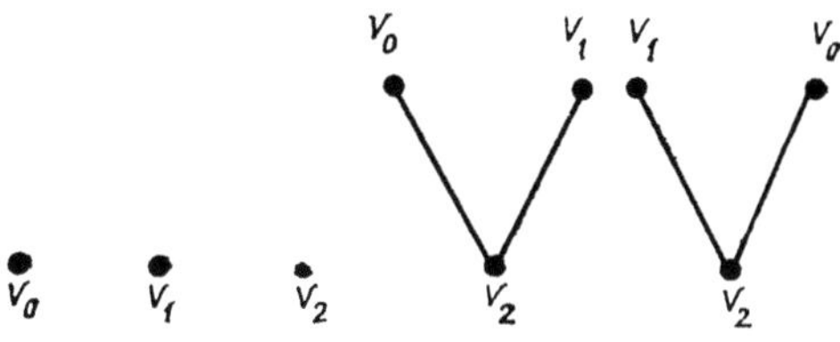

Fig. 7.1

We will denote the set of all trees belonging to the basis $v_0, v_1, \ldots, v_{n-1}$ by $\mathcal{T}$, the subset of trees whose root corresponds to v_k by $\mathcal{T}_k$, the set of end-vertices of a tree ω by $E(\omega)$. For one-vertex trees the end vertex coincides with the root. One-vertex trees can be identified with the basis elements $v_0, v_1, \ldots v_{n-1}$; we will denote these trees simply $0, 1, \ldots, n - 1$.

Theorem 7.3.4 *The minimal evolutionary spectrum of a genetic algebra* $(\mathcal{A}, \sigma)$ *with train roots* $\tau_1, \ldots \tau_{n-1}$ *is contained in the set of numbers of the form*

$$\lambda_\omega = \prod_{w \in E(\omega)} 2\tau_{i(w)}, \qquad (7.3.2)$$

$(\omega \in \mathcal{T})$ *where* $i(w)$ *is the index of the basis element corresponding to vertex* w.

Proof Note that the quotient algebra $(\mathcal{A}_k, \sigma_k)$ is associated with the family of trees $T^{(k)} = T_0 \cup T_1 \cup \cdots \cup T_{k-1}$. Thus it is possible to use induction on k, as usual. For $k = 1$ the assertion is trivial. For the induction step from k to $k + 1$ it is sufficient to adjoin to the minimal evolutionary spectrum of the quotient algebra $(\mathcal{A}_k, \sigma_k)$ the following: λ_k and the roots of all the polynomials annihilating the products $\xi_{t,j}\xi_{t,i}$, $(\xi_{t,0} \equiv 1)$ that actually are included in

$$f_k(\xi_{t,1}, \ldots, \xi_{t,k-1}) = \sum_{j,i=0}^{k-1} \lambda_{ji,k}\xi_{t,j}\xi_{t,i}, \tag{7.3.3}$$

i.e. those products for which $\lambda_{ji,k} \neq 0$. By the induction hypothesis the sequence $\{\xi_{t,j}\}_{t=0}^{\infty}$ is annihilated by the polynomial whose roots lie in the set of the numbers of the form (7.3.2) corresponding to the trees from T_j. The sequence $\{\xi_{t,i}\}_{t=0}^{\infty}$ behaves analogously. The product $\xi_{t,j}\xi_{t,i}$ is annihilated by the quasitensor product of the above polynomials; the roots of this product are products of roots of one polynomial and roots of the other one. Each root of the first (second) polynomial is determined by some tree $\omega_{j\alpha} \in T_j(\omega_{i\beta} \in T_i)$; the product of the roots is determined by the tree formed by gluing those trees to the branches starting from the vertex corresponding to the element v_k (Fig. 7.2). $\square$

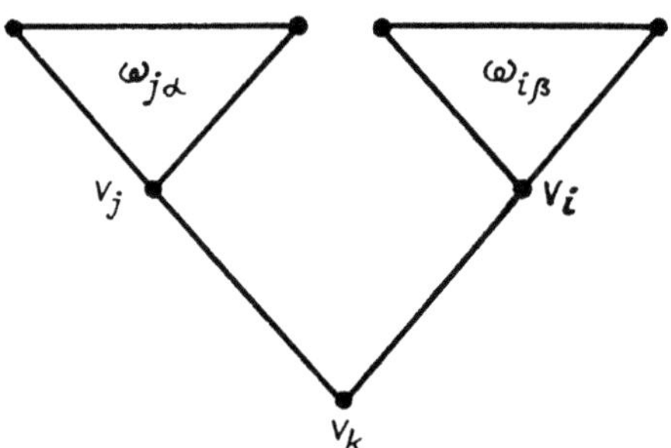

Fig. 7.2

For the algebra whose family of trees was shown on Fig. 7.1, the minimal evolutionary spectrum turns out to be $\{1, 2\tau_1, 2\tau_2\}$ and the maximal evolutionary spectrum also contains $(2\tau_1)^2$.

The minimal evolutionary spectrum is incidentally characterized by the following proposition.

Lemma 7.3.5 *Let Λ be a finite set of points on the complex plane such that each trajectory has the form*

$$x_t = \sum_{\lambda \in \Lambda} \theta_\lambda(t) \lambda^t, \qquad (7.3.4)$$

where $\theta_\lambda(t)$ are the polynomials in t, and for each $\lambda \in \Lambda$ there exists a trajectory for which $\theta_\lambda \neq 0$. The Λ coincides with minimal evolutionary spectrum.

Proof The degrees of the polynomials θ_λ are uniformly bounded on all the trajectories because the space of trajectories is finite dimensional, by Theorem 7.3.1, and the decomposition (7.3.4) is unique. Let $\max(\deg \theta_\lambda) = \nu_\lambda - 1$. The polynomial with the zero divisor $\sum_{\lambda \in \Lambda} \nu_\lambda \lambda$ annihilates all the trajectories. Hence the minimal evolutionary spectrum is contained in the set Λ. Moreover, if the point $\mu \in \Lambda$ is not in the spectrum then there exists a polynomial ψ with roots in the set $\Lambda \backslash \{\mu\}$ annihilating all the trajectories. It also annihilates all the addends in (7.3.3). (We can arbitrarily increase the multiplicity of the roots of the polynomial ψ if necessary). Therefore it annihilates $\theta_\mu(t)\mu^t$ for any trajectory, including the one mentioned in the condition of the theorem. But then $\psi(\mu) = 0$, which is a contradiction. $\square$

Corollary 7.3.6 *The evolutionary spectrum of a Reiersöl algebra coincides with its minimal evolutionary spectrum.*

Proof It follows from (6.5.7) and the linear independence of the normalized measures of nonequilibrium that for any evolutionary root $\lambda_{K_1|\cdots|K_m}$ there exists an initial state p_0 such that $C_{K_1|\cdots|K_m}(p_0) \neq 0$. The result applies to algebras with simple roots, but the conclusion is always valid because the roots of the polynomial continuously depend on the coefficients. $\square$

Later we will need to know the sufficient condition for the minimal evolutionary polynomial to contain no multiple roots.

Theorem 7.3.7 *If the minimal evolutionary polynomial contains no multiple roots then*

$$\lambda_k \neq \lambda_\omega, \qquad (\omega \in T^{(k)} \backslash \{k\}, \; 1 \leq k \leq n - 1). \qquad (7.3.5)$$

Proof If the minimal evolutionary polynomial of a quotient algebra $(\mathcal{A}_k, \sigma_k)$ contains no multiple roots then the sequence (7.3.3) has no secular terms, i.e. is annihilated by the polynomial ψ with no multiple roots.

But $\xi_{t,k}$ has the property if and only if $\lambda_k = 2\tau_k$ coincides with no root of the polynomial ψ (i.e. in the absence of resonance in the equation (7.2.3)), which is the assertion. $\square$

Hence the *minimal evolutionary spectrum of the generic genetic algebra is simple.*

7.4   The Explicit Evolutionary Formula

Any trajectory in a genetic algebra can be represented as

$$x_t = \sum_{\omega \in T} C_\omega \lambda_\omega^t, \qquad (7.4.1)$$

where C_ω may be polynomials in t, but when the inequalities (7.3.5) hold (and thus for a generic algebra), C_ω are constants. Note that in (7.4.1) the exponents λ_ω^t may be repeated because different trees may correspond to the same values of λ_ω as in the situation represented in Fig. 7.3, where $\lambda = \mu_1^2$. Therefore the coefficients C_ω are not uniquely defined, and here we will speak of *choosing* for C_ω explicit expressions satisfying (7.4.1).

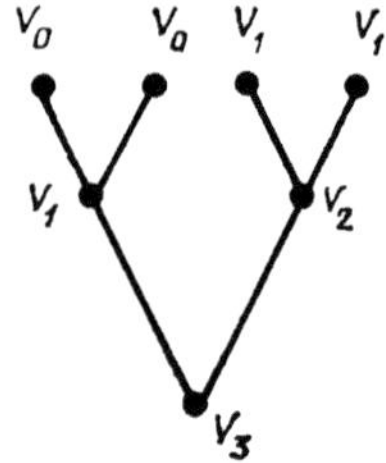

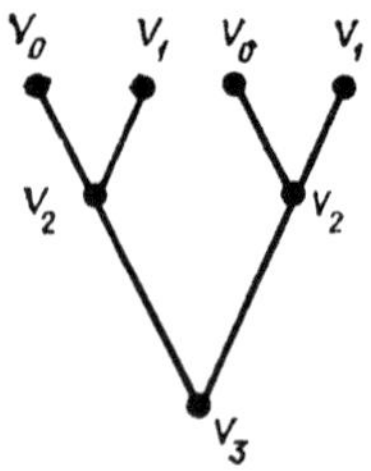

Fig. 7.3

As in the derivation of the explicit formula of evolution for the recombination process (section 6.5), one has to consider separately "diagonal" and "non-diagonal" coefficients. This separation is achieved by the following lemma, which can be proven by induction (cf. the proof of Theorem 7.3.4).

Lemma 7.4.1 *The trajectory (7.4.1) has the coordinate form*

$$\xi_{t,i} = \sum_{\omega \in T_i} C_{\omega,i} \lambda_\omega^t = C_{ii} \lambda_i^t + \sum_{\omega \in T_i \setminus \{i\}} C_{\omega,i} \lambda_\omega^t.$$

For a tree ω we denote the set of non-end vertices by $I(\omega)$. For a vertex $w \in I(\omega)$ we will denote by $\omega(w)$ the subtree in ω whose root is w. Let w', w'' be the successors of the vertex w. We define

$$P(w) = \prod_{w \in I(w)} \lambda_{i(w'),i(w''),i(w)},$$

($P(w) = 1$, if w has only one vertex). For an initial state x_0 and a tree ω we define

$$\pi(x_0, \omega) = \prod_{w \in E(\omega)} \xi_{0,i(w)}.$$

We will say that subtrees $\phi_1, \ldots, \phi_r$ of the tree ω form a *partition* of it, if 1) each vertex of ω belongs to a subtree of this system; 2) the subtrees do not have common branches. Subtrees with a single vertex are not considered. We will denote the set of partitions of the tree ω by $Z(\omega)$. For $z = \{\phi_1, \ldots, \phi_r\} \in Z(\omega)$ we write $|z| = r$. For a given partition z and a given vertex $w \in I(\omega)$ there exists a unique subtree ϕ_k for which this vertex is non-terminal. We will denote the subtree of ϕ_k whose root is w by $\psi_z(\omega)$ (Fig. 7.4).

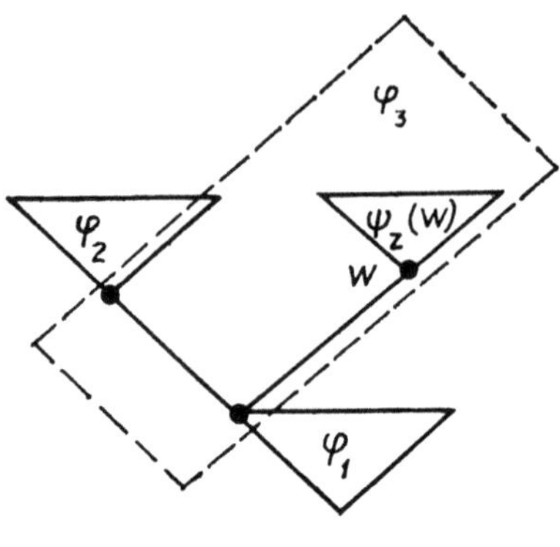

Fig. 7.4

Theorem 7.4.2 *The following choice of coefficients C_ω satisfies the expansion (7.4.1):*

$$C_{ii} = \sum_{\omega \in T_i} \pi(x_0, \omega) P(\omega) \sum_{z \in Z(\omega)} (-1)^{|z|} \prod_{w \in I(\omega)} \frac{1}{\lambda_{\psi_z(w)} - \lambda_{i(w)}}, \qquad (7.4.2)$$

$$C_{\omega,k} = P(\omega) \prod_{w \in I(\omega)} \frac{1}{\lambda_{\omega(w)} - \lambda_{i(w)}} \prod_{u \in E(\omega)} C_{i(u),i(u)}(\omega \in T_k \backslash \{i\}, \ 1 \le k \le n-1).$$

$$(7.4.3)$$

Proof Consider, as usual, the system of evolutionary equations in coordinates:

$$\xi_{t+1,k} = \lambda_k \xi_{t,k} + \sum_{i,j=0}^{k-1} \lambda_{ji,k} \xi_{t,j} \xi_{t,k} \quad (1 \le k \le n - 1). \qquad (7.4.4)$$

Substituting here the first of the expansion of Lemma 7.4.1, we get, instead of the double sum, the expression $\sum_{\omega \in T_k} \lambda_{k'k'',k} C_{\omega',k'} C_{\omega'',k''} \lambda_\omega^t$, where k', k'' are the indices of the basis elements corresponding to the successors of the root in the tree ω; ω', ω'' are the subtrees starting from these vertices. (The "non-commutativty" of trees adopted earlier is significant here, for then the two symmetric terms in (7.4.4), with the coefficient $\lambda_{ji,j}$ and with the coefficient $\lambda_{ij,j}$, are taken into consideration.) For these equations to hold, it would suffice that $C_{\omega,k} = (\lambda_{k'k'',k} C_{\omega',k'} C_{\omega'',k''})/(\lambda_\omega - \lambda_k)(\omega \in T_k)$. This holds if the $C_{\omega,k}$s are chosen as in (7.4.3), and the choice of the diagonal coefficient C_{ii} does not matter. It remains to satisfy the initial condition

$$C_{kk} = \xi_{k,0} - \sum_{\omega \in \mathcal{T}_k \setminus \{k\}} C_{\omega,k}. \tag{7.4.5}$$

Let h_k be the maximal depth of the trees from the family $\mathcal{T}_k$. The proof is by induction on h_k. If $h_k = 0$ then (7.4.5) reduces to $C_{kk} = \xi_{k,0}$, and the same gives (7.4.2) (the set of partitions $Z(k)$ has one element in this case). Suppose that the assertion holds for all $h_j < h$, and let $h_k = h$ for some k. For all the $C_{\omega,k}$ in (7.4.5) the diagonal coefficients $C_{i(u),i(u)}$ that enter the definition of $C_{\omega,k}$ in (7.4.3) are such that $h_{i(u)} < h$ and hence (7.4.2) holds for these diagonal coefficients. Indeed, if $u \in E(\omega)$, $\omega \in \mathcal{T}_k \setminus \{k\}$, then by attaching any tree from $\mathcal{T}_{i(u)} \setminus \{i(u)\}$ to the tree ω at the vertex u we obtain a deeper tree in $\mathcal{T}_k \setminus \{k\}$. Therefore $h_{i(u)} < h_k = h$. This conclusion holds even when $\mathcal{T}_{i(u)} \setminus \{i(u)\} = \emptyset$, for in this case $h_{i(u)} = 0$.

Performing the substitution described above, we obtain from (7.4.5)

$$C_{ii} = \xi_{i,0} - \sum_{\omega \in \mathcal{T}_k \setminus \{k\}} \prod_{w \in I(\omega)} \frac{P(\omega)}{\lambda_{\omega(w)} - \lambda_{i(w)}}$$

$$\prod_{u \in E(\omega)} \sum_{\epsilon \in \mathcal{T}_{i(u)}} \pi(x_0, \epsilon) P(\epsilon) \sum_{z \in Z(\epsilon)} (-1)^{|z|} \prod_{w \in I(\epsilon)} \frac{1}{\lambda_{\psi_z(w)} - \lambda_{i(w)}}.$$

This transforms into the desired expression if we consider the one-to-one correspondence between the non-trivial partitions of the tree $\omega \in \mathcal{T}_k \setminus \{k\}$ and the pairs (ω, z), where $z \in Z(\epsilon)$, $\epsilon \in \mathcal{T}_{i(u)}$, $u \in E(\omega)$. $\square$

Chapter 8

General Quadratic Evolutionary Operators

8.1 The Set of Equilibrium States

Let V be a quadratic evolutionary operator on the simplex $\Delta^{n-1} \subset \mathbf{R}^n$. What can be said about the set F_V of its fixed points (equilibrium states)? It is not empty by Brouwer's theorem, can be finite (the extremal example is a constant evolutionary operator sending the entire simplex into one point) or infinite (the extremal example is the identity operator). In any case F_V is a semi-algebraic manifold. What is its topology? What is its dimension? Can it be disconnected? What is the largest number of connected components it can have? At present these and other analogous questions are little investigated. We give a sketch of this topic.

In general, for a smooth map V on a manifold, a fixed point x is called *non-degenerate* if 1 is not an eigenvalue for the tangent linear map at x, $d_x V$, or, equivalently, if $d_x V - I$ is nonsingular. The utility of nondegeneracy comes from the implicit function theorem which implies that a non-degenerate fixed point x is isolated, i.e. there is a neighborhood of x on which $V(\tilde{x}) \neq \tilde{x}$ except for $\tilde{x} = x$. For the simplex Δ^{n-1} the tangent space is the hyperplane $\mathbf{R}_0^n = \{x : s(x) = \sum_i \xi_i = 0\}$.

A fixed point is called *degenerate* otherwise.

The evolutionary operator V has the coordinate form $\xi'_j = g_j(\xi_1, \ldots, \xi_n)$ $\equiv \sum_{i,k=1}^n p_{ik,j}\xi_i\xi_k$, where

$$p_{ki,j} = p_{ik,j} \geq 0, \quad \sum_{j=1}^{n} p_{ik,j} = 1 \quad (1 \leq i, k \leq n). \tag{8.1.1}$$

The coordinates $\xi_1, \ldots, \xi_n$ in the simplex are subject to the constraints $\xi_j \geq 0$ $(1 \leq j \leq n)$, $s(x) \equiv \sum_{j=1}^{n} \xi_j = 1$. At a point $x \in \Delta$ the tangent map $d_x V$ is the restriction to $\mathbf{R}_0^n$ of the linear map with Jacobian matrix $(\frac{\partial g_j}{\partial \xi_k})_{j,k=1}^{n}$. At $x \in \Delta$ with coordinate $\xi_1, \ldots, \xi_n$,

$$\frac{\partial g_j}{\partial \xi_k} = 2 \sum_{i=1}^{n} \xi_i p_{ik,j}. \tag{8.1.2}$$

Thus, $d_x V$ is the restriction to $\mathbf{R}_0^n$ of $2M_x$, where M_x is the multiplication operator in the evolutionary algebra. It is clear from the equality $Vx = x^2$ as well.

Lemma 8.1.1 *For a matrix $A = (a_{jk})_{j,k=1}^{n}$ define the associated linear operator on $\mathbf{R}^n$ by $A(x)_j = \sum_{k=1}^{n} a_{jk}\xi_k$ and the dual by $A^*(f)_k = \sum_{j=1}^{n} f_j a_{jk}$. Then $\mathbf{R}_0^n$ is an invariant subspace for A_0 if and only if the linear functional s with coordinates $(1, \ldots, 1)$ is an eigenvector for A^*, or, equivalently*

$$\sum_{j=1}^{n} a_{j1} = \sum_{j=1}^{n} a_{j2} = \cdots = \sum_{j=1}^{n} a_{jn}. \tag{8.1.3}$$

Assuming such invariance then the restriction of A to $\mathbf{R}_0^n$ is nonsingular if and only if

$$\begin{vmatrix} a_{11} & \cdots & a_{1,n} \\ \vdots & \ddots & \vdots \\ a_{n-1,1} & \cdots & a_{n-1,n} \\ 1 & \cdots & 1 \end{vmatrix} \neq 0. \tag{8.1.4}$$

Proof Clearly, (8.1.3) is true if and only if s is an eigenvector for A^* and the common sum is the eigenvalue. Also (8.1.3) obviously implies $A(\mathbf{R}_0^n) \subset \mathbf{R}_0^n$. Conversely, let $\{e_k\}_1^n$ be the canonical basis of $\mathbf{R}^n$ and $x_k = e_k - e_n$. $A(x) \in \mathbf{R}_0^n$ says: $\sum_{j=1}^{n}(a_{jk} - a_{jn}) = 0$. So invariance of $\mathbf{R}_0^n$ implies (8.1.3).

On Δ we use $\xi_1, \ldots, \xi_{n-1}$ as independent coordinates, with $\xi_n = -\sum_{i=1}^{n-1} \xi_i$. With respect to this coordinate system the matrix of A on $\mathbf{R}_0^n$ is $(a_{jk} - a_{jn})_{jk=1}^{n-1}$. By using elementary column operations to convert the last row to

$(0, \ldots, 0, 1)$ in (8.1.4) we see that the determinant coincides with that of the restriction of A. $\square$

Remark 1 In (8.1.4) the first $n - 1$ rows of the matrix A can be replaced by any $n - 1$ rows.

Remark 2 Observe that the proof showed that the left side of (8.1.4) is exactly the determinant of the restriction of A to $\mathbf{R}_0^n$. We can denote this by $\det_0 A$, defined only when the subspace $\mathbf{R}_0^n$ is A invariant. If $\det_0 A \neq 0$ then rank $A \geq n - 1$ and rank $A = n - 1$ if and only if there exists a row whose elements are equal.

Theorem 8.1.2 *For $x \in \Delta$ the multiplication operator M_x has matrix $(\sum_{i=1}^{n} \xi_i p_{ik,j})_{j,k=1}^{n}$ which is stochastic, i.e. s is an eigenvector for M_x^* with eigenvalue 1. M_x has an eigenvector $\tilde{x} \in \Delta$ with eigenvalue 1 and the entire spectrum of M_x is contained in the unit circle.*

A point x is fixed, $Vx = x$ if and only if x is an eigenvector of M_x with eigenvalue 1. Such a fixed point is non-degenerate if and only if $\frac{1}{2}$ is not an eigenvalue of M_x, i.e. if

$$\det_0(2M_x - I) \neq 0. \tag{8.1.5}$$

Proof For $x \in \Delta$ the matrix of M_x is stochastic by (8.1.1), i.e. it is nonnegative and the common value in (8.1.3) is 1, for $a_{jk} = \sum_{i=1}^{n} \xi_i p_{ik,j}$. Thus, M_x is a nonnegative matrix with dominant eigenvalue 1. The existence of $\tilde{x}$ and the spectral property follow from the Perron-Frobenius theory of such operators.

As $V(x) = M_x(x)$, x is a fixed point of V precisely when $M_x(x) = x$. As the linearization $d_x V$ is $2M_x$, such a fixed point is non-degenerate when (8.1.5) holds, i.e. when $\frac{1}{2}$ is not an eigenvalue of M_x. $\square$

Remark If M_x is indecomposable then the eigenvector $\tilde{x} \in \Delta$ is unique. Hence if x is a fixed point then $\tilde{x} = x$.

Example Consider $V = A|\Delta$ where A is a stochastic linear operator, i.e. the evolutionary algebra is A-induced. Then $M_x = \frac{1}{2}(A + s \otimes Ax)$ and M_x is reduced to $\frac{1}{2}A$ by the restriction on $\mathbf{R}_0^n$. So $\det_0(2M_x - I) = \det_0(A - I)$ does not depend on x. It is not equal to zero if and only if the eigenvalue 1 of A is simple. It is exactly the case of the fixed point being unique (otherwise F_V is infinite). Thus in linear situation non-degeneracy is equivalent to the uniqueness of fixed point. In particular, if A is indecomposable then it has the unique fixed point and the latter is non-degenerate.

By (8.1.1) the space of evolutionary operators can be identified with the $n(n + 1)/2$-th Cartesian product of $(n - 1)$-dimensional simplices. It comes with a natural topology induced by the metric

$$\rho(V^{(1)}, V^{(2)}) = \max_{ik,j} |p_{ik,j}^{(1)} - p_{ik,j}^{(2)}|$$

or some other equivalent metric.

Theorem 8.1.3 *The set of evolutionary operators all of whose fixed points are non-degenerate is open and everywhere dense in the space $\mathcal{E}$ of evolutionary operators.*

Thus in the general case the set F_V is finite. It follows from the preceding chapters that evolutionary operators arising in genetics usually have continuously many equilibrium states and thus are not "typical".

Proof Suppose $x \in F_V$ is degenerate. By Theorem 8.1.2

$$\begin{vmatrix} \frac{\partial g_1}{\partial \xi_1} - s & \cdots & \frac{\partial g_1}{\partial \xi_{n-1}} & \frac{\partial g_1}{\partial \xi_n} \\ \vdots & \ddots & \vdots & \vdots \\ \frac{\partial g_{n-1}}{\partial \xi_1} & \cdots & \frac{\partial g_{n-1}}{\partial \xi_{n-1}} - s & \frac{\partial g_{n-1}}{\partial \xi_n} \\ 1 & \cdots & 1 & 1 \end{vmatrix} = 0, \tag{8.1.6}$$

$$\xi_j s - \sum_{i,k=1}^{n} p_{ik,j} \xi_i \xi_k = 0 \quad (1 \le j \le n-1). \tag{8.1.7}$$

This is a system of n homogeneous equations with n unknowns $\xi_1, \ldots, \xi_n$ ($s = \sum_i \xi_i$) and x is its non-trivial solution. Thus the resultant of the system $R(\{p_{ik,j}\})$ becomes zero. The equation $R = 0$ defines an algebraic variety in $\mathcal{E}$ containing all operators having at least one degenerate fixed point. It is nowhere dense, unless all the operators satisfy this equation. But $R \ne 0$ for the operator $\xi_j' = \xi_j^2$ $(1 \le j \le n-1)$, $\xi_n' = s^2 - \sum_{j=1}^{n-1} \xi_j^2$ because the system (8.1.6), (8.1.7) has only the trivial complex solution in this case. $\square$

We will call an operator all of whose fixed points are non-degenerate *a non-degenerate operator*. The set of non-degenerate operators is open in $\mathcal{E}$ because the limit of sequence of operators, each of which has a degenerate point, obviously has such point. The set F_V of fixed points of a non-degenerate operator is finite. How large can it be? Because all the fixed points satisfy $\xi_j = \sum_{i,k=1}^{n} p_{ik,j} \xi_i \xi_k$ $(1 \le j \le n-1)$, $\sum_{i=1}^{n} \xi_i = 1$ by Bézout's Theorem

$$|F_V| \le 2^{n-1}. \tag{8.1.8}$$

If moreover the boundary of the simplex $\partial\Delta^{n-1}$ contains no fixed points then the vector field $Vx - x$ on the boundary is directed inside the simplex and hence its index is 1. On the other hand, the index of every critical point of the field (i.e. fixed point of V) is ± 1 by nondegeneracy. By the Hopf Index Theorem, the index of the field on the boundary is equal to the sum of indices of the critical points inside the simplex. Hence the following theorem has been proven.

Theorem 8.1.4 *If V is a non-degenerate evolutionary operator with $F_V \cap \partial\Delta = \emptyset$ then $|F_V|$ is odd.*

There is a simple sufficient condition for the absence of boundary fixed points.

Lemma 8.1.5 *If the matrix of diagonal coefficients $(p_{ii,j})_{i,j=1}^n$ of evolutionary operator is indecomposable then $F_V \cap \partial\Delta = \emptyset$.*

This lemma follows from Theorem 3.8.4 and its Corollary 3.8.5.
We note that the matrix $(p_{ii,j})$ obviously is indecomposable if

$$p_{ii,j} > 0 \quad (1 \le i, j \le n). \tag{8.1.9}$$

The condition (8.1.9) is equivalent to

$$\mathrm{Im}V \subset \mathrm{Int}\Delta^{n-1}. \tag{8.1.10}$$

Indeed, (8.1.9) can be written as $Ve_i \in \mathrm{Int}\Delta^{n-1}$ ($1 \le i \le n$), where $\{e_i\}_1^n$ is the canonical basis in $\mathbf{R}^n$, so the condition is necessary. It is also sufficient because $Vx \ge \sum_{i=1}^n \xi_i^2 Ve_i$ ($x \in \Delta^{n-1}$).

Corollary 8.1.6 *If for a non-degenerate operator the matrix $(p_{ii,j})$ is indecomposable then $|F_V|$ is odd.*

Being odd, the estimate (8.18) is decreased by one. This fact can be proved without assuming that the operator is non- degenerate.

Corollary 8.1.7 *If $F_V \cap \partial\Delta^{n-1} = \emptyset$ (in particular, if the matrix $(p_{ii,j})$ is indecomposable) and $|F_V| < \infty$ then*

$$|F_V| \le 2^{n-1} - 1. \tag{8.1.11}$$

Proof If $|F_V| = 2^{n-1}$ then the multiplicity of every fixed point is one, all these points are non-degenerate and Theorem 8.1.4 leads to contradiction. $\square$

A sufficient condition that there be a unique fixed point is that $F_V \cap \partial \Delta^{n-1} = \emptyset$ and the determinant on the left side of (8.1.6) does not become zero inside the simplex. Indeed, the latter implies that all the interior fixed points are non-degenerate and the Jacobian of V has the same sign at these points and the Index Theorem implies uniqueness.

Conjecture 8.1.8 *If all the coefficients of an evolutionary operator V are positive (and hence $V \in \text{Int } \mathcal{E}$),*

$$p_{ik,j} > 0 \quad (1 \le i,j,k \le n), \tag{8.1.12}$$

then $|F_V| < \infty$.

Condition (8.1.12) does not necessarily imply that there is a unique fixed point, but by Corollary 8.1.7, the estimate (8.1.11) holds. In particular, if $n = 3$ then $|F_V| \le 3$. We will see that this estimate is sharp.

Example Consider an operator V_ϵ in Δ^2 depending on a small parameter $\epsilon > 0$:

$$\xi_1' = (1 - 4\epsilon)\xi_1^2 + 2\epsilon\xi_2^2 + 10\epsilon\xi_3^2 + 4\epsilon\xi_1\xi_2 + (1 + 4\epsilon)\xi_1\xi_3 + 8\epsilon\xi_2\xi_3,$$

$$\xi_2' = 2\epsilon\xi_1^2 + (1 - 3\epsilon)\xi_2^2 + \epsilon\xi_3^2 + (\frac{1}{2} + 2\epsilon)\xi_1\xi_2 + 2\epsilon\xi_1\xi_3 + (1 - 12\epsilon)\xi_2\xi_3,$$

$$\xi_3' = 2\epsilon\xi_1^2 + \epsilon\xi_2^2 + (1 - 11\epsilon)\xi_3^2 + (\frac{3}{2} - 6\epsilon)\xi_1\xi_2 + (1 - 6\epsilon)\xi_1\xi_3 + (1 + 4\epsilon)\xi_2\xi_3.$$

Here all the coefficients are positive for $\epsilon < \frac{1}{12}$. Moreover for $\epsilon < \frac{9 - 5\sqrt{2}}{124}$ there are exactly three fixed points. This example also shows that F_V can be disconnected.

At present the conjecture has been proven for $n \le 4$. The proof is given below. It uses only the fact that the matrix of diagonal coefficients is indecomposable, so a stronger conjecture is possible.

The case $n \le 3$ is quite elementary. Also it need not be considered for a formal reason: if the conjecture (of either form above) holds for $n + 1$ then it holds for n. This follows from the "descent" construction which we will now describe.

Consider the family of maps $l_\tau : \Delta^{n-1} \to \Delta^n$, $0 \le \tau \le 1$, defined by $l_\tau(\xi_1, \ldots, \xi_{n-1}, \xi_n) = (\xi_1, \ldots, \xi_{n-1}, \tau\xi_n, (1 - \tau)\xi_n)$. The map $h : \Delta^n \to \Delta^{n-1}$

defined as $h(\xi_1, \ldots, \xi_n, \xi_{n+1}) = (\xi_1, \ldots, \xi_{n-1}, \xi_n + \xi_{n+1})$ is the left inverse for all the l_τ i.e. $hl_\tau = \mathrm{id}$ and hence $l_\tau h|\mathrm{Im}\, l_\tau = \mathrm{id}$.

An operator V on simplex Δ^{n-1} defines a family of operators $V_\tau = l_\tau V h$ in Δ^n. This can be illustrated by a commutative diagram:

$$
\begin{array}{ccc}
 & V_\tau & \\
\Delta^n & \longrightarrow & \Delta^n \\
\downarrow h & & \uparrow l_\tau \\
\Delta^{n-1} & \longrightarrow & \Delta^{n-1} \\
 & V &
\end{array}
$$

Obviously $V_\tau l_\tau = l_\tau V$. Hence l_τ maps fixed points of V to fixed points of V_τ. Moreover l_τ is injective. Hence the following lemma holds.

Lemma 8.1.9 *If, for a certain τ, $|F_{V_\tau}| < \infty$ then $|F_V| < \infty$ (and moreover $|F_V| \le |F_{V_\tau}|$).*

Remark Evolutionary operators V_τ and V are internal kin (see section 3.9).

The matrix $P(\tau) = (p_{ii,j}(\tau))_{i,j=1}^n$ of diagonal coefficients of the operators V_τ is easily computed through the matrix $P = (p_{ii,j}^{n+1})_{i,j=1}$ of diagonal coefficients of the operator V as follows. (The simplex Δ^{n-1} is canonically included in Δ^n, i.e. if $\{e_1, \ldots, e_n, e_{n+1}\}$ is a basis Δ^n then $\{e_1, \ldots, e_n\}$ is a basis in Δ^{n-1}.)

$$
p_{ii,j}(\tau) = \left\{
\begin{array}{ll}
p_{ii,j} & i \le n; j \le n - 1, \\
\tau p_{ii,n} & i \le n; j = n, \\
(1 - \tau)p_{ii,n} & i \le n; j = n + 1, \\
p_{nn,j} & i = n + 1; j \le n - 1, \\
\tau p_{nn,n} & i = n + 1; j = n, \\
(1 - \tau)p_{nn,n} & i = n + 1; j = n + 1
\end{array}
\right.
$$

This implies that if P is indecomposable and $0 < \tau < 1$ then $P(\tau)$ is also indecomposable. Also the condition (8.1.12) is preserved when going from V to V_τ (the same is true for intermediate condition (8.1.9)).

So if the conjecture is true for $n + 1$ then it is true for n. Thus, either it is true for all n or there exists a dimension n_0 such that the conjecture is true for $n \le n_0$ and false for $n > n_0$.

Theorem 8.1.10 *If the matrix of diagonal coefficients $(p_{ii,j})$ of an evolutionary operator V is indecomposable then for $n \le 4$, $|F_V| < \infty$.*

As we already know, it is sufficient to give the proof for the case $n = 4$. It applies some elementary algebraic geometry. We will also need the following lemma.

Lemma 8.1.11 *For any evolutionary operator V the quadratic map $R = V - I$ is such that*

$$[\text{Im } R] = J^\perp, \tag{8.1.13}$$

where J is the subspace of invariant linear forms of the corresponding evolutionary algebra.

Proof The map R extends to the entire space as a quadratic map $\tilde{R}x = x^2 - s(x)x$ which "measures" the extent to which the algebra is not the unit algebra. Obviously $[\text{Im } R] = [\text{Im } \tilde{R}]$. A linear form f annihilates $\tilde{R}x$ for all x if and only if $f(x^2) = s(x)f(x)$, i.e. $f \in J$. $\square$

Going directly to the proof of Theorem 8.1.10 we note that in this case $\dim J = 1$, i.e. $J = [s]$. Then (8.1.13) implies that the subspace $[\text{Im } R]$ is 3-dimensional and coincides with $\ker s$. Hence $Rx = \sum_{i=1}^3 \phi_i(x)u_i$, where u_1, u_2, u_3 is a basis in $\ker s$ and ϕ_1, ϕ_2, ϕ_3 are quadratic forms. These forms are linearly independent, for otherwise $\dim[\text{Im } R] < 3$.

The set F_V is defined by the equation $Rx = 0$, $x \in \Delta^3$, or in this case by the system of scalar equations $\phi_i(x) = 0$, $(x \in \Delta^3, 1 \leq i \leq 3)$. Suppose that F_V is infinite. Then in a 3-dimensional complex projective space $\mathbf{C}P^3$ the quadric given by the equation $\phi_3(z) = 0$ has an infinite intersection with the curve

$$\phi_1(z) = 0, \quad \phi_2(z) = 0. \tag{8.1.14}$$

Suppose this curve is irreducible. Then all of it lies on the surface $\phi_3(z) = 0$. From the classification of the curves of the form (8.1.14) (the intersection of two quadrics in $\mathbf{C}P^3$) and from Hilbert's Nullstellensatz it follows that the quadratic form ϕ_3 is a linear combination of forms ϕ_1 and ϕ_2, a contradiction.

If the curve (8.1.14) is reducible then it consists, by the above classification, of a finite number of parts lying in two- dimensional planes. We will show that this also leads to contradiction. Consider the intersection of the curve with a basic simplex Δ^3; it is infinite by assumption. Hence some 2-dimensional plane π contains an infinite part of the intersection. Because the boundary of the simplex contains no fixed points, π has internal points common with the simplex. Hence it intersects the simplex in a triangle or a quadrilateral, from whose vertices one can pick a triple a, b, c such that either a, b, c lie on three edges of the simplex coming from one vertex of the simplex or a, b lie on two edges coming from one vertex and c lies on an edge not containing this vertex.

The part of the intersection of the curve (8.1.14) with the plane π that lies in the simplex is an ellipse. Indeed, it satisfies the equations $\phi_1(x) = 0$, $s(x) = 1$, $\psi(x) = c$ (the last two linear equations define the plane π), i.e. it is a part of a real curve of second order that fits inside the simplex and is infinite. The possibility of replacing ϕ_1 by ϕ_2 shows that the quadratic forms ϕ_1 and ϕ_2 are proportional on the plane π.

The plane π is the affine hull of the triple a, b, c i.e. $x = \alpha a + \beta b + \gamma c$, $(\alpha + \beta + \gamma = 1)$. Substituting this expression in the equation $\xi_1 s(x) = \sum_{i,k=1}^{4} p_{ik,1} \xi_i \xi_k$, equivalent to the equation $\phi_1(x) = 0$ we get the equation

$$A\alpha^2 + B\beta^2 + C\gamma^2 + 2D\alpha\beta + 2E\alpha\gamma + 2F\beta\gamma = 0 \tag{8.1.15}$$

with coefficients $A = \sum_{ik=1}^{4} p_{ik,1} a_i a_k - a_1$, $D = \sum_{ik=1}^{4} p_{ik,1} a_i b_k - (a_1 + b_1)/2, \ldots$. Let a, b, c lie on edges coming from vertex e_4 correspondingly to vertices e_1, e_2, e_3. Then $a_2 = a_3 = 0$, $b_1 = b_3 = 0$, $c_1 = c_2 = 0$. In the equation (8.1.15) $B, C, F > 0$ (since $b_1 = c_1 = 0$) i.e. in this case the coefficients at $\beta^2, \gamma^2, \beta\gamma > 0$. If we rewrite the equation $\phi_2(x) = 0$ in the same way, we will have positive coefficients at $\alpha^2, \gamma^2, \alpha\gamma$. Hence the proportion coefficient is positive. Then in each equation all coefficients are positive and it has no positive solution, i.e. the curve does not intersect the interior of the triangle abc. The same contradiction arises in the second case of placing a, b, c on the boundary of the simplex (if the intersection $\pi \cap \Delta^3$ is a quadrilateral then it should be split into two triangles).

This section ends with the estimate of the dimension of F_V related to Theorem 8.1.10.

Theorem 8.1.12 *If the matrix of diagonal coefficients of an evolutionary operator* $V : \Delta^{n-1} \to \Delta^{n-1}$ *is indecomposable then* dim $F_V \le n - 4$ *for* $n \ge 4$.

This estimate may be rather weak since if the corresponding conjecture is proved then dim $F_V = 0$.

To prove the theorem we will use the "ascent" construction dual to the "descent" construction. Precisely, given $W : \Delta^n \to \Delta^n$ using maps l_τ and h as above we define the family $W_\tau : \Delta^{n-1} \to \Delta^{n-1}$, $0 \le \tau \le 1$, letting

$W_\tau = hWl_\tau$, i.e. the following diagram

$$
\begin{array}{ccc}
& W & \\
\Delta^n & \longrightarrow & \Delta^n \\
\uparrow l_\tau & & \downarrow h \\
\Delta^{n-1} & \longrightarrow & \Delta^{n-1} \\
& W_\tau &
\end{array}
$$

commutes. If $x \in F_W \cap \operatorname{Im} l_\tau$ then $W_\tau hx = hWl_\tau hx = hWx = hx$, for $l_\tau h|\operatorname{Im} l_\tau = \operatorname{id}$, i.e. $hx \in F_{W_\tau}$.

If the matrix of diagonal coefficients of operator W is indecomposable then so is the matrix of diagonal coefficients of W_τ for all τ. This is verified by directly computing the coefficients, as with descent.

Proof of Theorem 8.1.12 We will use induction on n. For $n = 4$ the assertion holds by Theorem 8.1.10. Going from n to $n+1$ we suppose that $W : \Delta^n \to \Delta^n$ satisfies the conditions of the theorem, but dim $F_W > (n+1)-4$, i.e. dim $F_W \geq n-2$. Then in some point $a = (a_1, \ldots, a_{n+1}) \in F_W$, $a_i > 0$, there exists a real- analytic map $\theta : \mathbf{D}^{n-2} \to \mathbf{R}^{n+1}$, where $\mathbf{D}^{n-2}$ is the unit $(n-2)$-dimensional open Euclidean ball, such that $\operatorname{Im} \theta \subset F_W$, $\theta(0) = a$, the derivative $\theta'(0)$ is injective, i.e. rank $\theta'(0) = n-2$. Suppose that in the matrix $\theta'(0)$ the last $n-2$ rows, i.e. 4 through $n+1$, are linearly independent. Apply the ascent construction for $\tau = a_n(a_n + a_{n+1})^{-1}$. The hyperplane $(1 - \tau)\xi_n - \tau\xi_{n+1} = 0$ passes through the point a. Consider its intersection with $\operatorname{Im} \theta$. Its complete preimage in the sphere $\mathbf{D}^{n-2}$ is given by the equation $(1-\tau)\theta_n(y)-\tau\theta_{n+1}(y) = 0$. Because the differentials $d\theta_n, d\theta_{n+1}$ are linearly independent at zero this equation is solvable in a neighborhood of zero with respect to one of the coordinates $\eta_1, \ldots, \eta_{n-1}$ of the point y, say $\eta_{n-2} = f(\eta_1, \ldots, \eta_{n-3})$. We note that the hyperplane in question lies in $\operatorname{Im} l_\tau$. Hence its intersection with F_W is mapped by h into the set F_{W_τ}. Suppose $ha = b$. Introduce the map $\phi = h\phi\tilde{f}$, where $\tilde{f}(\eta_1, \ldots, n_{n-3}) = (\eta_1, \ldots, \eta_{n-3}, f(\eta_1, \ldots, \eta_{n-3}))$. It is analytic in the neighborhood of zero which it maps into F_{W_τ} and $\phi(0) = b$. The derivative $\phi'(0)$ is injective because h' is injective on $\operatorname{Im} (\theta'\tilde{f}')$. Indeed, on ker h' we have $d\xi_i = 0$ $(1 \leq i \leq n - 1)$, $d\xi_n + d\xi_{n+1} = 0$ and $(1 - \tau)d\xi_n = \tau d\xi_{n+1}$ on $\operatorname{Im} (\theta'\tilde{f}')$. Hence dim $F_{W_\tau} \geq n - 3$ which contradicts the induction hypothesis. $\square$

8.2 Contracting Quadratic Operators

An extremal example of a quadratic contraction is the constant operator. In this case the coefficients $p_{ik,j}$ do not depend on i and k. This suggests that for a sufficiently small scattering of coefficient for every fixed j the quadratic operator will be a contraction. This remark can be expressed as a precise theorem.

The *Lipschitz constant* of an operator V is

$$l(V) = \sup_{x \neq y} \|Vx - Vy\|/\|x - y\|$$

, where $\|\cdot\|$ is some norm in $\mathbf{R}^n$. If this norm can be chosen so that $l(V) < 1$ then V will be a strict contraction in this norm with the consequences: unique fixed point, convergence of all trajectories to this point, exponential rate of convergence. Unless otherwise specified, we will use the l_1-norm in the canonical basis $\{e_i\}_{i=1}^n$ defined as $\|z\| = \sum_{i=1}^n |\xi_i|$ for $z = \sum_{i=1}^n \xi_i e_i$.

Theorem 8.2.1 $l(V) = \max_{i_1, i_2, k} \sum_{j=1}^n |p_{i_1 k, j} - p_{i_2 k, j}|.$

The proof will require two lemmas, the first of which is of a general character.

Lemma 8.2.2 *Let Δ be a convex n-dimensional compact in $\mathbf{R}^n$, $F : \Delta \to \Delta$ be a smooth map. Then (for any norm) $l(F) = l'(F) \equiv \max_{x \in \Delta} \|d_x F\|.$*

Proof The Mean Value Theorem yields the estimate $\|Fx - Fy\| \leq l'(F)\|x - y\|$, i.e. $l(F) \leq l'(F)$. On the other hand, Δ is the closure of its interior so for any $\epsilon > 0$ there exists $x_0 \in \mathrm{Int}\Delta$ in which $\|d_{x_0}F\| > l'(F) - \epsilon$. Hence there exists a vector z such that $\|z\| = 1$, $\|(d_{x_0}F)(z)\| > l'(F) - \epsilon$. From the definition $F(x_0 + \tau z) - F(x_0) = \tau d_{x_0}Fz + o(\tau)$ as $\tau \to 0$, we get $l(F)\tau > (l'(F) - \epsilon)\tau - o(\tau)$, which in the limit gives $l(F) \geq l'(F) - \epsilon$ and hence $l(F) \geq l'(F)$. $\square$

Lemma 8.2.3 *Let a matrix $A = (a_{ji}^n)_{j,i=1}$ satisfy (8.1.3). Then*

$$\|A|\mathbf{R}_0^n\| = \frac{1}{2} \max_{i_1 \neq i_2} \sum_{j=1}^n |a_{ji_1} - a_{ji_2}|. \tag{8.2.1}$$

Proof We rewrite (8.2.1) as $\|A|\mathbf{R}_0^n\| = \max_{i_1 \neq i_2} \|A((e_{i_1} - e_{i_2})/2)\|$. To prove it, it suffices to show that $\epsilon_{i_1 i_2} = (e_{i_1} - e_{i_2})/2$, $(1 \leq i_1, i_2 \leq n; i_1 \neq i_2)$ is the set of all extremal points of the intersection of unit ball $\|z\| \leq 1$ with the hyperplane $\mathbf{R}_0^n$.

Let

$$\epsilon_{i_1 i_2} = \frac{1}{2}(\sum_{i=1}^n \alpha_i e_i + \sum_{i=1}^n \beta_i e_i), \tag{8.2.2}$$

where $\sum_{i=1}^{n} \alpha_i = \sum_{i=1}^{n} \beta_i = 0$ and $\sum_{i=1}^{n} |\alpha_i|, \sum_{i=1}^{n} |\beta_i| \leq 1$. Then $\alpha_{i_1} + \beta_{i_1} = 1, \alpha_{i_2} + \beta_{i_2} = -1$ and $\alpha_i + \beta_i = 0$ for $i \neq i_1, i_2$. Hence $\alpha_{i_1} + \beta_{i_1} - \alpha_{i_2} - \beta_{i_2} = 2$ while $|\alpha_{i_1}| + |\beta_{i_1}| + |\alpha_{i_2}| + |\beta_{i_2}| \leq 2$. Hence $\alpha_{i_1}, \beta_{i_1} \geq 0$, $\alpha_{i_2}, \beta_{i_2} \leq 0$ and $|\alpha_{i_1}| + |\beta_{i_1}| + |\alpha_{i_2}| + |\beta_{i_2}| = 2$. Hence $\alpha_i = \beta_i = 0$ for $i \neq i_1, i_2$ and $\alpha_{i_1} = \beta_{i_1} = -\alpha_{i_2} = -\beta_{i_2} = \frac{1}{2}$ and the representation (8.2.2) is trivial. Thus all the $\epsilon_{i_1 i_2}$ are extremal.

If a point $z \in \mathbf{R}_0^n$, $(\|z\| \leq 1)$ is not on the convex hull of the set $\{\epsilon_{i_1 i_2}\}$ then it is separated from it by some linear form f, i.e. $f(z) > 0$, $f(\epsilon_{i_1 i_2}) \leq 0$. Since $\epsilon_{i_2 i_1} = -\epsilon_{i_1 i_2}, f(\epsilon_{i_1 i_2}) = 0$. Hence $f(e_1) = \cdots = f(e_n)$, i.e. f is proportional to s. But then $f(z) = 0$, a contradiction. $\square$

Proof of Theorem 8.2.1 For the operator V in the simplex Δ^{n-1} the derivative is $d_x V = 2M_x = 2\sum_{k=1}^{n} x_k M_k$, where $M_k = M_{e_k}$ is the multiplication map with matrix $(p_{ik,j})_{i,j=1}^{n}$. By Lemma 8.2.2, $l(V) = 2\max_{x \in \Delta^{n-1}} \|M_x\| = 2\max_k \|M_k\|$ and by Lemma 8.2.3, $\|M_k\| = \frac{1}{2}\max_{i_1,i_2} \sum_{k=1}^{n} |p_{i_1 k,j} - p_{i_2 k,j}|$. $\square$

Corollary 8.2.4 *An evolutionary operator V is a strict contraction if and only if* $\max_{i_1,i_2,k} \sum_{j=1}^{n} |p_{i_1 k,j} - p_{i_2 k,j}| < 1$.

For operators with positive coefficients there is a multiplicative estimate of the distance from the operator to the constant one. Let $\mu(V) = \max_{i_1,i_2,k,j}(p_{i_1 k,j}/p_{i_2 k,j})$, and let $m(V) = \max_{i_1,i_2,k} \sum_{j=1}^{n} |p_{i_1 k,j} - p_{i_2 k,j}|$ which is in Theorem 8.2.1.

Lemma 8.2.5 $m(V) \leq 2(\mu(V) - 1)/(\mu(V) + 1)$.

Proof If $\alpha, \beta > 0$ and $\mu = \max(\alpha/\beta, \beta/\alpha)$ then obviously

$$|\alpha - \beta| = \frac{\mu - 1}{\mu + 1}(\alpha + \beta).$$

Hence

$$|p_{i_1 k,j} - p_{i_2 k,j}| \leq (\mu(V) - 1)(\mu(V) + 1)(p_{i_1 k,j} + p_{i_2 k,j}).$$

It remains to sum these inequalities over j, keeping in mind that

$$\sum_{j=1}^{n} p_{i_1 k,j} = \sum_{j=1}^{n} p_{i_2 k,j} = 1.$$

$\square$

Corollary 8.2.6 $l(V) \leq 2(\mu(V) - 1)/(\mu(V) + 1)$.

Corollary 8.2.7 *If $\mu(V) < 3$ then the evolutionary operator V is a strict contraction.*

8.3   An Example of Irregular Trajectory Behavior

Consider in the 2-dimensional simplex $\Delta \equiv \Delta^2$ (in a 1-dimensional simplex, the limit set of every trajectory in either a point or a cycle of order 2 due to non-negativity of all the coefficients) an evolutionary operator V defined by

$$
\begin{aligned}
\xi_1' &= \xi_1^2 + 2\xi_1\xi_2, \\
\xi_2' &= \xi_2^2 + 2\xi_2\xi_3, \\
\xi_3' &= \xi_3^2 + 2\xi_1\xi_3.
\end{aligned} \tag{8.3.1}
$$

It has four fixed points: the center of the simplex $C = (\frac{1}{3}, \frac{1}{3}, \frac{1}{3})$ and the vertices $M_1 = (1,0,0)$, $M_2 = (0,1,0)$ and $M_3 = (0,0,1)$. C is a repelling focus and $M_i s$ are hyperbolic saddles. The edge $M_1 M_2$ is an attracting branch of the point M_1 and the edge $M_1 M_3$ its repelling branch. The operator V can also be written as

$$
\begin{aligned}
\xi_1' &= \xi_1(1 + \xi_2 - \xi_3), \\
\xi_2' &= \xi_2(1 + \xi_3 - \xi_1), \\
\xi_3' &= \xi_3(1 + \xi_1 - \xi_2).
\end{aligned} \tag{8.3.2}
$$

In this form V is sometimes referred to as the "paper-rock- scissors" dynamic of game dynamics.

Because of the inequality of geometric mean and arithmetic mean we have $\xi_1'\xi_2'\xi_3' \leq \xi_1\xi_2\xi_3$. The inequality is strict if the point $x = (\xi_1, \xi_2, \xi_3) \in \hat{\Delta} \equiv$ Int $\Delta \backslash \{C\}$.   Consider the function $P(x) = \xi_1\xi_2\xi_3$.   Obviously, $\max_{x \in \Delta} P(x) = \frac{1}{27}$ is achieved only at C. Let $\{x_t\}_{t=0}^\infty$ be a trajectory, $x_0 \in \hat{\Delta}$. Because $\hat{\Delta}$ is invariant, $P(x_{t+1}) < P(x_t)$ and there exists $\lim_{t \to \infty} P(x_t) \equiv \lambda < \frac{1}{27}$. We will show that $\lambda = 0$. Indeed, if $\lambda > 0$ then the product $(1 + \xi_{t,2} - \xi_{t,3})(1 + \xi_{t,3} - \xi_{t,1})(1 + \xi_{t,1} - \xi_{t,2})$ approaches 1, its upper bound, because of the inequality of means. Then the differences of the type $\xi_{t,1} - \xi_{t,2}$ approach zero from which it follows that $x_t \to C$. This contradicts $\lambda < \frac{1}{27}$.

The equality $\lambda = 0$ shows that *all limit points of a trajectory lie on $\partial\Delta$.* We will now show that the *set of limit points is infinite.*

A trajectory cannot converge to a vertex because if for example in the neighborhood of the vertex $M_1 = (1,0,0)$ defined by the inequality $\xi_1 > \xi_2$ we had $\xi_3' > \xi_3$ for $\xi_3 > 0$ which would prevent $\xi_{t,3} \to 0$. Nor can the limit set consist only of vertices because the vertices are fixed and so Lemma 1.2.1 applies. Therefore, the limit set contains an interior point of some edge. But the limit set is invariant, hence it contains the entire trajectory starting at this point. This trajectory is infinite: for example, the restriction of the map to edge $M_1 M_2$ has the form $\xi_1' = \xi_1(2 - \xi_1)$ which implies that if $0 < \xi_1 < 1$ then $\xi_1 < \xi_1' < 1$. Therefore the limit set is infinite.

Thus any trajectory $x_t = V^t x$ starting at $x \in \hat{\Delta}$ diverges.

Theorem 8.3.1 *For all $x \in \hat{\Delta}$ and the operator V defined by (8.3.1), the limit of averages $\lim_{t \to \infty}(x + Vx + \cdots + V^t x)/(t+1)$ does not exist.*

Proof The triangle Δ is dissected by heights into six parts. Consider their closures: $F_1 = \{x | \xi_1 \geq \xi_2 \geq \xi_3\}, \ldots, F_6 = \{x | \xi_2 \geq \xi_1 \geq \xi_3\}$. (Fig 8.1)

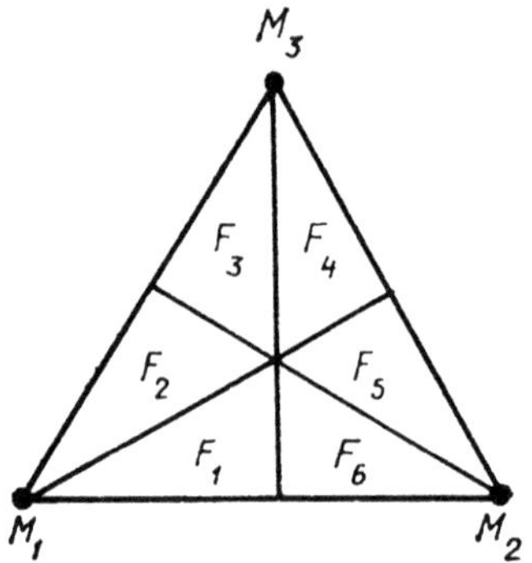

Fig. 8.1

We will show that $x_t = V^t x$ goes around $F_1 \to F_2 \to F_3 \to F_4 \to F_5 \to F_6 \to F_1$. Indeed, if $x \in F_1$ then (8.3.2) implies $\xi_1' \geq \xi_1, \xi_2' \leq \xi_2, \xi_3' \geq \xi_3$. Also $\xi_3' \leq \xi_1'$ (this is easily seen from (8.3.1)). Hence either $\xi_1' \geq \xi_2' \geq \xi_3'$ and $x' \in F_1$ or $\xi_1' \geq \xi_3' \geq \xi_2'$ and $x' \in F_2$. Analogously if $x \in F_2$ then either $x' \in F_2$ or $x' \in F_3$, etc.

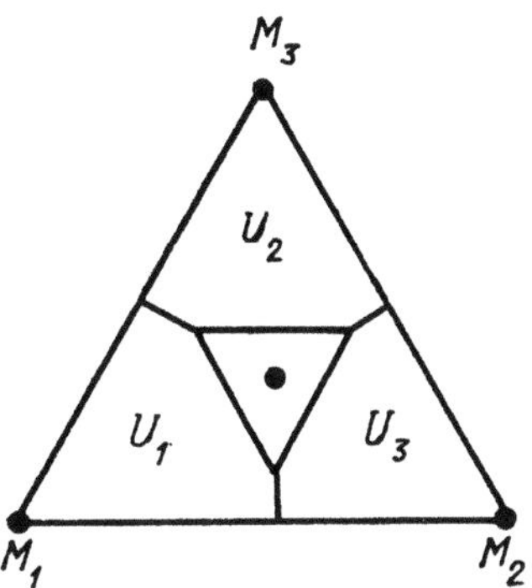

Fig. 8.2

Delete from the simplex a triangular neighborhood of C (Fig. 8.2). This defines the neighborhoods U_1, U_2, U_3 of vertices M_1, M_2, M_3. Obviously U_1, U_2, U_3 are convex and the intersection of their closures is empty. For t large enough the point x_{t+1} goes from F_6 to U_1, stays there for some time $\tau(t)$, then goes to F_3, etc. Let us estimate the time $\tau(t)$.

Let $x \in F_6 \cap \mathrm{Int}\ \Delta$, $x_t \in U_1\ (1 \le t \le \tau)$, $x_{\tau+1} \in F_3$. Then $\xi_{\tau+1,3} = \xi_3' \prod_{t=1}^{\tau} \xi_{t+1,3}/\xi_{t,3} = \xi_3' \prod_{t=1}^{\tau} (1 + \xi_{t,1} - \xi_{t,2}) < 2^\tau \xi_3'$. Hence $2^\tau P(x') > \xi_1' \xi_2' \xi_{\tau+1,3}$. But in F_1 the largest coordinate is ξ_1 and in F_3, ξ_3. Hence $\xi_1' \ge \frac{1}{3}$, $\xi_{\tau+1,3} \ge \frac{1}{3}$. Also $\xi_2' = \xi_2^2 + 2\xi_2\xi_3 \ge \xi_1\xi_2 \ge \frac{1}{3}(\xi_1^2 + 2\xi_1\xi_2) = \frac{1}{3}\xi_1' \ge 1/9$. Hence $2^\tau P(x') > 1/81$. replacing x by x_t we get

$$2^{\tau(t)} P(x_{t+1}) > \frac{1}{81} \tag{8.3.3}$$

for t before x went from F_6 to U_1. Most of the time the trajectory is arbitrarily close to M_1, M_2 or M_3. Indeed, if a trajectory is on an edge then it converges to the corresponding vertex and only its finite part lies outside the neighborhood of this vertex. Thus for any neighborhood U of the vertex set and any $\epsilon > 0$ there exists a neighborhood N of $\partial\Delta$ such that for a trajectory that enters the neighborhood N the ratio of time outside U to the total time is less than ϵ.

Let us show that $P(x_t)$ decreases faster than any geometric progression. Evidently, $P(x') = P(x)Q(x)$, where Q is a polynomial. On the edge M_1, M_2 $Q(x) = 2(2 - \xi_1)\xi_1(1 - \xi_1)$. Analogous relations hold for the other two edges.

Hence for any $\eta > 0$ there exists a neighborhood U of the vertex set such that $P(x') \leq \eta P(x)$ for $x \in U$.

For any t let $t = t_1 + t_2$, where t_1 is the time spent by x_t in U and t_2 is the remaining time. Then $P(x_t) \leq \eta^{t_1} P(x)$. Hence $\limsup_{t \to \infty} \sqrt[t]{P(x_t)} \leq \eta$ since $t_1 t^{-1} \to 1$ as $t \to \infty$ by the above. Because η is arbitrary, $\lim_{t \to \infty} \sqrt[t]{P(x_t)} = 0$. Now (8.3.3) implies that $\tau(t) t^{-1} \to \infty$ as $t \to \infty$. But then the center of mass $(x + Vt + \cdots + V^t x)/(t+1)$ at the time the point leaves U_1 is arbitrarily close to the closure $\bar{U}_1$. Hence if the limit of averages exists, it lies in $\bar{U}_1$. By the same argument it lies in $\bar{U}_2$ and $\bar{U}_3$ which contradicts the property $\bar{U}_1 \cap \bar{U}_2 \cap \bar{U}_3 = \emptyset$. $\square$

Finally we remark that it is not known whether the limit set of arbitrary trajectory starting at $\hat{\Delta}$ coincides with all of $\partial \Delta$.

Chapter 9

Selection Dynamics

9.1 Evolutionary Selection Equations for an Autosomal Multiallele Locus

With $A_1, \ldots, A_m$ the alleles for the locus, let $\lambda_{ik}(= \lambda_{ki})$ be the survival coefficient of a given zygote $A_i A_k \equiv A_k A_i$. The symmetric matrix $\Lambda = (\lambda_{ik})$ is called the *population fitness matrix*. By definition, all $\lambda_{ik} \geq 0$. We further suppose that all $\lambda_{ii} > 0$, i.e. there are no lethal homozygotes.

During the reproductive stage we get the state

$$x'_{ik} = \frac{\lambda_{ik} \bar{x}_{ik}}{\sum_{j,l} \lambda_{jl} \bar{x}_{jl}}$$

(see (1.2.5) and (1.2.8)), where $\bar{x}_{ik}$ are the probabilities of the zygotes at fertilization with the convention $\bar{x}_{ik} = \bar{x}_{ki}$ so that $\bar{x}_{ik} = p_i p_k \ (1 \leq i, k \leq n)$ where p_i, $1 \leq i \leq m$, are the gene probabilities for the previous generation $p_i = \sum_k x_{ik}$ (with the similar convention $x_{ik} = x_{ki}$, see (2.1.11), (2.1.12)).

At the gamete level we have

$$p'_i = p_i \frac{W_i(p)}{W(p)} \tag{9.1.1}$$

where $W_i(p) = \sum_{k=1}^{m} \lambda_{ik} p_k = \frac{1}{2} \frac{\partial W}{\partial p_i}$ is the *mean fitness* of the allele A_i,

$$W(p) = \sum_{i,k=1}^{m} \lambda_{ik} p_i p_k = \sum_{i=1}^{m} p_i W_i(p) \tag{9.1.2}$$

is the *mean fitness of the population* at a state $p = (p_1, \ldots, p_m)$. Obviously $W(p) > 0$ because all $\lambda_{ii} > 0$.

The equation (9.1.1) defines the evolutionary *Fisher selection operator* $F : \Delta \to \Delta$, where $\Delta \equiv \Delta^{m-1}$. We note that every face of the simplex Δ is invariant, i.e. alleles not present in the initial state do not appear during evolution. The interior of every face is also invariant because $\lambda_{ii} > 0$, i.e. alleles present in the initial state do not disappear during evolution. (They may disappear in the limit as $t \to \infty$).

Let us study the set of equilibrium states. From (9.1.1) we get the equilibrium equations

$$p_i[W_i(p) - W(p)] = 0 \qquad\qquad 1 \le i \le m. \qquad (9.1.3)$$

This system of equations is easily reduced to a union of a set of linear systems. Let $M \subset \{1, \ldots, m\}$, $M \ne \emptyset$. Denote the smallest element of M by $i(M)$. Consider the system of linear equations

$$\left. \begin{array}{rcl} W_i(p) - W_{i(M)}(p) & = & 0 \ \ i \in M, i > i(M), \\ p_i & = & 0 \ \ i \notin M, \\ \sum_{i \in M} p_i & = & 1 \end{array} \right\}. \qquad (9.1.4)$$

Every solution of this system lying in simplex Δ (i.e. nonnegative) is an equilibrium since (9.1.4) implies (9.1.3) by (9.1.2). Conversely, every equilibrium state p is obtained thus with $M = \text{supp}(p)$. The set of equilibrium states is "piecewise linear" since it is a union of intersections of simplex Δ with the panes defined by equations (9.1.4) We note that the singleton set $M = \{k\}$ corresponds to the vertex e_k. The vertices of the simplex Δ are *monomorphic* equilibrium states, i.e each one of them contains a single allele. Any other equilibrium state contains more than one allele, i.e. is a *polymorphism*. The equations (9.1.4) show that all the alleles contained in an equilibrium polymorphism have the same mean fitness (equal to the mean fitness of the population). An equilibrium containing all the alleles is called *interior*. Otherwise it is a *boundary* equilibrium, i.e. lies in $\partial\Delta$.

The number of linear systems of the form (9.1.4) is $2^m - 1$. The set of solutions of each system is convex. Therefore the following theorem holds.

Theorem 9.1.1 *If the set of equilibrium states is finite then it contains no more than $2^m - 1$ points.*

Proof In this case every face contains no more than one equilibrium state. $\square$

This upper estimate is achieved. For example for the identity matrix Λ only the centers of the faces are equilibria. On the other hand, the number of equilibrium states is no less than m taking into account the vertices. This low estimate is achieved as well.

Example Suppose that $\lambda_{ii} \equiv \lambda_i$ $(1 \leq i \leq m)$ and $\lambda_{ik} \equiv \mu_{k-1}$ $(1 \leq i < k \leq m)$. Also suppose that the λ's and μ's *alternate*: $\lambda_1 < \mu_1 < \lambda_2 < \cdots < \mu_{m-1} < \lambda_m$. Biologically it means that each allele A_k $(k > 1)$ dominates over all A_i for $i < k$, the heterozygote A_iA_k is intermediate between the homozygotes A_iA_i and A_kA_k with respect to the fitness and the fitness of A_kA_k increases $(1 \leq k \leq m)$. We will call this the *alternation model.* If the system (9.1.5) has a solution $p \gg 0$ then $(\mu_1 - \lambda_1)p_1 + (\lambda_2 - \mu_1)p_2 = 0$ conradicts to positivity of the left side. It remains to note that every principal sub-matrix of the given matrix has the same form.

It will be interesting to know for an arbitrary $r, m < r < 2^m - 1$ if there exists such a matrix Λ that the number of equilibrium states equals r. More precisely, can we arrange $r - m$ such states into given $r - m$ faces different from vertices? It is an open question.

Let p be an interior equilibrium. This is a positive solution of a system of equations

$$\left.\begin{array}{rcl} \sum_{k=1}^m \lambda_{ik}p_k &=& W \quad (1 \leq i \leq m) \\ \sum_{k=1}^m p_k &=& 1 \end{array}\right\} \tag{9.1.5}$$

with the additional unknown W. Suppose that the fitness matrix is nondegenerate. Then $p_k = \Lambda_k W / \det \Lambda$ where

$$\Lambda_k = \begin{vmatrix} \lambda_{11} & \cdots & 1 & \cdots & \lambda_{1m} \\ \vdots & \ddots & \vdots & \ddots & \vdots \\ \lambda_{m1} & \cdots & 1 & \cdots & \lambda_{mm} \end{vmatrix}$$

($1s$ are in the kth column). Obviously all $\Lambda_k \neq 0$, have the same sign and $\sum_{k=1}^m \Lambda_k = \det \Lambda / W$. This allows us to eliminate W:

$$p_k = \frac{\Lambda_k}{\sum_{i=1}^m \Lambda_i}. \tag{9.1.6}$$

Conversely, if $\sum \Lambda_i \neq 0$ then (1.1.6) is a solution of the system (1.1.5) with $W = \det \Lambda / \sum \Lambda_i$.

Theorem 9.1.2 *Assume the fitness matrix is nondegenerate. If an interior equilibrium exists then all the determinants Λ_k are different from zero and have the same sign. Conversely, this condition implies the existence of a unique interior equilibrium (given by formula (1.1.6)).*

Corollary 9.1.3 *If in the fitness matrix all the principal minors are nonzero then the set of equilibrium states if finite*

Equilibrium states have a simple but important "variation" characterization.

Theorem 9.1.4 *A state p^* is an equilibrium if and only if it is a critical point for the restriction of mean fitness $W(p)$ to the minimal face containing p^*.*

Proof It suffices to consider an interior equilibrium, in which case the critical point condition has the form $\frac{\partial W}{\partial p_i} = \alpha$, $\sum_{k=1}^{m} p_k = 1$, where α is the Lagrange multiplier, and this is equivalent to (9.1.5). $\square$

Corollary 9.1.5 *Mean fitness is constant on every connected component of the set of equilibrium states.*

Corollary 9.1.6 *All local extrema of restrictions of function W to the interiors of faces of simplex Δ are equilibria.*

9.2   Stability of Equilibrium States. Fundamental Theorem

The following lemma is useful for the investigation of local stability of equilibrium states.

Lemma 9.2.1 *Let p^* be an equilibrium and assume, for concreteness, that $p_i^* > 0$ for $1 \leq i \leq r$, $p_i^* = 0$ for $r + 1 \leq i \leq m$. Let $y_i = p_i - p_i^*$ be local coordinates, so that $y \in \mathbf{R}_0^m$. Then the evolutionary operator is linearly approximated as*

$$y_i' \approx y_i + \frac{p_i^*}{W(p^*)} \sum_{k=1}^{m} \lambda_{ik} y_k - \frac{2p_i^*}{W(p^*)} \sum_{k=r+1}^{m} (W_k(p^*) - W(p^*))y_k \quad (i \leq r),$$

$$y_i' \approx \frac{y_i W_i(p^*)}{W(p^*)} \quad (i > r).$$

Proof Compute the differential of the Fisher operator at a point $p \in \Delta$:

$$\frac{\partial p_i'}{\partial p_k} = \frac{1}{W(p)}(\delta_{ik}W_i(p) + \lambda_{ik}p_i) - \frac{2p_iW_i(p)W_k(p)}{W^2(p)}. \tag{9.2.1}$$

At the point p^* by equilibrium equations we have

$$\left(\frac{\partial p_i'}{\partial p_k}\right)_{p^*} = \begin{cases} \delta_{ik} + \frac{\lambda_{ik}p_i^*}{W_k(p^*)} - \frac{2p_i^*W_k(p^*)}{W(p^*)} & (i \leq r), \\ \delta_{ik}\frac{W_i(p^*)}{W(p^*)} & (i > r). \end{cases}$$

This proves the $(i > r)$ part of the Lemma and for $i \leq r$ we note that

$$\sum_{k=1}^{m} y_k = 0 \tag{9.2.2}$$

because of which

$$\sum_{k=1}^{m} W_k(p^*)y_k = \sum_{k=1}^{m}(W_k(p^*) - W(p^*))y_k = \sum_{k=r+1}^{m} (W_k(p^*) - W(p^*))y_k.$$

$\square$

Corollary 9.2.2 *The spectrum of the linear approximation of the evolutionary operator at the equilibrium state P^* consists of values*

$$\omega_i = \frac{W_i(p^*)}{W(p^*)} \qquad (i > r) \tag{9.2.3}$$

and eigenvalues of the matrix $L_1^{(r)} = (\delta_{ik} + p_i^*\lambda_{ik}/W(p^*))_{i,k=1}^{r} = I + L^{(r)}$ on the invariant subspace $\mathbf{R}_0^r$ of $\mathbf{R}^r$ where

$$\sum_{k=1}^{r} y_k = 0. \tag{9.2.4}$$

Proof The matrix of the linear approximation has the block form

$$\begin{pmatrix} L_1^{(r)} & * \\ 0 & D \end{pmatrix},$$

where D is a diagonal matrix with elements (9.2.3). (9.2.2) for the block $L_1^{(r)}$ takes the form (9.2.4) and for the block D is immaterial. There are no other dependencies among $y_1, \ldots, y_m$. $\square$

Remark Note that 1 is an eigenvalue for $L^{(r)}$ with left eigenvector $(1, 1, \ldots, 1)$ in $\mathbf{R}^r$ and right eigenvector $(p_1^*, \ldots, p_r^*)$. The former shows that the subspace $\mathbf{R}_0^r$ defined by (9.2.4) is invariant (cf. Lemma 8.1.1), and the determinant of the restriction is the same as the determinant of $L^{(r)}$ which is $\det(\lambda_{ik})_{i,k=1}^r$ times $p_1^* \cdots p_r^* / W(p^*)^r$.

Corollary 9.2.3 *The spectrum of linear approximation of evolutionary process at any equilibrium state is real and nonnegative.*

Proof The eigenvalues of $L^{(r)}$ are real because it is symmetrizable and they lie in the circle $|\lambda| \leq 1$ because the matrix is stochastic (by columns). So the eigenvalues of $L_1^{(r)}$ lie in the interval $[0,2]$. $\square$

An important class of equilibrium states is formed by the so called *hyperbolic equilibria*, i.e. the points for which the spectrum of linear approximation contains no point with absolute value 1. In this case this reduces to the absence of point $\lambda = 1$ in the spectrum.

Corollary 9.2.4 *An equilibrium state p^* is hyperbolic if and only if $W_i^* \neq W^*$ $(i > r)$ and $\det(\lambda_{ik})_{i,k=1}^r \neq 0$.*

Let us now establish sufficient conditions for asymptotic stability.

Theorem 9.2.5 *If p^* is an equilibrium state such that*

$$W_i(p^*) < W(p^*) \quad (i > r), \tag{9.2.5}$$

and the quadratic form $W_r(\xi) = \sum_{i,k=1}^r \lambda_{ik} \xi_i \xi_k$ has the type $(1, r-1)$ (i.e. inertia indices $\nu_+ = 1$, $\nu_- = r - 1$) then the state is asymptotically stable, i.e. it is stable local attractor.

Proof The matrix $L^{(r)}$ symmetrizes to $(\sqrt{p_i^* p_k^*} \lambda_{ik} / W^*)$ and hence its eigenvalues have the same signs as eigenvalues of the form $W_r(\xi)$. As remarked above, the matrix $L^{(r)}$ has an eigenvalue $\lambda = 1$ with eigenvector $(p_1^*, \ldots, p_r^*)$ not lying in the invariant subspace (9.2.4). Hence all other eigenvalues of $L^{(r)}$ are negative and the spectrum of $L_1^{(r)}$ is in $[0,1)$. $\square$

Remark Inequalities (9.2.5) mean that the alleles not present in the given state have smaller mean fitness than those present.

Example The vertice e_k is always an equilibrium. By Corollary 1.2.2 the spectrum at the state e_k is $\{\lambda_{ik} / \lambda_{kk}\}_{i \neq k}$. Thus it is asymptotically stable if

$\lambda_{ik} < \lambda_{kk}$ for all $i \neq k$, i.e. all heterozygotes containing the allele A_k have lesser fitness than the homozygote $A_k A_k$.

A known sufficient condition for the form $W_r(\xi)$ to have type $(1, r-1)$ is for the determinant inequalities

$$\lambda_{11} > 0, \quad \begin{vmatrix} \lambda_{11} & \lambda_{12} \\ \lambda_{21} & \lambda_{22} \end{vmatrix} < 0, \dots, \quad (-1)^r \begin{vmatrix} \lambda_{11} & \cdots & \lambda_{1r} \\ \vdots & \ddots & \vdots \\ \lambda_{r1} & \cdots & \lambda_{rr} \end{vmatrix} < 0, \qquad (9.2.6)$$

to hold. The first one is automatic. Inequalities (9.2.6) are necessary if the determinants are nonzero.

A formal converse of Theorem 9.2.5 is false (see Section 2.4) but the following weaker converse is obvious.

Theorem 9.2.6 *If an equilibrium state p^* is stable then $W_i(p^*) \leq W(p^*)$ $(i > r)$ and the positive inertia index of the quadratic form $W_r(\xi)$, $\nu_+ = 1$.*

Section 9.4 will give the necessary and sufficient conditions for stability and asymptotic stability.

Now we will study the behavior of trajectories "as a whole" with arbitrary initial state $p^{(0)}$. The basis for this in selection theory is the following Fundamental Theorem.

Theorem 9.2.7 *During the evolutionary process the mean fitness of the population increases (and does not change in equilibrium).*

It is desirable to refine the Fundamental Theorem to get a lower bound for the increment of mean fitness on one step. Let p be a state and p' be the next state defined by the evolutionary equations (9.1.1). Considering $W_i(p)$, the mean fitness of the gene A_i as random variable with probability distribution p_i we get mean value $W(p) = \sum p_i W_i(p)$ and the variance

$$\sigma^2(p) = \sum_{i=1}^{m} p_i (W_i(p) - W(p))^2.$$

Theorem 9.2.8 $W(p') - W(p) \geq C\sigma^2(p)$, *where C is a positive constant.*

Proof It follows from (9.1.1) that

$$W(p')W^2(p) = \sum_{i,k=1}^{m} \lambda_{ik} p_i p_k W_i(p) W_k(p)$$

$$= \sum_{ij,k=1}^{m} \lambda_{ik}\lambda_{jk}p_ip_jp_kW_i(p)$$

$$= \sum_{ij,k=1}^{m} \lambda_{ik}\lambda_{jk}p_ip_jp_k\frac{1}{2}(W_i(p) + W_j(p))$$

$$\geq \sum_{ij,k=1}^{m} \lambda_{ik}\lambda_{jk}p_ip_jp_k\sqrt{W_i(p)W_j(p)} = \sum_{k=1}^{m}p_k(\sum_{i=1}^{m}\lambda_{ik}p_i\sqrt{W_i(p)})^2$$

$$\geq (\sum_{i,k=1}^{m} \lambda_{ik}p_ip_k\sqrt{(W_i(p))})^2 = (\sum_{i=1}^{m}p_iW_i^{\frac{3}{2}}(p))^2$$

by the convexity of τ^2. Because $\tau^{\frac{3}{2}}$ is also convex, $W(p')W^2(p) \geq W^3(p)$, i.e. $W(p') \geq W(p)$, which proves the Fundamental Theorem. The refinement of Theorem 9.2.8 is based ont he fact that if a function $\phi(\tau)$ is twice differentiable on some finite interval and $\phi''(\tau) \geq \mu > 0$ then for any random variable ξ with values in this interval we average the estimate $\phi(\tau) \geq \phi(M\xi) + \phi'(M\xi)(\tau - M\xi) + (\mu/2)(\tau - M\xi)^2$ to get

$$M\phi(\xi) \geq \phi(M\xi) + (\mu/2)V\xi$$

where M and V are mean and variance.

If we let $\phi(\tau) = \tau^{\frac{3}{2}}$, $(0 \leq \tau \leq \lambda$, where $\lambda = \max_{i,k}\lambda_{ik})$ and take ξ to be the allelic mean fitness $W_i(p)$ then $M\xi = W(p)$, $V\xi = \sigma^2(p)$ and $\sum_{i=1}^{m}p_iW_i^{\frac{3}{2}}(p) \geq W^{\frac{3}{2}}(p)+(\mu/2)\sigma^2(p)$, where $\mu = \frac{3}{4}\lambda^{-\frac{1}{2}}$. Hence $W(p')W^2(p) \geq W^3(p) + \mu W^{\frac{3}{2}}(p)\sigma^2(p)$, from which $W(p') \geq W(p) + C\sigma^2(p)$, where $C = \mu\lambda^{-\frac{1}{2}} = \frac{3}{4}\lambda^{-1}$. $\Box$

In further estimates we introduce other positive constants and denote them all C or c.

Corollary 9.2.9 $W(p\prime) = W(p)$ *if and only if p is an equilibrium.*

Proof If $W(p') = W(p)$ then $\sigma^2(p) = 0$ and (9.1.3) follows. The converse is trivial. $\Box$

The variance $\sigma^2(p)$ can thus be viewed as a measure of disequilibrium of the population state In the next section we will get upper bounds using $\sigma^2(p)$ for other quantities measuring disequilibrium.

Finally we use the Fundamental Theorem to obtain a criterion of asymptotic stability for an interior equilibrium.

Theorem 9.2.10 *Let p^* be an interior equilibrium. Then the following statement are equivalent: 1) p^* is asymptotically stable; 2) p^* is an isolated local maximum of mean fitness $W(p)$; 3) the quadratic form $W(y)$ on the subspace $\mathbf{R}_0^n$ where $s(y) \equiv \sum_{k=1}^{m} y_k = 0$, is negative definite; 4) the quadratic form W has the type $(1, m-1)$.*

Proof 1) $\Rightarrow$ 2): Let U be a neighborhood of p^* such that all the trajectories starting in U converge to p^*. Then if $p \in U$, $p \neq p^*$, then $W(p) < W(p') \leq W(p^*)$ by the Fundamental Theorem.

2) $\Rightarrow$ 3): If $y = p - p^*$ then because p^* is a critical point for W (Theorem 9.1.4)

$$W(p) = W^* + W(y); \tag{9.2.7}$$

3) $\Rightarrow$ 4): For any $x \in \mathbf{R}^m$ we decompose $x = s(x)p^* + y$ $(s(y) = 0)$ from which

$$W(x) = s^2(x)W(p^*) + W(y). \tag{9.2.8}$$

Because $W(p^*) > 0$ and $W(y)$ is negative definite, W on the whole space has the type $\nu_+ = 1, \nu_- = m - 1$.

4) $\Rightarrow$ 1) by Theorem 9.2.5. $\square$

Remark If p^* is a local attractor then p^* is asymptotically stable.

Corollary 9.2.11 *An interior equilibrium is asymptotically stable if and only if it is the unique absolute maximum of mean fitness W.*

Proof If the form $W(y)$ for $s(y) = 0$ is negative definite then (9.2.7) implies that $W(p) < W(p^*)$ for all $p \in \Delta$, $p \neq p^*$. $\square$

Corollary 9.2.12 *If an interior equilibrium exits and is asymptotically stable then all equilibrium states are asymptotically stable in their minimal faces.*

In this case the set of equilibrium states is finite.

Proof The negative definite quadratic form $W(y)$, $s(y) = 0$, remains so when further restricted to the intersection with coordinate subspaces. $\square$

The criterion of stability of an interior equilibrium analogous to Theorem 9.2.10 is a "degenerate" analog of conditions 2)-4). We delay its study until Section 9.4.

9.3   Variance Estimates of Disequilibrium

Introduce in the space $\mathbf{R}^m$ with the simplex Δ the l_1-metric $||x|| = \sum_{i=1}^m |x_i|$, $x = (x_1, \ldots, x_m)$. Rewriting (9.1.1) as

$$p_i' - p_i = p_i(W_i(p) - W(p))/W(p)$$

we get $||p' - p|| = (1/W(p)) \sum_{i=1}^m p_i |W_i(p) - W(p)|$, from which by Cauchy inequality

$$||p' - p|| \le C\sigma(p), \tag{9.3.1}$$

where $C = \min_\Delta W(p))^{-1}$. Certainly, it is valid for any norm (with a constant depending on the choise of norm).

The following estimates apply to a neighborhood of an equilibrium state $p^* = (p_1^*, \ldots, p_m^*)$. Consider the set of indices $I = \{i | W_i(p^*) = W(p^*)\}$ and its complement J. It follows from (9.1.3) that $p_i^* = 0$ for $i \in J$. Hence $I \ne \emptyset$ and p^* lies in the face Γ_I of the simplex Δ whose vertices correspond to the indices in I. (Γ_I may not coincide with the minimal face containing p^*). Equivalently, we have $\mathrm{supp}(p^*) \subset I$. Observe that in the previous section we ordered the alleles so that $\mathrm{supp}(p^*) = \{1, \ldots, r\}$.

Lemma 9.3.1 $\left.\frac{\partial(\sigma^2)}{\partial p_k}\right|_{p^*} = (W_k(p^*) - W(p^*))^2$. *In particular, these derivatives are equal to zero for $k \in I$.*

Proof We have

$$\frac{\partial(\sigma^2(p))}{\partial p_k} = (W_k(p) - W(p))^2 + 2\sum_{i=1}^m p_i(W_i(p) - W(p))(\lambda_{ik} - 2W_k(p)).$$

But $\sum_{i=1}^m p_i(W_i(p) - W(p)) = 0$. Hence

$$\frac{\partial(\sigma^2)}{\partial p_k} = (W_k(p) - W(p))^2 + 2\sum_{i=1}^m \lambda_{ik} p_i(W_i(p) - W(p)). \tag{9.3.2}$$

In equilibrium, by definition of I and J and by (9.1.3) this reduces to (9.3.2). $\square$

Now consider any point $p \in \Delta$ and consider its orthogonal projection $\bar{p}$ on face Γ_I. Its coordinates $\bar{p}_k$ are $p_k + (1/|I|)\sum_{j\in J} p_j$ for $k \in I$ and 0 for $k \in J$. We remark that

$$\sum_{k\in I} |p_k - \bar{p}| = \sum_{k\in J} |p_k - \bar{p}_k| = \sum_{k\in J} p_k. \tag{9.3.3}$$

Lemma 9.3.2 *In a certain neighborhood of p^* the following estimate holds:* $||p - \bar{p}|| \leq C(\sigma^2(p) - \sigma^2(\bar{p}))$.

Proof Consider the function $D(\tau) = \sigma^2(\bar{p} + \tau(p - \bar{p}))$ for $\tau \in [0, 1]$. By the Mean Value Theorem we have

$$\sigma^2(p) - \sigma^2(\bar{p}) = D(1) - D(0) = \sum_{k=1}^{m} \frac{\partial(\sigma^2(p))}{\partial p_k}\Big|_{p(\theta)}(p_k - \bar{p}_k),$$

where $p(\theta) = \bar{p} + \theta(p - \bar{p})$ for some $\theta \in (0, 1)$. Let $c = \frac{1}{4}\min_{k \in J}(W_k(p^*) - W(p^*))^2$. By Lemma 1.3.1 there exists a ball $||p - p^*|| < \epsilon$ on which $\frac{\partial(\sigma^2(p))}{\partial p_k} \geq 3c$ for $k \in J$ and $|\frac{\partial(\sigma^2(p))}{\partial p_k}| \leq c$ for $k \in I$. On this ball

$$\sum_{k \in J} \frac{\partial(\sigma^2(p))}{\partial p_k}\Big|_{p(\theta)}(p_k - \bar{p}_k) = \sum_{k \in J} \frac{\partial(\sigma^2(p))}{\partial p_k}\Big|_{p(\theta)}p_k \geq 3c \sum_{k \in J} p_k.$$

On the other hand

$$|\sum_{k \in I} \frac{\partial(\sigma^2(p))}{\partial p_k}\Big|_{p(\theta)}(p_k - \bar{p}_k)| \leq c \sum_{k \in I} |p_k - \bar{p}_k| = c \sum_{j \in J} p_j.$$

Hence $\sigma^2(p) - \sigma^2(\bar{p}) \geq 2c \sum_{k \in J} p_k$. By (9.3.3), $\sum_{k \in J} p_k = \frac{1}{2} \sum_{k=1}^{m} |p_k - \bar{p}_k| = \frac{1}{2}||p - \bar{p}||$ which proves the assertion. $\square$

Corollary 9.3.3 *In a neighborhood of p^*, $\sigma(\bar{p}) < \sigma(p)$ for $p \notin \Gamma_I$.*

Let $v = \bar{p} - p^*$ denote the radius-vector of the point $\bar{p}$ with respect to p^*. Obviously $v_i = 0$ for $i \in J$.

Lemma 9.3.4 $\sigma^2(\bar{p}) = \sum_{i \in I} \bar{p}_i W_i^2(v) - W^2(v)$.

Proof Obviously $W(\bar{p}) = W(p^*) + W(v) + 2\sum_{i=1}^{m} v_i W_i(p^*)$. But $v_i = 0$ for $i \in J$ and $W_i(p^*) = W(p^*)$ for $i \in I$. Hence $\sum_{i=1}^{m} v_i W_i(p^*) = W(p^*)\sum_{i \in I} v_i = W(p^*)\sum_{i=1}^{m} v_i = 0$. Hence simply

$$W(\bar{p}) = W(p^*) + W(v). \tag{9.3.4}$$

At the same time $W_i(\bar{p}) = W_i(p^*) + W_i(v)$ because W_i is a linear function. Hence $W_i(\bar{p}) - W(\bar{p}) = W_i(v) - W(v)$ and $\sigma^2(\bar{p}) = \sum_{i \in I} \bar{p}_i(W_i(v) - W(v))^2$. Further, $\sum_{i \in I} \bar{p}_i W_i(v) = \sum_{i=1}^{m} p_i^* W_i(v) + \sum_{i=1}^{m} v_i W_i(v) = \sum_{i,k=1}^{m} \lambda_{ik} p_i^* v_k + \sum_{i,k=1}^{m} \lambda_{ik} v_i v_k = \sum_{i=1}^{m} v_i W_i(p^*) + W(v) = W(v)$ which implies the assertion. $\square$

Denote $f(v) = \sum_{i \in I} |v_i| W_i^2(v)$.

Corollary 9.3.5 *In a neighborhood of p^*, $\sigma^2(\bar{p}) \geq f(v) - W^2(v)$.*

Proof If $p_i^* = 0$ then $\bar{p}_i = v_i = |v_i|$ and if $p_i^* > 0$ then $\bar{p}_i = p_i^* + v_i > |v_i|$ for $|v| < \frac{1}{2}p_i^*$. $\square$

$f(v)$ is a homogeneous function of third degree. $W^2(v)$ is a homogeneous polynomial of fourth degree that becomes zero when $f(v) = 0$. It is not *a priori* clear that near the manifold $f(v) = 0$ the difference $f(v) - W^2(v)$ is nonnegative and Corollary 9.3.5 has content for small v; we will show this now.

Lemma 9.3.6 *Let $Q(z)$ $(z \in \mathbf{R}^\nu)$ be a quadratic form, denote $P(z) = \sum_{k=1}^\nu |z_k|Q_k^2(z)$, where $Q_k(z) = \frac{1}{2}\frac{\partial Q}{\partial z_k}$. Then*

$$|Q(z)| \leq CP^{\frac{2}{3}}(z). \tag{9.3.5}$$

Proof The inequality (9.3.5) is homogeneous and it suffices to prove it on the unit Euclidean sphere $S = \{z : ||z|| = 1\}$.

If $P(z) \neq 0$ for $z \neq 0$ then (9.3.5) holds with $C = \max_z |Q(z)P^{-\frac{2}{3}}(z)|$.

For the general case we will use induction on the dimension ν. The case $\nu = 1$ is trivial. For the induction step from $\nu - 1$ to ν it suffices to show that for every $z \in S$ there exits $\delta > 0$, $K > 0$ such that for a coordinate hyperplane H transversal to z

$$|Q(z+y)| \leq KP^{\frac{2}{3}}(z+y) \tag{9.3.6}$$

for $||y|| < \delta$, $y \in H$. Indeed, (9.3.6) because of homogeneity becomes (9.3.5) in a neighborhood of z in S. Because S is compact it suffices to check (9.3.5) locally.

If $P(z) \neq 0$ then (9.3.6) holds by continuity. Suppose $P(z) = 0$. We may assume that $Q_k(z) \neq 0$ for $1 \leq k \leq l$ and $Q_k(z) = 0$ for $l < k \leq \nu$. Then $z_k = 0$ for $1 \leq k \leq l$ and hence $Q(z) = \sum_{k=1}^\nu z_k Q_k(z) = 0$. Therefore $Q(z+y) = 2\sum_{k=1}^l y_k Q_k(z) + Q(y)$. If $y \in H$ then by induction hypothesis $|Q(y)| \leq CP(y)^{\frac{2}{3}}$ (one y-coordinate is zero) and $|Q(z+y)| \leq C_1(\sum_{k=1}^l |y_k| + P(y)^{\frac{2}{3}})$.

On the other hand

$$P(z+y) = \sum_{k=1}^l |y_k|Q_k^2(z+y) + \sum_{k=l+1}^\nu |z_k + y_k|Q_k^2(y).$$

Let $\delta > 0$ be such that if $\|y\| < \delta$ then $|Q_k(y)| < \frac{1}{2}|Q_k(z)|$ for $1 \le k \le l$ (hence $|Q_k(z+y)| > \frac{1}{2}|Q_k(z)|$) and $|z_k + y_k| > |y_k|$ for $k > l$, $z_k \ne 0$. Then for some $c > 0$ we have $P(z+y) \ge c(\sum_{k=1}^{l} |y_k| + \sum_{k=l+1}^{\nu} |y_k|Q_k^2(y))$.

Denote $\alpha = \sum_{k=1}^{l} |y_k|$, $\beta = \sum_{k=l+1}^{\nu} |y_k|Q_k^2(y)$ and $\gamma = \sum_{k=1}^{l} |y_k|Q_k^2(y)$. The problem reduces to the estimate $\alpha + (\beta + \gamma)^{\frac{2}{3}} = O((\alpha + \beta)^{\frac{2}{3}})$ for small $\alpha, \beta, \gamma \ge 0$ and $\gamma = O(\alpha)$. This estimate is clear. $\square$

Corollary 9.3.7 *For small v we have $f(v) - W^2(v) \ge 0$.*

Proof By the lemma $C^2 f(v)^{\frac{4}{3}} \ge W^2(v)$. For $\|v\|$ small enough, $C^2 f(v)^{\frac{1}{3}} \le 1$ and so $f(v) \ge W^2(v)$. $\square$

The next lemma is of utmost importance in our chain of estimates but it is easily proved now.

Lemma 9.3.8 *In a neighborhood of p^* we have $|W(p^*) - W(p)| \le C\sigma^{\frac{4}{3}}(p)$.*

Proof We have

$$|W(p) - W(\bar{p})| \le C_1\|p - \bar{p}\| \le C_2\sigma^2(p) \le C_3\sigma^{\frac{4}{3}}(p)$$

by Lemma 9.3.2. Now by Corollary 9.3.3, it suffices to prove that $|W(p^*) - W(\bar{p})| \le C\sigma^{\frac{4}{3}}(\bar{p})$ and this, by (9.3.4) and Corollaries 9.3.5 and 9.3.7 reduces to the estimate

$$|W(v)| \le C(f(v) - W^2(v))^{\frac{2}{3}} \tag{9.3.7}$$

for small v. Denote $Q = W(v)$, $P = f(v)$. By Lemma 9.3.6, $|Q| \le MP^{\frac{2}{3}}$, M is a constant. The inequality (9.3.7) is easily brought to the form

$$|Q|^3 + C^3(2PQ^2 - Q^4) \le C^3 P^2. \tag{9.3.8}$$

But $|Q|^3 + 2C^3 PQ^2 \le (M^3 + 2C^3 M^2 P^{\frac{1}{3}})P^2$. Therefore for (9.3.8) to hold it suffices that $M^3 + 2C^3 M^2 P^{\frac{1}{3}} \le C^3$.

Because P is small for small v, the constant C exists. $\square$

The situation is much simpler and Lemma 9.3.8 can be made much stronger if $p_i^* \ne 0$ for $i \in I$, i.e. $p^* \in \text{Int } \Gamma_I$. We will call such equilibrium points *regular*. (Biologically this describes a polymorphism that contains all the alleles having mean fitness equal to the mean fitness of the entire population.) All hyperbolic equilibria are regular Corollary 9.2.4.

Lemma 9.3.9 *On a neighborhood of regular point p^* we have*

$$|W(p^*) - W(p)| \le C\sigma^2(p).$$

Proof We have to show that $|W(v)| \leq C\sigma^2(\bar{p})$. Instead of the inequality of Corollary 9.3.5 we can now use the exact expression of Lemma 9.3.4 and the fact that for sufficiently small $v = \bar{p} - p^*$ we have $\bar{p}_i > \frac{1}{2}p_i^* > 0$ for $i \in I$. Hence it suffices to prove the estimate $|W(v)| \leq C \sum_{i \in I} p_i^* W_i^2(v)$. This follows from the inequality $|(Tx, x)| \leq K\|Tx\|^2$, where T is arbitrary self-adjoint operator on Euclidean space and K is a constant. $\square$

9.4 Convergence to Equilibrium for Selection in an Autosomal Multiallele Locus

The Fundamental Theorem together with the accumulated estimates allows us to use the techniques of the general theory of relaxation processes.

A *relaxation process* with respect to a real function $\phi(z)$, $z \in \mathbf{R}^\nu$ is a sequence $\{z^{(t)}\}_{t=0}^\infty$ on which $\phi(z^{(t+1)}) \leq \phi(z^{(t)})$. When we emphasize the initial process $\{z^{(t)}\}_{t=0}^\infty$, we call such a function $\phi(z)$ a *Lyapunov function* for the process.

By the Fundamental Theorem each trajectory $p^{(t+1)} = Fp^{(t)}$ (where F is the Fisher selection operator given by (9.1.1)) is a relaxation process with respect to the function $-W(p)$. Because the sequence is monotone and bounded, there exists $W^* = \lim_{t\to\infty} W(p^{(t)})$. If Ω is the limit set of the trajectory then $W|\Omega = W^*$.

Lemma 9.4.1 Ω *is a closed, connected invariant set consisting entirely of equilibria and upon which the function W is constant.*

Proof Because F is continuous, the set Ω is invariant, i.e. $F\Omega \subset \Omega$ and $W|\Omega$ is constantly W^*. Hence $W(p') = W(p)$ for $p \in \Omega$ and hence $p' = p$ by Corollary 9.2.9. As Ω consists entirely of equilibria, it is connected by Lemma 1.2.1. $\square$

Corollary 9.4.2 *If the limit set of a trajectory is finite then the trajectory converges.*

Corollary 9.4.3 *If the set of equilibrium states of a population is finite then all the trajectories converge.*

In particular, this happens (by Corollary 9.1.3) in the case when in the fitness matrix all main minors are nonzero.

Example In the alternation model (see Section 9.1) only the vertices $\{e_k\}_1^m$ are equilibria. Moreover, the state e_m is asymptotically stable. At the state e_k for $k < m$ the invariant subspaces $[e_1,\dots,e_{k-1}]$ and $[e_{k+1},\dots,e_m]$ are stable and unstable respectively. So if a trajectory lies in Int Δ then it cannot converge to e_m. The same behavior is observed in every face of Δ. Thus every trajectory converges to the homozygote state with maximal fitness (taking into account only the alleles which are available at the origin).

We will now use the variance estimates to prove a general result totally superseding the preceding elementary facts.

Theorem 9.4.4 *All the trajectories for the Fisher dynamics converge.*

Thus under selection a one-locus autosomal multiallele population always approaches equilibrium.

Proof Consider a limit point p^* of the given trajectory. Let $\Delta^{(t)} = W(p^*) - W(p^{(t)})$. Obviously $\Delta^{(t)}$ monotonically approaches zero.

By Lemma 9.4.1, p^* is an equilibrium. By Lemma 9.3.8 there is a neighborhood $U_\delta = \{p : \|p - p^*\| < \delta\}$ on which $|W(p^*) - W(p)| \leq C\sigma^{\frac{4}{3}}(p)$.

Let $p^{(t)} \in U_\delta$, Then by (9.3.1), Theorem 9.2.8 and the definition of U_δ

$$
\begin{aligned}
\|p^{(t+1)} - p^{(t)}\| \leq C_1\sigma(p^{(t)}) &= C_1\sigma^2(p^{(t)})(\sigma^{\frac{4}{3}}(p^{(t)}))^{-\frac{3}{4}} \\
&\leq C_2(W(p^{(t+1)}) - W(p^{(t)}))(W(p^*) - W(p^{(t)}))^{-\frac{3}{4}} \\
&= C_2(\Delta^{(t)} - \Delta^{(t+1)})(\Delta^{(t)})^{-\frac{3}{4}} \\
&\leq C_2 \int_{\Delta^{(t+1)}}^{\Delta^{(t)}} \tau^{-\frac{3}{4}} d\tau \\
&= 4C_2(\sqrt[4]{\Delta^{(t)}} - \sqrt[4]{\Delta^{(t+1)}}).
\end{aligned}
\tag{9.4.1}
$$

We will show that if $p^{(t)} \in U_{\delta/2}$, $\Delta^{(t)} < (\delta/8C_2)^4$ then $p^{(t+k)} \in U_\delta$ for $k \geq 0$ by induction on k. Indeed, if $p^{(t)}, \ldots, p^{(t+k-1)} \in U_\delta$ then

$$
\|p^{(t+k)} - p^{(t)}\| \leq \sum_{j=1}^{k} \|p^{(t+j)} - p^{(t+j-1)}\|
$$

$$
\leq 4C_2(\sqrt[4]{\Delta^{(t)}} - \sqrt[4]{\Delta^{(t+k)}} \leq 4C_2\sqrt[4]{\Delta^{(t)}} < \delta/2.
\tag{9.4.2}
$$

Now we can analogously write $\|p^{(t+k)} - p^{(t+l)}\| \leq 4C_2\sqrt[4]{\Delta^{(t+l)}}$ for $k > l \geq 0$ and since $\Delta^{(t+l)} \to 0$ as $l \to \infty$, the trajectory converges. $\square$

Let us study the rate of convergence. First we note that it follows from (9.4.1) that $\sum_{t=0}^{\infty} \|p^{(t+1)} - p^{(t)}\| < \infty$. In this sense the convergence is absolute.

Theorem 9.4.5 $\|p^{(t)} - p^*\| = O(t^{-\frac{1}{2}})$.

Proof It follows from (9.4.1) that $\|p^{(t)} - p^*\| = O(\sqrt[4]{\Delta^{(t)}})$. Therefore, it suffices to prove that

$$
\Delta^{(t)} = O(t^{-2}).
\tag{9.4.3}
$$

We have (by Lemma 9.3.8 and Theorem 9.2.8, again) $\Delta^{(t)} = O(\sigma^{\frac{4}{3}}(p^{(t)})) = O([W(p^{(t+1)})-W(p^{(t)})]^{\frac{2}{3}}) = O((\Delta^{(t)}-\Delta^{(t+1)})^{\frac{2}{3}})$ from which $(\Delta^{(t)}-\Delta^{(t+1)})(\Delta^{(t)})^{-\frac{3}{2}}$

$C > 0$ and $\int_{\Delta^{(t+1)}}^{\Delta^{(t)}} \tau^{-\frac{2}{3}} d\tau \geq C$, i.e. $(\Delta^{(t+1)})^{-\frac{1}{2}} - (\Delta^{(t)})^{-\frac{1}{2}} \geq C/2$. Adding up we get $(\Delta^{(t)})^{-\frac{1}{2}} - (\Delta^{(0)})^{-\frac{1}{2}} \geq Ct/2$ which implies (9.4.3). $\square$

Thus the convergence is quite slow. We note that a somewhat faster estimate $O(t^{-1})$ is possible, for example in selection of dominant allele versus recessive one (see section 2.4.) It is desirable to know when the rate of convergence is fast (exponential).

Theorem 9.4.6 *A trajectory $\{p^{(t)}\}_{t=0}^{\infty}$ converges to p^* exponentially, i.e. with estimate $\|p^{(t)} - p^*\| = O(\kappa^t)$, where $0 < \kappa < 1$, if and only if $W_i(p^*) \neq W(p^*)$ for all i such that $p_i^* = 0$ and $p_i^{(0)} \neq 0$. In particular, these conditions are fulfilled for a regular point p^*.*

Proof $\Rightarrow$: In coordinate form the estimate becomes

$$|p_i^{(t)} - p_i^*| = O(\kappa^t). \tag{9.4.4}$$

For the indices i under consideration it follows from (9.1.1) that

$$\lim_{t \to \infty} p_i^{(t+1)} / p_i^{(t)} = \frac{W_i(p^*)}{W(p^*)}$$

from which

$$\lim_{t \to \infty} \sqrt[t]{p_i^{(t)}} = \frac{W_i(p^*)}{W(p^*)}.$$

If $p_i^* = 0$ then $p_i^{(t)} = O(\kappa^t)$. Hence $\lim_{t \to \infty} \sqrt[t]{p_i^{(t)}} \leq \kappa$. Hence $W_i(p^*) \leq \kappa W(p^*)$ and $W_i(p^*) \neq W(p^*)$.

$\Leftarrow$: If $p_j^{(0)} = 0$ for some j then $p_j^{(t)} = 0$ for $t \geq 0$ and $p_j^* = 0$. Thus (9.4.4) holds in jth coordinate trivially for any κ and the system (9.1.1) reduces to the same system of smaller dimensions. Thus we can assume that $p_i^{(t)} \neq 0$ for all i and all t. Then the hypothesis of the theorem reduces to regularity of the equilibrium point p^*. But for a regular point the estimate of Lemma 9.3.8 is replaced by a stronger estimate of Lemma 9.3.9 which correspondingly strengthens the estimate (9.4.1):

$$\|p^{(t+1)} - p^{(t)}\| \leq C_1 \sigma^2(p^{(t)})(\sigma^2(p(t)))^{-\frac{1}{2}} \leq C_2(\Delta^{(t)} - \Delta^{(t+1)})(\Delta^{(t)})^{-\frac{1}{2}}$$

$$\leq C_2 \int_{\Delta^{(t+1)}}^{\Delta^{(t)}} \tau^{-\frac{1}{2}} d\tau = 2C_2(\sqrt{\Delta^{(t)}} - \sqrt{\Delta^{(t+1)}}).$$

Hence $||p^{(t)} - p^*|| = O(\sqrt{\Delta^{(t)}})$. At the same time $\Delta^{(t)} = O(\sigma^2(p^{(t)}))$, and so by Theorem 9.2.8 again, $\Delta^{(t)} \leq C(\Delta^{(t)} - \Delta^{(t+1)})$ and hence $\Delta^{(t+1)} \leq \kappa^2 \Delta^{(t)}$ where $\kappa = \sqrt{1 - 1/C}$. Thus $\Delta^{(t)} = O(\kappa^{2t})$ and $||p^{(t)} - p^*|| = O(\kappa^t)$ $\square$

Slightly modifying the proof of Theorem 9.4.4 we obtain the criterion of local stability (cf. Theorem 9.2.6).

Theorem 9.4.7 *An equilibrium state p^* is stable if and only if it is a local maximum of the mean fitness $W(p)$.*

Proof $\Rightarrow$ follows easily from the Fundamental Theorem. Indeed, suppose the state p^* is stable. Let U be its neighborhood so small that all equilibrium states $p \in U$ lie in the same connected component of the set of equilibrium states. Let the neighborhood $U_0 \subset U$ be such that if $p^{(0)} \in U_0$ then the trajectory starting at $p^{(0)}$ lies in U. Then $W(p) \leq W(p^*)$ for $p \in U_0$. Otherwise if $W(p) > W(p^*)$ then, for any limit point $\bar{p}$ of the trajectory starting at p, $W(\bar{p}) > W(p^*)$ contradicting Corollary 9.1.5.

$\Leftarrow$: Let p^* be a local maximum of $W(p)$. Take $\delta > 0$ so small that in the neighborhood U_δ, $|W(p^*) - W(p)| \leq C\sigma^{\frac{4}{3}}(p)$ and $W(p) \leq W(p^*)$, $W(Fp) \leq W(p^*)$. Consider a trajectory $\{p^{(t)}\}_{t=0}^\infty$. If $p^{(t)} \in U_\delta$ then the estimate (9.4.1) holds. Now if $p^{(0)} \in U_\epsilon$ for some $0 < \epsilon < \delta/2$ and ϵ is so small that $\Delta^{(0)} < (\delta/8C_2)^4$ then the entire trajectory lies in U_δ since if $p^{(0)}, \ldots p^{(k-1)} \in U_\delta$ then analogously to (9.4.2) $||p^{(k)} - p^{(0)}|| < \delta/2$ from which $||p^{(k)} - p^*|| < \delta/2 + \epsilon < \delta$, i.e. $p^{(k)} \in U_\delta$. $\square$

Corollary 9.4.8 *An equilibrium state p^* is asymptotically stable if and only if it is an isolated local maximum of mean fitness $W(p)$.*

We will now obtain a criterion of stability for an interior equilibrium analogous to Theorem 9.2.10.

Theorem 9.4.9 *Let p^* be an interior equilibrium. Then the following statements are equivalent: 1) p^* is stable; 2) p^* is a local maximum of mean fitness $W(p)$; 3) quadratic form $W(y)$ is negative semidefinite on the subspace $\mathbf{R}_0^n$ where $s(y) = 0$; 4) quadratic form W has a positive inertia index $\nu_+ = 1$.*

Proof 1) $\Rightarrow$ 2) by Theorem 9.4.7. 2) $\Rightarrow$ 3) $\Rightarrow$ 4) as in the proof of Theorem 9.2.10. Finally 4) $\Rightarrow$ 1) because if $\nu_+ = 1$ then the decomposition (9.2.8) implies that $W(y) \leq 0$ for $s(y) = 0$ and then by (9.2.7) $W(p) \leq W(p^*)$ i.e. p^* is a local maximum (in fact absolute) of $W(p)$. $\square$

Corollary 9.4.10 *An interior equilibrium p^* is stable if and only if it is an absolute maximum of the mean fitness $W(p)$.*

We will now investigate the structure of domains of attraction of equilibrium states. If p^* is an equilibrium then its domain of attraction $Z(p^*)$ consists of initial states p such that the trajectory starting at p converges to p^*. For simplicity we will assume that all $\lambda_{ik} > 0$, i.e. there are no lethal zygotes.

Lemma 9.4.11 *The Fisher operator F is an orientation preserving diffeomorphism of the simplex Δ.*

Proof The differential of the map F (naturally extended to the entire space $\mathbf{R}^m$ except for the cone $W(p) = 0$) is given by (9.2.1). Consider the matrix $\tilde{L}^{(m)} = (W_i \delta_{ik} + \lambda_{ik} p_i)_{i,k=1}^m$ and let us prove that it is nonsingular.

Suppose for concreteness that $p_1, \ldots, p_r > 0$, $p_{r+1} = \cdots = p_m = 0$. Then $\det \tilde{L}^{(m)} = W_{r+1} \cdots W_m \det \tilde{L}^{(r)}$. But $\det \tilde{L}^{(r)} \neq 0$ for in this matrix the diagonal dominates (the columns): $\sum_{i \neq k} \lambda_{ik} p_i = W_k - \lambda_{kk} p_k < W_k$ for $1 \leq k \leq r$. It is also obvious that $W_i = \sum_{k=1}^r \lambda_{ik} p_k > 0$ for $1 \leq i \leq m$. Hence $\det \tilde{L}^{(m)} \neq 0$.

The matrix of the differential of map F is obtained from $\tilde{L}^{(m)}$ by one-dimensional perturbation $(-2p_i W_i W_k W^{-1})$ and multiplication by W^{-1}. Therefore its rank is $\geq m - 1$. In fact it is equal to $m - 1$ because the matrix annihilates the vector $(p_1, \ldots, p_m)$. (F is degenerate, its image lies in the hyperplane $s(x) = 1$.) Therefore on the invariant hyperplane $s(dp) = 0$ the differential is nondegenerate and F is a local diffeomorphism of the simplex Δ. It preserves orientation because at vertex e_k the Jacobian equals $\lambda_{kk}^{-m} \prod_{i=1}^m \lambda_{ik} > 0$.

We will prove that F is bijective by induction on dimension using the invariance of faces. If $q \in \text{Int } \Delta$ then the complete preimage $F^{-1}q \subset \text{Int } \Delta$ and this set is finite. The vector field $Fp - q$, on the boundary of the simplex is directed outward because $Fp \in \partial \Delta$ for $p \in \Delta$. Hence the index is 1. The indices of all the singular points are also 1. By the Index Theorem the number of singular points, i.e. the preimages of q, is 1 (cf. the proof of Theorem 8.1.4). $\square$

Corollary 9.4.12 *The domain of attraction of an asymptotically stable state $p*$ is open (in the simplex) and connected.*

Proof A sufficiently small neighborhood U_δ of p^* is in $Z(p^*)$. Hence $Z(p^*) = \cup_{t=0}^\infty F^{-t}U_\delta$. But all the $F^{-t}U_\delta$ are open, connected and have a common point p^*. $\square$

Corollary 9.4.13 *The domain of attraction of a hyperbolic equilibrium state p^* is a smooth connected submanifold of the simplex Δ. Its dimension equals to the number of eigenvalues of differential of F at the point p^* in the interval $(0,1)$.*

Proof The intersection $Z(p^*) \cap U_\delta$ for δ small enough has the required properties by Hadamard-Perron Invariant Manifold Theorem $\square$

We will call the operator F *hyperbolic* if al of its fixed points are hyperbolic. The set of equilibrium states of a hyperpbolic operator F is finite. Let $p^{(1)}, \ldots p^{(l)}$ be the asymptotically stable ones.

Corollary 9.4.14 Δ *decomposes as* $\Delta = Z(p^{(1)}) \cup \cdots \cup Z(p^{(l)}) \cup D$. *where D is a smooth submanifold whose dimension is less than $m-1$.*

Proof By Theorem 9.4.4, the entire simplex decomposes into domains of attraction of equilibrium states. We denote by D the union of domains of attraction of the states that are not asymptotically stable ad invoke Corollary 9.4.l3. $\square$

The operator F is hyperbolic in the generic case.

Theorem 9.4.15 *The set $\mathcal{H}$ of hyperbolic operators is open and dense in the space of all Fisher operators.*

Proof The fact that $\mathcal{H}$ is open is obvious. To show that it is dense, let F be Fisher operator with fitness matrix Λ. We assume that all main minors of Λ are nonzero (we can achieve it by adding to each λ_{ii} an arbitrarily small ϵ). Then by Theorem 9.1.2, each face of the simplex Δ contains no more than one fixed point. Let $p^* = (p_1^*, \ldots, p_r^*, 0, \ldots, 0)$ be one of them. If p^* is not a hyperbolic point, then $W_i(p^*) = W(p^*)$ for some $i > r$. Let $W_{r+1}(p^*) = \cdots = W_j(p^*) = W(p^*)$ and $W_i(p^*) \neq W(p^*)$ for $i > j$. Disturb the matrix Λ letting $\tilde{\lambda}_{ik} = \lambda_{ik} + \epsilon$ $(\epsilon > 0)$, $\tilde{\lambda}_{ki} = \tilde{\lambda}_{ik}$ for $r < i \leq j, 1 \leq k \leq r$ and not changing the other elements. Then $W_i(p^*)$ for $r < i \leq j$ increase by ϵ and the other $W_i(p^*)$ do not change. This point p^* will be fixed (the only one in the face $p_{r+1} = \cdot = p_m = 0$ for ϵ small enough) and hyperbolic. Going to the next nonhyperbolic point we preserve the previous ones as hyperbolic for small enough disturbance (although these points can move a little). $\square$

The dynamical picture is much simplified if an asymptotically stable interior equilibrium exists. Let us consider a slightly more general situation.

Theorem 9.4.16 *If the quadratic form W is negative semidefinite on $\mathbf{R}_0^n$, i.e. $W(y) \leq 0$ when $s(y) = 0$, then any trajectory starting interior to the simplex converges to a point where the absolute maximum of mean fitness $W(p)$ is achieved.*

This global *maximization principle* is much stronger than the maximization along a trajectory implied by the Fundamental Theorem, but unlike the latter it only holds under special conditions.

Proof Suppose that in equilibrium state p^*, with $p_i^* > 0$ for $1 \leq i \leq r$ and $p_i^* = 0$ for $r < i \leq m$, the inequalities

$$W_i(p^*) \leq W(p^*) \qquad (i > r) \qquad (9.4.5)$$

hold. We will show that $W(p)$ achieves the absolute maximum at this point. If $y = p - p^*$ is the local coordinate vector then

$$W(p) - W(p^*) = 2 \sum_{i=1}^{m} W_i(p^*)y_i + W(y)$$

$$= 2 \sum_{i=r+1}^{m} (W_i(p^*) - W(p^*))y_i + W(y).$$

Here, as usual, we used the equilibrium equation $W_i(p^*) = W(p^*)$ for $i \leq r$ and the constraint $s(y) \equiv \sum_{i=1}^{m} y_i = 0$. Applying (9.4.5), the inequalities $y_i \geq 0$ for $i > r$ and $W(y) \leq 0$ we see that $W(p) \leq W(p^*)$ for any $p \in \Delta$.

Now suppose some trajectory $\{p^{(t)}\}_0^\infty$ converges to the point p^* where the absolute maximum is not achieved. Then there exists index $i > r$ for which $W_i(p^*) > W(p^*)$. We will show that $p^{(0)} \in \partial\Delta$. We may assume the entire trajectory lies in such a neighborhood of p^* that $W_i(p) \geq cW(p)$ for some constant $c > 1$. Then $p_i^{(t+1)} \geq cp_i^{(t)}$ and $p_i^{(t)} \geq c^t p_i^{(0)}$ and hence $p_i^{(0)} = 0$. $\square$

Remark Inequality (9.4.5) is necessary for a local maximum (cf. Theorem 9.2.6).

Corollary 9.4.17 *If p^* is an asymptotically stable interior equilibrium then its domain of attraction coincides with the interior of the simplex, $Z(p^*) =$ Int Δ.*

Proof Because the faces of the simplex are invariant, we have $Z(p^*) \subset$ Int Δ. We apply Theorem 9.4.16 and Corollary 9.2.11 to get the reverse inclusion. $\square$

Under the same conditions it follows from Corollary 9.2.12 that if p^* is any other equilibrium state then $Z(p^*) = \mathrm{Int}\ \Gamma$, where Γ is the minimal face containing p^*. All these states are asymptotically stable in their faces and unstable in the whole. In particular, all the vertices are unstable.

Thus *in the presence of an asymptotically stable interior equilibrium evolution leads to the maximum value of mean fitness possible for the collection of alleles present at the beginning of the process.*

If the operator F is hyperbolic then, as the decomposition of Corollary 9.4.14 shows, every trajectory not lying on the exceptional set D converges to some (depending on the trajectory) local maximum of $W(p)$. However in the nonhyperbolic situation there may be an equilibrium state in which there is no local maximum of $W(p)$ (i.e. it is not stable) but its domain of attraction contains interior points. In this sense the state is dynamically observable, despite its instability.

Example Let the fitness matrix have the form

$$
\begin{pmatrix}
\lambda + a_{11} & \lambda + a_{12} & \lambda \\
\lambda + a_{21} & \lambda + a_{22} & \lambda \\
\lambda & \lambda & \lambda
\end{pmatrix}
$$

where $a_{11} > 0 > a_{22} > a_{12}\ (= a_{21})$ and $\lambda > |a_{12}|$. Then the allele A_3 is dominant over A_1, A_2; homozygote $A_1 A_1$ is better (in fitness) than homozygote $A_3 A_3$ better than homozygote $A_2 A_2$, better than heterozygote $A_1 A_2$. We have $W_1 = \lambda + a_{11} p_1 + a_{12} p_2$, $W_2 = \lambda + a_{21} p_1 + a_{22} p_2$ and $W_3 = \lambda$. They are positive, hence $W > 0$. The equilibrium states are the vertices e_1, e_2, e_3 of the simplex $\Delta = \Delta^2$ and the point p^* on the edge $[e_3, e_2]$ defined by equation $(a_{11} - a_{21})p_1 = (a_{22} - a_{12})p_2 = 0$. This equation in the simplex defines a line segment $[e_3, p^*]$ that splits the simplex into two triangles $\Delta_1 \ni e_1$ and $\Delta_2 \ni e_2$ (Fig. 9.1). e_1 is asymptotically stable, e_2 is a hyperbolic saddle, p^* is a repelling point. All this is established by linear approximation. But at e_3 the linear approximation is the identity operator, so it becomes necessary to take the nonlinear corrections into account. At least e_3 is unstable because on the edge $[e_1, e_3]$, the point e_3 repels.

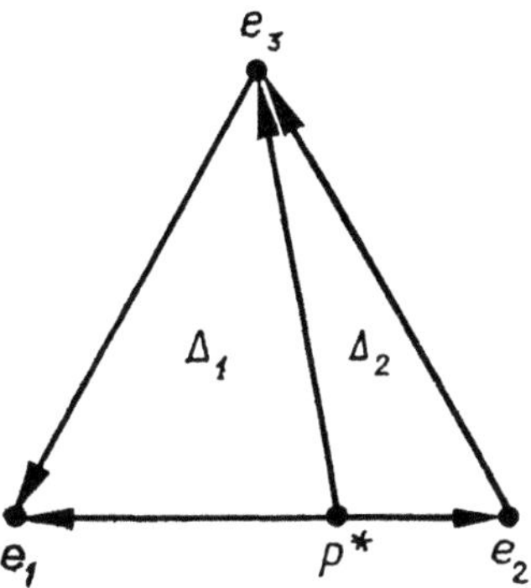

Fig. 9.1

Because p^* is a repelling point and $Z(e_2) = (p^*, e_2]$, a trajectory that starts at $p^{(0)}$, $p_3^{(0)} > 0$ may converge only to e_1 or e_3. But the triangle Δ_2 is invariant, for if $(a_{11} - a_{21})p_1 - (a_{22} - a_{12})p_2 < 0$ then $W_1 < W_2$ and $W((a_{11} - a_{21})p_1' - (a_{22} - a_{12})p_2') < W_1((a_{11} - a_{21})p_1 - (a_{22} - a_{12})p_2) < 0$. The set $M = \Delta_2 \backslash [p^*, e_2]$ is also invariant, since if $p_3 > 0$ then $p_3' > 0$. If $p^{(0)} \in M$ then the trajectory converges to e_3 because $e_1 \notin \bar{M}$. It is also possible to show that $Z(e_1) = N = \Delta_1 \backslash [p^*, e_3]$ but this is immaterial.

Mean fitness $W(p)$ achieves the absolute maximum at e_1 and has no other local maxima.

9.5 Evolutionary Selection Equations in a System of Autosomal Multiallel Loci

In this section we will use the language and notation of chapter 6. In particular, $\Delta \equiv \Delta(\Gamma)$ denotes the simplex generated by gametes. We will begin by deriving the evolutionary equation. Assume that the survival coefficient of a zygote $\zeta = g \circ h$ depends only on its gametes g and h and denote it $\lambda(g, h)(= \lambda(h, g))$. We also assume that $\lambda(g, g) > 0$ for all g.

If the gamete fund of a generation is in the state p then when the next generation appears after a round of random mating, the distribution of gamete pairs is $p \otimes p$, i.e. $\mathrm{Pr}\{g \circ h\} = p(g)p(h)$. By the reproductive stage after selection has occurred, it has become $\lambda(g,h)p(g)p(h)/W(p)$, where $W(p) = \sum_{g,h} \lambda(g,h)p(g)p(h)$ is the mean fitness of the population. Obviously $W(p) > 0$ on the entire simplex of gamete states.

The next state p' arises from meiosis. Hence (cf. (6.2.14)):

$$p'(g) = W^{-1}(p) \sum_{U|V} r(U|V) \sum_h \lambda(g_U h_V, h_U g_V) p(g_U h_V) p(h_U g_V). \qquad (9.5.1)$$

As in section 9.1 the mean fitness of gamete g is

$$W_g(p) = \sum_h \lambda(g,h)p(h). \qquad (9.5.2)$$

And, as before,

$$W(p) = \sum p(g) W_g(p). \qquad (9.5.3)$$

In the equation (9.5.1) it is convenient to isolate the terms not depending on the linkage distribution (contrast (6.2.19)):

$$p'(g) = \frac{p(g)W_g(p)}{W(p)} + \frac{1}{W(p)} \sum_{U|V \neq L} r(U|V) D_{U|V;g}(p), \qquad (9.5.4)$$

where,

$$D_{U|V;g}(p) = \sum_h (\lambda(g_U h_V, h_U g_V) p(g_U h_V) p(h_U g_V) - \lambda(g,h)p(g)p(h)). \qquad (9.5.5)$$

According to (9.5.4) the evolutionary operator can be written as $F + R$, where the operator F is, formally speaking, Fisher, i.e. the dynamic of a one-locus population in which the gametes play the roles of alleles. The operator R accounts for recombination. If all the loci under consideration are completely linked, i.e. $r(L) = 1$ then $R = 0$.

We will later need the following lemma. Recall the $a_i(g)$ is the gene at the i-th locus in the gamete g.

Lemma 9.5.1 *For any allele a at the i-th locus ($i = i(a)$)*

$$\sum_{g:a_i(g)=a} D_{U|V;g}(p) = 0.$$

Proof In the sum

$$\sum_{a_i(g)=a} \sum_h \lambda(g_U h_V, h_U g_V) p(g_U h_V) p(h_U g_V) \qquad (9.5.6)$$

we make the substitution

$$g_U h_V = \gamma \qquad h_U g_V = \chi. \qquad (9.5.7)$$

It is invertible ($g = \gamma_U \chi_V, h = \chi_U \gamma_V$) and if for concreteness $i \in U$ then the condition $g_i = a$ is equivalent to the condition $a_i(\gamma) = a$. As the result, (9.5.6) becomes $\sum_{a_i(\gamma)=a} \sum_\chi \lambda(\gamma, \chi) p(\gamma) p(\chi) \equiv \sum_{a_i(g)=a} \sum_h \lambda(g, h) p(g) p(h)$. It remains to invoke the definition of $D_{U|V;g}(p)$. $\square$

Now recall the gene probabilities in state p (defined by the map F^i of section 6.2):

$$f_a(p) = \sum_{a_i(g)=a} p(g). \qquad (9.5.8)$$

Obviously $\sum_{a \in \Gamma_i} f_a(p) = 1$ ($1 \le i \le l$) where Γ_i is the set of all the alleles at the i-th locus. From (9.5.4) and Lemma 9.5.1 it follows that

$$f_a(p') = \frac{1}{W(p)} \sum_{a_i(g)=a} p(g) W_g(p). \qquad (9.5.9)$$

These equations define the dynamics of gene probabilities. We note however that the system (9.5.9) is not closed but is only a projection of the closed system (9.5.1). Thus, the change in the gene probabilities at locus i does depend on the probabilities of all gametes.

We will now consider the conditions under which the Fundamental Theorem holds.

Lemma 9.5.2 *If* $\mathrm{Im}R \subset \ker W$, *then the Fundamental Theorem holds:* $W(p') \ge W(p)$, *and equality is equivalent to the system of equations*

$$p(g)(W_g(p) - W(p)) = 0, \qquad (9.5.10)$$

describing equilibrium with respect to the operator F.

Proof We have $W(p') = W(Fp + Rp) = W(Fp)$. By the Fundamental Theorem for one locus population, $W(Fp) \ge W(p)$ and the equality is equivalent to (9.5.10). $\square$

Corollary 9.5.3 *If the survival coefficients $\lambda(g,h)$ satisfy*

$$\lambda(k, g_U h_V) + \lambda(k, h_U g_V) = \lambda(k, g) + \lambda(k, h) \qquad (9.5.11)$$

for all partitions $U|V$ then the Fundamental Theorem holds and the equality is achieved if and only if (9.5.10) holds.

Proof The inclusion Im $R \subset$ ker W can be represented as

$$\sum_g \lambda(k, g) \sum_{U|V \neq L} r(U|V) D_{U|V;g}(p) = 0,$$

i.e. by (9.5.5)

$$\sum_{U|V \neq L} r(U|V) \sum_{g,h} \lambda(k,g)\{\lambda(g_U h_V, h_U g_V)p(g_U h_V)p(h_U g_V)$$

$$-\lambda(g,h)p(g)p(h)\} = 0,$$

which, after substitution (9.5.7) and relabeling, becomes

$$\sum_{U|V \neq L} r(U|V) \sum_{g,h} (\lambda(k, g_U h_V) - \lambda(k,g))\lambda(g,h)p(g)p(h) = 0.$$

But the inner sum is zero, which follows immediately from (9.5.11) after symmetrizing on g and h. $\square$

The condition (9.5.11) is satisfied, in particular, by *additive* selection in which the survival coefficient of each zygote is additively defined by single locus survival coefficients of allele pairs:

$$\lambda(g,h) = \sum_{i=1}^{l} \lambda_i(a_i(g), a_i(h)). \qquad (9.5.12)$$

The fact that (9.5.11) holds follows from the obvious formula

$$\lambda(k, g_U h_V) = \sum_{i \in U} \lambda_i(a_i(k), a_i(g)) + \sum_{i \in V} \lambda_i(a_i(k), a_i(h)).$$

Thus the following theorem holds.

Theorem 9.5.4 *The Fundamental Theorem holds for additive selection with equality achieved if and only if (9.5.10) holds.*

Unless otherwise specified we will assume additive selection in the future.

Corollary 9.5.5 *Let $\{p^{(t)}\}_{t=0}^{\infty}$ be a trajectory of a population, Ω its limit set. Then Ω is a closed set, invariant for the evolutionary operator $F + R$, and upon which W is constant at $W^* = \lim_{t\to\infty} W(p^{(t)})$. Furthermore, $p(g)(W_g(p) - W^*) = 0$ for each $p \in \Omega$.*

The latter follows from the fact that for $p \in \Omega$, we have $p' \in \Omega$ and $W|\Omega = \text{const} = W^*$. Thus *all limit points of the trajectory are equilibria for F.* In particular, all *equilibrium states of the population are equilibria for F.*

Theorem 9.5.4 follows also from the next remark, which is of independent interest.

Theorem 9.5.6 *Under additive selection, the mean fitnesses of the gametes and of the entire population depend only on the gene probabilities.*

Proof Substituting (9.5.12) into (9.5.2) and considering (9.5.8) we get $W_g(p) = \sum_{i=1}^{l} \sum_{a\in\Gamma_i} \lambda_i(a_i(g), a) f_a(p)$, from which, by (9.5.3) and (9.5.8) $W(p) = \sum_{i=1}^{l} \sum_{a,\bar{a}\in\Gamma_i} \lambda_i(a, \bar{a}) f_a(p) f_{\bar{a}}(p).$ $\square$

Finally, we have the corresponding refinement of the Fundamental Theorem for additive selection.

Theorem 9.5.7 $W(p') - W(p) \geq C\sigma^2(p)$, *where*

$$\sigma^2(p) = \sum_g p(g)(W_g(p) - W(p))^2 \tag{9.5.13}$$

is the variance of gametic mean fitness $W(p)$ and C is a positive constant.

Proof $W(p') - W(p) = W(Fp) - W(p) \geq C\sigma^2(p)$ by Theorem 9.2.8. $\square$

9.6 Convergence to Equilibrium under Additive Selection for a System of Autosomal Multiallele Loci

The assumption of additive selection first allows us to use relaxation techniques to prove the existence of limits of gene probabilities. These probabilities can be described by a vector $f \equiv f(p) = \{f_a(p)\}_{1 \leq i \leq l; a \in \Gamma_i}$.

Lemma 9.6.1 *The following estimate holds*

$$\|f(p') - f(p)\| \leq C\sigma(p). \tag{9.6.1}$$

Proof p' is the outcome of selection plus recombination applied to p, i.e. $p' = F(p) + R(p)$. Define $\tilde{p} = F(p)$, the result of selection alone regarding the loci as totally linked. (9.3.1) says that $\|\tilde{p} - p\|$ is $O(\sigma(p))$. Because the association $p \mapsto f(p)$ is the restriction of a linear map we have $\|f(\tilde{p}) - f(p)\|$ is $O(\|\tilde{p} - p\|)$. But by (9.5.8) and (9.5.9) $f(p') = f(\tilde{p})$. So $\|f(p') - f(p)\|$ is $O(\sigma(p))$. $\square$

The following variance estimate follows from Lemma 9.3.8 and Theorem 9.5.6.

We will call the vector $\bar{f}$ of the values of gene probabilities *equilibrial* if there exists an equilibrium $\bar{p} \in \Delta$ for the operator F (i.e. satisfying (9.5.10)) such that $f(\bar{p}) = \bar{f}$. By Theorem 9.5.6 mean fitness at p depends only on $f(p)$ and so we can write $W(f)$ for $W(p)$ when $f = f(p)$. Similarly, $W_g(f) = W_g(p)$ with $f = f(p)$.

Lemma 9.6.2 *An equilibrial vector $\bar{f}$ of gene probabilities has a neighborhood N such that for all $p \in \Delta$ with $f(p) \in N$, the following estimate holds:*

$$|W(\bar{f}) - W(p)| \leq C\sigma^{\frac{4}{3}}(p). \tag{9.6.2}$$

Proof Let $\bar{p}$ be such that $\bar{f} = f(\bar{p})$. If $\bar{p}$ is an equilibrium for F then by Lemma 9.3.8 there exists a neighborhood on which the estimate (9.6.2) holds. If $\bar{p}$ is not an equilibrium then $\sigma(\bar{p}) \neq 0$ and we can choose a neighborhood $U(\bar{p})$ on which $\sigma(p) > \frac{1}{2}\sigma(\bar{p}) > |W(\bar{p}) - W(p)|$ so that (9.6.2) holds there as well. Because the set $\{\bar{p}|\bar{p} \in \Delta, f(\bar{p}) = \bar{f}\}$ is compact, we can cover it with finitely many sets $U(\bar{p})$ (with union U) and by using the maximum of the associated constants C, (9.6.2) holds on U. Finally, by compactness of $\Delta \backslash U$ there is an open set N such that $\bar{f} \in N$ and $f^{-1}(N) \subset U$. $\square$

Now let p^* be a limit point of a sequence $\{p^{(t)}\}_{t=0}^{\infty}$. By Corollary 9.5.5 it is an equilibrium for F and hence the vector $f^* = f(p^*)$ is equilibrial. If $f^{(t)} = f(p^{(t)})$ lies in the neighborhood N of f^* in which (9.6.2) holds, then from (9.6.2), (9.6.1) and Theorem 9.5.7 we get the estimate (cf. (9.4.1)) $\|f^{(t+1)} - f^{(t)}\| \le C(\sqrt[4]{\Delta^{(t)}} - \sqrt[4]{\Delta^{(t+1)}})$ where $\Delta^{(t)} = W(p^*) - W(p^{(t)})$. From here, in a way analogous to section 9.4, we have the following theorem.

Theorem 9.6.3 *Let $\{p^{(t)}\}_{t=0}^{\infty}$ be a population trajectory. Then there exists a limit $f^* = \lim_{t\to\infty} f(p^{(t)})$.*

Corollary 9.6.4 *There exist limits of the gametic mean fitnesses $W_g^* = \lim_{t\to\infty} W_g(p^{(t)})$.*

Corollary 9.6.5 *If $W_g^* \ne W^*$ for some gamete g, then $p^*(g) = 0$ for all $p^* \in \Omega$, the limit set.*

On the set Ω (obviously invariant for the evolutionary operator, $F + R$) the function W is constantly W^* and it cannot be used to study the dynamic. But it turns out that the entropy $H(p) = -\sum_g p(g)\ln p(g)$ grows when moving in Ω as in Theorem 6.3.5. Be aware, however, that our goal is to trivialize this observation. Having proved it we will use it to show that no motion occurs, i.e. Ω is a single point.

For every partition $U|V$ we introduce the auxiliary two-locus population in which the loci are the classes U, V the genes are all possible subgametes g_U, g_V and the probability of crossing over is 1 (cf. Section 6.3). (In the case of the trivial partition $L|\emptyset$ this population becomes a one-locus population with evolutionary operator F.) The evolutionary equation of such population is

$$p'_{U|V}(g) = \frac{1}{W(p_{U|V})}(p_{U|V}(g)W_g(p_{U|V}) + D_{U|V;g}(p_{U|V})). \tag{9.6.3}$$

by (9.5.4), and so the evolutionary operator of the original population is

$$p'(g) = \sum_{U|V} r(U|V)p'_{U|V}(g) \tag{9.6.4}$$

with the gametes in original and auxiliary populations naturally identified.

Consider the action of the auxiliary operator (9.6.3) for $p \in \Omega$. Let u be an allele of "locus" U, v an allele of "locus" V, $f_u(p) = \sum_{g_U=u} p(g)$,

$f_v(p) = \sum_{g_V = v} p(g)$ be gene probabilities in the auxiliary population. We have

$$\lambda(g, h) = \lambda_U(g_U, h_U) + \lambda_V(g_V, h_V) \tag{9.6.5}$$

where for example $\lambda_U(g_U, h_U) = \sum_{i \in U} \lambda_i(a_i(g), a_i(h))$. Hence $W_g(p) = \sum_u \lambda_U(g_U, u) f_u(p) + \sum_v \lambda_V(g_V, v) f_v(p)$.

If $p \in \Omega$ then

$$p'_{L|\emptyset}(g) = p(g)$$

$$p'_{U|V}(g) = p(g) + \frac{1}{W^*} D_{|U|V;g}(p) \tag{9.6.6}$$

$$= \frac{1}{W^*}(f_{g_U}(p) \sum_u \lambda_U(g_U, u) p(u g_V) + f_{g_V}(p) \sum_v \lambda_V(g_V, v) p(g_U v))$$

for $U|V \neq L$.

Without loss of generality we will assume $r(U|V) \neq 0$ for any nontrivial partition $U|V$. We will also assume that all one-locus survival coefficients are positive.

Lemma 9.6.6 *If $p \in \Omega$ and $W_g^* \neq W^*$ then for any partition $U|V$ we have* $p'_{U|V} = 0$, $f_{g_U}(p) f_{g_V}(p) = 0$.

For the trivial partition the latter becomes $p(g) = 0$ which holds by Corollary 9.6.5.

Proof It follows from Corollary 9.6.5 and the invariance of Ω that $p'(g) = 0$. But $r(U|V) \neq 0$ for nontrivial $U|V$. Hence by (9.6.4), $p'_{U|V}(g) = 0$. For the trivial partition this equality follows from (9.6.6). Hence the first equality is established. From here and (9.6.6) it follows that $f_{g_U}(p) \sum_u \lambda_U(g_U, u) p(u g_V) = 0$, from which either $f_{g_U}(p) = 0$ or $p(u g_V) = 0$ for all u, i.e. $f_{g_V}(p) = 0$. $\square$

It is clear now that in (9.6.6) we may assume $W_g^* = W^*$, $f_{g_U}(p) > 0$, $f_{g_V}(p) > 0$ for any g since by Lemma 9.6.6 and preservation of gene probabilities we may disregard the genes g_U or g_V for which this is false. We will assume that this has been done.

Lemma 9.6.7 *For any g, h and $p \in \Omega$*

$$\sum_u \lambda_U(g_U, u) f_u(p) = \sum_u \lambda_U(h_U, u) f_u(p)$$

(and similarly vor V).

Proof This follows from the equality $W^*_{h_U g_V} = W^* = W^*_g$. $\square$
We will now turn to the proof of the H-theorem.

Lemma 9.6.8 *If $p \in \Omega$ then $H(p'_{U|V}) \geq H(p_{U|V})$ and the equality occurs if and only if $p(g) = f_{g_U}(p) f_{g_V}(p)$ for all g.*

Proof For simplicity we will use i, k to index the U-subgametes, i.e. the elements of Γ_U, and j, l to index the V-subgametes. So gamete g with $g_U = i$, $g_V = j$ is represented by the ordered pair ij. But now we need to write f_i^U and f_j^V instead $f_i \equiv f_{g_U}$ and $f_j \equiv f_{g_V}$. In this notation, the equation (9.6.6) has the matrix form

$$p'_{ij} = \frac{1}{W^*}(f_i^U \sum_k \lambda_U(i,k)p_{kj} + f_j^V \sum_l \lambda_V(j,l)p_{jl}), \qquad (9.6.7)$$

where f_i^U, f_j^V, $\lambda_U(i,k)$, $\lambda_V(j,l)$ are positive. Denote $\alpha = 1/W^* \sum_i \lambda_U(i,k)f_k^U$, $\beta = 1/W^* \sum_j \lambda_V(j,l)f_j^V$. By Lemma 9.6.7, these numbers do not depend on k and l and by $W_g^* = W^*$ and (9.6.5) we have $\alpha + \beta = 1$. Let $s_{ik} = \lambda_U(i,k)f_i^U/(\alpha W^*)$, $t_{jl} = \lambda_V(j,l)f_j^V/(\beta W^*)$ and consider the positive stochastic (by columns) matrices $S = (s_{ik})$, $T = (t_{jl})$ and the matrix $P = (p_{ij})$. Then (9.6.7) can be rewritten as $P' = \alpha SP + \beta PT$, and $\sum_j p_{ij} = f_i^U$, $\sum_i p_{ij} = f_j^V$. It remains to show $H(\alpha SP + \beta PT) \geq H(P)(\equiv - \sum p_{ij}\ln p_{ij})$ and the equality occurs if and only if $p_{ij} = f_i^U f_j^V$, i.e. $P = f^U \otimes f^V$. Because the entropy H is strictly concave and $\alpha + \beta = 1$, $\alpha, \beta > 0$ it suffices to prove the following general fact.

Lemma 9.6.9 *Let S be an $n \times n$ stochastic (by columns) matrix, and x be a stochastic column vector such that $Sx = x$. If $S \gg 0$ then on set of $n \times m$ matrices $P = (p_{ij}) \geq 0$ satisfying $Pe = x$ (e is a column of m ones), the entropy $H(P)$ does not decrease as P maps to SP, i.e. $H(SP) \geq H(P)$. The equality occurs if and only if $P = xy$, where y is a stochastic row vector.*

Proof Let π_x be the set of all nonnegative stochastic (by columns) matrices S such that $Sx = x$. This set is a convex polyhedron. For every matrix P satisfying the stated conditions we define the concave function $h_P(S) = H(SP) - H(P)$ on the set π_x.

For the special case when $x_i = \frac{1}{n}$ for $1 \leq i \leq n$ the set π_x consists of bistochastic matrices whose extreme points are the permutation matrices. For a permutation matrix S it is clear that $H(SP) = H(P)$ and so the result for every positive S follows from concavity of entropy. For general x we lack such a simple description of the extreme points.

If $P = xy$ then $h_P = 0$. Suppose $P \neq xy$, where y is the vector (row) of column sums of P. Consider the $n \times n$ matrix

$$X = \begin{pmatrix} x_1 & \cdots & x_1 \\ \vdots & \ddots & \vdots \\ x_n & \cdots & x_n \end{pmatrix}.$$

Then $XP = xy$. But $H(xy) > H(P)$. This and concavity of h_P imply that $H(SP) > H(P)$ for $S \in \text{Int } \pi_x$ and $P \neq xy$ if and only if the derivative of h_P at the points $\bar{S} \in \pi_x$ at which $h_P(\bar{S}) = 0$ is nonnegative along vectors pointing into the polyhedron π_x.

Suppose $h_P(\bar{S}) = 0$ for some $\bar{S} \in \pi_x$, i.e. $H(\bar{S}P) = H(P)$. Then the derivative $h'_P(\bar{S})$ along the vector $S - \bar{S}$ is $-H(P) - \sum_{ij}(SP)_{ij}\ln((\bar{S}P)_{ij})$. It remains to show that $\psi_P(S) \equiv \sum_{ij}(SP)_{ij}\ln((\bar{S}P)_{ij}) \leq -H(P)$. This is a linear programming problem because the function $\psi_P(S)$ is linear and $\psi_P(\bar{S}) = -H(P)$. Consider its dual problem of minimizing the function $\phi_x(z) = \sum_{i=1}^{n}(x_i z_i + z_{n+i})$ with restrictions $x_i z_k + z_{n+i} \geq \sum_{t=1}^{m} p_{it}\ln((\bar{S}P)_{kt})$ for $1 \leq i, k \leq n$. We need to find such solution z^* of the dual problem that $\phi_x(z^*) = -H(P)$. Let $z_i^* = \ln x_i$, $z_{n+i}^* = \sum_{t=1}^{m} p_{it} \ln p_{it} - x_i \ln x_i$, for $1 \leq i \leq n$. We will denote $\bar{S}P$ by $\bar{P}$ and we will prove that the point z^* is admissible for the dual problem, i.e. the inequality

$$x_i \ln x_k + \sum_{t=1}^{m} p_{it} \ln p_{it} - x_i \ln x_i \geq \sum_{i=1}^{m} p_{it} \ln \bar{p}_{kt} \quad (1 \leq i, k \leq n).$$

This is equivalent to the inequality

$$\ln(\frac{x_k}{x_i}) \geq \frac{1}{x_i} \sum_{t=1}^{m} p_{it} \ln (\frac{\bar{p}_{kt}}{p_{it}}),$$

or

$$\ln(\frac{1}{x_i} \sum_{t=1}^{m} \bar{p}_{kt}) \equiv \ln(\sum_{t=1}^{m} \frac{p_{it}}{x_i} \frac{\bar{p}_{kt}}{p_{it}}) \geq \frac{1}{x_i} \sum_{t=1}^{m} p_{it} \ln(\frac{\bar{p}_{kt}}{p_{it}}).$$

The latter inequality is obvious because logarithm is concave. It is easy to check that $\phi_x(z^*) = -H(P)$. $\square$

To formulate the H-theorem for the original population, let

$$p^*(g) = \prod_{i=1}^{l}(\lim_{t \to \infty} f_{a_i(g)}(p^{(t)}))$$

(cf. Theorem 9.6.3).

Theorem 9.6.10 *If $p \in \Omega$, $p \neq p^*$ then $H(p') > H(p)$.*

Proof By equation (9.6.4), the concavity of the entropy and Lemma 9.6.8 we get

$$H(p') \geq \sum_{U|V} r(U|V) H(p'_{U|V}) \geq \sum_{U|V} r(U|V) H(p_{U|V}) = H(p).$$

The equality $p = p^*$ holds only if $p(g) = f_{g_U}(p) f_{g_V}(p)$ for all gametes g and all partitions $U|V$. $\square$

From Theorem 9.6.10 we derive the main result of this section.

Theorem 9.6.11 *Under additive selection, every trajectory in the population converges.*

Let $V : X \to X$ be a continuous mapping of a compact set X, $\{x_t\}_0^\infty$ be a trajectory ($x_{t+1} = V x_t$), Ω be the set of limit points, and $\bar{x} \in \Omega$. A function $\phi : \Omega \to \mathbf{R}$ is called a *strong Lyapunov function (s.L.f)* on Ω, if $\phi(Vx) < \phi(x)$ for $x \in \Omega$, $x \neq \bar{x}$. As Theorem 9.6.10 says that $(-H)$ is an s.L.f. the result follows from the following general lemma.

Lemma 9.6.12 *If there exists a s.L.f on the limit set Ω of the trajectory $\{x_t\}_0^\infty$ then the trajectory converges.*

Proof Note first that if $y_0 \in \Omega$, the Lyapunov function ϕ is constant on the limit point set of the orbit $\{y_t\}_0^\infty$ and so all the trajectories in Ω converge to $\bar{x}$. It follows also that $V\bar{x} = \bar{x}$ and $\bar{\phi} \equiv \phi(\bar{x}) = \min_\Omega(\phi) < \phi(y)$ for $y \in \Omega$, $y \neq \bar{x}$.

For any $\epsilon > 0$ define the open neighborhood $U_\epsilon = \{y \in \Omega : \phi(y) > \bar{\phi} - \epsilon\}$ containing $\bar{x}$. For any $y \in \Omega$ the trajectory $\{V^t(y)\}_0^\infty$ eventually enters U_ϵ. It then remains there. So there is an open neighborhood $G(y)$ of y and an integer T_y such that $V^t(G(y)) \subset U_\epsilon$ for $t = T_y$ and hence for all $t \geq T_y$. Cover Ω with finitely many such $G(y)$'s and let T_ϵ be the maximum of the associated T_y's. Then $V^T(\Omega) \subset U_\epsilon$. But as the limit point set of an orbit in X, Ω is not only V invariant, i.e. $V(\Omega) \subset \Omega$ but, in fact, $V(\Omega) = \Omega$ for if the subsequence $\{x^{t_i}\}$ converges to $y \in \Omega$ then any limit point $\bar{y}$ of $\{x^{t_i-1}\}$ satisfies $\bar{y} \in \Omega$ and $V\bar{y} = y$.

Because V is surjective on Ω we have $\Omega = V^t(\Omega) \subset U_\epsilon$ for all $\epsilon > 0$. Hence $\Omega \subset \cap_{\epsilon>0} U_\epsilon$, i.e. ϕ is constantly $\bar{\phi}$ on Ω. This implies $\Omega = \{\bar{x}\}$. $\square$

The previous discussion also gives rise to the following theorem.

Theorem 9.6.13 *Under additive selection the equilibrium states are exactly those which give equilibrium both for a population with the same linkage distribution but without selection and for a population with the same selection but complete linkage.*

If the selection is not additive, then even for the two-locus two-allele case the trajectories may diverge.

For the two-locus two-allele case we use the notation of Section 2.2, numbering the four gamete types $1 \equiv AB$, $2 \equiv Ab$, $3 \equiv aB$, $4 \equiv ab$. The linkage distribution is described by the single probability $r = r(U|V)$ where $U|V$ is the unique nontrivial partition of the two loci. In the notation of Section 2.2 we are thus assuming that $r = \rho$. The only nontrivial crossing over occurs between double heterozygote pairs: 14 or 23. In particular, in the sum (9.5.5) for each $g \equiv i = 1, 2, 3, 4$ there is only one nonzero term and we compute $D_{U|V;i}(p) = -\lambda\epsilon_i D$ where $\lambda \equiv \lambda_{14} = \lambda_{23}$ (we assume the two double heterozygotes have a common fitness value); $\epsilon_1 = \epsilon_4 = +1$, $\epsilon_2 = \epsilon_3 = -1$; and $D = p_1 p_4 - p_2 p_3$ is the measure of disequilibrium (cf. (2.2.5)).

So the evolutionary operator is defined by

$$p_i' = \frac{1}{W(p)}(p_i W_i(p) - r\lambda D\epsilon_i). \qquad (9.6.8)$$

Example We consider a particular fitness matrix and exploit its symmetric character:

$$\Lambda = \begin{pmatrix} \sigma^2 & \sigma & \sigma & 1 \\ \sigma & \sigma^2 & 1 & \sigma \\ \sigma & 1 & \sigma^2 & \sigma \\ 1 & \sigma & \sigma & \sigma^2 \end{pmatrix} \quad (0 < \sigma < 1).$$

Lemma 9.6.14 *If $p_1 = p_4$ then $W_1(p) = W_4(p)$ and $p_1' = p_4'$. Similarly, $p_2 = p_3$ implies $W_2(p) = W_3(p)$ and $p_2' = p_3'$. In particular, the line segment defined by $p_A = p_a = p_B = p_b = \frac{1}{2}$ is an invariant set for the evolutionary operator.*

Proof $W_1 = W_4$ follows by multiplying the first and fourth rows of Λ by the column vector p. Since $\epsilon_1 = \epsilon_4$, $p_1' = p_4'$ follows from (9.6.8). By (2.2.7) the pair of equations $p_1 = p_4$ and $p_2 = p_3$ is equivalent to the pair $p_A = p_a$ and $p_B = p_b$. But $p_A + p_a = 1 = p_B + p_b$. $\square$

On the invariant segment the equations analogous to (2.2.9) can be written as

$$4p_i = 1 + 4D\epsilon_i \tag{9.6.9}$$

and so we see that on the segment $z \equiv p_1 - p_3 - p_3 + p_4 (= \sum_i p_i \epsilon_i) = 4D$ and we can rewrite (9.6.9) in vector form

$$4p = e + z\epsilon \tag{9.6.10}$$

where $e_i = 1$ for $1 \leq i \leq 4$.

Theorem 9.6.15 *Restricted to the invariant segment the one dimensional evolutionary operator is given by*

$$z' = \frac{2z(1 + \sigma^2 - 2r)}{(1 + \sigma)^2 + z^2(1 - \sigma)^2} \tag{9.6.11}$$

Proof Observe that $\Lambda e = (1 + \sigma)^2 e$ and $\Lambda \epsilon = (1 - \sigma)^2 \epsilon$. Also $e^T \epsilon = \epsilon^T e = 0$ and $\epsilon^T \epsilon = e^T e = 4$. From (9.6.10) we compute

$$W_i(p) = (\Lambda p)_i = \frac{(1 + \sigma)^2 + z(1 - \sigma)\epsilon_i}{4} \tag{9.6.12}$$

$$W(p) = p^T \Lambda p = \frac{(1 + \sigma)^+ z^2(1 - \sigma)}{4}$$

Now by definition of z ($\lambda = 1$ in this case); $Wz' = p_1 W_1 - p_2 W_2 - p_3 W_3 + p_4 W_4 - rD(\epsilon_1 - \epsilon_2 - \epsilon_3 + \epsilon_1)$. Since $4D = z$, we can write this as

$$Wz' = \tilde{p}^T \Lambda p - rz \tag{9.6.13}$$

where $\tilde{p}_i = p_i \epsilon_i$ or from (9.6.10) $4\tilde{p} = \epsilon + ze$. Hence, $4\tilde{p}^T \Lambda p = (1 + \sigma)^2 z + (1 - \sigma)^2 z = 2(1 + \sigma)^2 z$. Substituting this and (9.6.12) into (9.6.13) we get (9.6.11). $\square$

Let p^* denote the barycenter of the simplex $p_i = \frac{1}{4}$ for $1 \leq i \leq 4$ which is equivalently described as the point on the segment when $z = 0 = D$.

Corollary 9.6.16 p^* *is an equilibrium for all values of r.*
1) For $0 \leq r < (1 - \sigma)^2/4$, p^* is repelling on the segment. Trajectories with initial values of z positive (resp. negative) are attracted monotonically to the asymptotically stable equilibrium p_+ (resp. p_-) defined by

$$z_\pm = \pm\sqrt{1 - \frac{4r}{(1 - \sigma)^2}} \tag{9.6.14}$$

for $r \geq (1 - \sigma)^2/4$, p^* is the only equilibrium on the segment.

2) For $(1 - \sigma)^2/4 \leq r < (1 + \sigma)^2/2$, p^* attracts all trajectories on the segment monotonically.

3) For $r = (1+\sigma)^2/2$, the evolutionary operator maps the entire segment to p^*.

4) For $(1 + \sigma)^2/2 < r \leq (3 + 2\sigma + 3\sigma^2)/4$, the evolutionary operator reverses orientation on the segment and each trajectory on the segment converges to p^* in an oscillatory manner.

5) For $(3 + 2\sigma + 3\sigma^2)/4 < r \leq 1$, p^* is a repelling equilibrium. There is a periodic orbit of period two consisting of the two points defined by

$$z = \frac{\pm\sqrt{4r - 3 - 2\sigma - 3\sigma^2}}{1 - \sigma} \tag{9.6.15}$$

The periodic orbit is attracting and so nearby orbits do not converge but instead approach the periodic orbit asymptotically.

Proof (9.6.11) can be written as $z' = Vz$ with

$$V(z) = \frac{Mz}{1 + Kz^2}, \qquad M = \frac{2(1 + \sigma^2 - 2r)}{(1 + \sigma)^2}, \qquad K = (\frac{1 - \sigma}{1 + \sigma})^2.$$

In particular, $V(0) = 0$ with derivative M at 0. With σ fixed M is a decreasing linear function of r and $M = 1, 0$ and -1 at the bifurcation values of $r : (1 - \sigma)^2/4, (1 + \sigma)^2/2$ and $(3 + 2\sigma + 3\sigma^2)/4$. Because $K > 0$ we have that $|V(z)| < |Mz|$ for $z \neq 0$, and so when $|M| \leq 1$, $z = 0$ attracts all trajectories.

When $M > 1$ the alternative fixed points, nonzero solutions of $V(z) = z$ are given by $z^2 = (M - 1)/K$ which is equivalent to (9.6.14).

$$\frac{dV}{dz} = \frac{M(1 - Kz^2)}{(1 + Kz^2)^2}$$

and so on the segment $-1 \leq z \leq 1$ the derivative has the same sign as M because $K < 1$. It easily follows graphically that when $M > 1$ the pair of nonzero equilibria attract all nonzero trajectories monotonically.

When $M = 0$, V degenerates to map the entire segment to 0. Once M is negative $V(z)$ has the opposite sign from z and so V reverses orientation. Thus, the approach of trajectories to 0 is oscillating when $0 > M \geq -1$.

When M moves below -1 a period doubling bifurcation occurs. p^* becomes repelling again and an orbit of period 2 appears as a pair of points above and below $z = 0$. We compute them by substituting to get

$$V^2(z) = \frac{M^2 z(1 + Kz^2)}{(1 + Kz^2)^2 + M^2 Kz^2} = \frac{m^2 z(1 + Kz^2)}{(1 + Kz^2)^2 + M^2(1 + Kz^2) - M^2}.$$

So for the nonzero solutions of $V^2(z) = z$ one solves $(1 + Kz^2)^2 = M^2$ whose two real solutions are given by (9.6.15). The derivative of V^2 is the product of two derivatives of V and so has the sign of M^2 (positive) for $|z| \leq 1$. By graphical analysis the two fixed points attract V^2-trajectories and so the periodic orbit attracts V-trajectories. $\square$

In addition to it symmetry observe that the fitness matrix Λ is multiplicative, i.e. $\log \lambda_{ij}$ is additive between loci.

The sample is notable both for the nonconvergence in case 5 and the degeneracy in case 3. Contrast the latter with Lemma 9.4.11.

9.7 Appendix. Characterization of Additive Selection

Let $X_1, \ldots, X_l$ be nonempty sets, G be an additive (hence abelian) group. A map $f : X_1 \times \cdots \times X_l \to G$ is called *additive* if it can be represented as $f(x_1, \ldots, x_l) = \sum_{k=1}^{l} f_k(x_k)$, where $f_k : X_k \to G$. For $l = 1$ any map is additive.

Theorem 9.7.1 *A map $f : X_1 \times \cdots \times X_l \to G$ is additive if and only if for each pair $i \neq j$*

$$\begin{aligned} f(\ldots, x_i', \ldots, x_j', \ldots,) &+ f(\ldots, x_i, \ldots, x_j, \ldots,) \\ - f(\ldots, x_i', \ldots, x_j, \ldots,) &- f(\ldots, x_i, \ldots, x_j', \ldots,) = 0 \end{aligned} \qquad (9.7.1)$$

(other arguments fixed).

Proof $\Rightarrow$ is clear. To prove $\Leftarrow$ we use induction on l. For $l = 1$ the assertion is trivial. Let $l \geq 2$ and suppose that the assertion holds for $l - 1$. Fix $x_0^1 \in X_1$ and consider the difference $f(x_1, x_2, \ldots x_l) - f(x_1^0, x_2, \ldots, x_l)$. By (9.7.1), it does not depend on $x_2, \ldots, x_l$, so $f(x_1, x_2, \ldots, x_l) = f(x_1^0, x_2, \ldots, x_l) + f_1(x_1)$, where $f_1 : X_1 \to G$. But $f(x_1^0, x_2, \ldots, x_l)$ as a function of $x_2, \ldots, x_l$ satisfies (9.7.1). By induction hypothesis it has the form $f(x_1^0, x_2, \ldots, x_l) = \sum_{k=2}^{l} f_k(x_k)$. $\square$

Corollary 9.7.2 *If a map $f : X_1 \times \cdots \times X_l \to G$ is additive with respect to every pair of arguments then it is additive.*

We know that additive selection satisfies the linear equation (9.5.11), i.e.

$$\lambda(k, g_U h_V) + \lambda(k, h_U g_V) = \lambda(k, g) + \lambda(k, h), \qquad (9.7.2)$$

ensuring that the Fundamental Theorem applies. This equation may have nonadditive solutions.

Example Consider the two-locus two-allele case. Then (9.7.2) becomes the system of equations, in the notation of the end of Section 9.6:

$$\lambda_{i1} + \lambda_{i4} = \lambda_{i2} + \lambda_{i3} \qquad (i = 1, 2, 3, 4). \qquad (9.7.3)$$

On the other hand any additive system also satisfies

$$\lambda_{14} = \lambda_{23} \qquad (9.7.4)$$

as the double heterozygotes both have the same fitness. Thus, for example, with $\alpha, \beta > 0$ and unequal and $\gamma = \frac{1}{2}(\alpha + \beta)$ the matrix

$$\Lambda = \begin{pmatrix} \alpha & \gamma & \gamma & \beta \\ \gamma & \beta & \alpha & \gamma \\ \gamma & \alpha & \beta & \gamma \\ \beta & \gamma & \gamma & \alpha \end{pmatrix}$$

satisfies $\Lambda \epsilon = 0$ which is equivalent to (9.7.3). (9.7.4) does not hold because the cross diagonal is not constant.

It turns out that additive selection is characterized among the solutions of the equation (9.7.2) by virtue of being *recombination-invariant*, i.e.

$$\lambda(k_U g_V, g_U k_V) = \lambda(k, g) \qquad (9.7.5)$$

for all partitions $U|V$ of the locus set L. We note that this requires not only definition (9.5.12) but also the symmetry of one-locus survival coefficients. Moreover, (9.7.5) implies (for $U = \emptyset$) that λ is a symmetric function. Incidentally, (9.7.2) says exactly that the function $\phi_k(g, h) = \lambda(k, g) + \lambda(k, h)$ is recombination-invariant.

Theorem 9.7.3 *Any recombination-invariant solution $\lambda(k, g)$ of (9.7.2) is additive.*

Proof By Corollary 9.7.2 the proof reduces to the two-locus case.

For the two-locus two-allele case recombination invariance is exactly (9.7.4), i.e. $\lambda(AB, ab) = \lambda(Ab, aB)$. The space of symmetric solutions of the system (9.7.3) with this additional condition has dimension 5. But the space of additive solutions has the same dimension. Indeed, these solutions depend on six parameters $\lambda(AA)$, $\lambda(aa)$, $\lambda(Aa)$, $\lambda(BB)$, $\lambda(bb)$, $\lambda(Bb)$, and the kernel of the parameterizing operator

$$\lambda(AB, AB) = \lambda(AA) + \lambda(BB),$$

$$\lambda(AB, Ab) = \lambda(AA) + \lambda(Bb)$$

$$\ldots$$

is one-dimensional: $\lambda(AA) = \lambda(aa) = \lambda(Aa) = -\lambda(BB) = -\lambda(bb) = -\lambda(Bb)$. Hence all recombination-invariant resolutions are additive.

Suppose now that one locus contains the alleles $A_1, \ldots, A_{m_1}$ and the other one contains $B_1, \ldots B_{m_2}$. Temporarily discard all but two alleles in each locus, say A_i, A_j, B_k and B_l. The resulting two-allele loci still satisfy (9.7.2) and are recombination-invariant. Therefore the corresponding restriction of the function λ has the form $\alpha_{ij,kl} + \beta_{kl,ij}$ where the first term is a function on genotype set $A_iA_i, A_jA_j, A_iA_j = A_jA_i$ and the second term on $B_kB_k, B_lB_l, B_kB_l = B_lB_k$. We will show that each term does not depend on the second pair of indices with error up to an additive constant.

Let $s, t \in \{i, j\}$, $r, q \in \{k, l\}$. Then $\lambda(A_sB_k, A_tB_k) = \alpha_{ij,kl}(A_sA_t) + \beta_{kl,ij}(B_kB_k)$. This implies that the function $\alpha_{ij,kl} - \alpha_{ij,kl'}$ does not depend on A_sA_t, i.e. is constant. The same is true for $\alpha_{ij,k'l'} - \alpha_{ij,kl'}$ and hence for $\alpha_{ij,k'l'} - \alpha_{ij,kl}$. The restriction $l' \neq k$ is immaterial since for $l' = k$ the resulting difference is $\alpha_{ij,k'k} - \alpha_{ij,kl} = \alpha_{ij,kk'} - \alpha_{ij,kl}$ (the symmetry with respect to each pair of indices is obvious). Thus $\alpha_{ij,kl} = \alpha_{ij} + a_{ij,kl}$, where $a_{ij,kl}$ is constant. Analogously, $\beta_{kl,ij} = \beta_{kl} + b_{kl,ij}$, where $b_{kl,ij}$ is constant. Hence $\lambda(A_sB_r, A_tB_q) = \alpha_{ij}(A_sA_t) + \beta_{kl}(B_rB_q) + c_{ij,kl}$, where $c_{ij,kl}$ is constant.

Let $\lambda_{ijk} = \lambda(A_iB_k, A_jB_k)$. Obviously $\lambda_{jik} = \lambda_{ijk} = \alpha_{ij}^0 + \beta_{kl}^0 + c_{ij,kl}$, where $\alpha_{ij}^0 = \alpha_{ij}(A_iA_j)$, $\beta_{kl}^0 = \beta_{kl}(B_kB_k)$. We eliminate $c_{ij,kl}$ obtaining

$$\lambda(A_sB_r, A_tB_q) = \alpha'_{ij}(A_sA_t) + \beta'_{kl}(B_rB_q) + \lambda_{ijk}, \qquad (9.7.6)$$

where $\alpha'_{ij} = \alpha_{ij} - \alpha_{ij}^0$, $\beta'_{kl} = \beta_{kl} - \beta_{kl}^0$. In particular,

$$\lambda(A_iB_k, A_iB_l) = \alpha'_{ij}(A_iA_i) + \beta'_{kl}(B_kB_l) + \lambda_{ijk}. \qquad (9.7.7)$$

Take any self-map $i \to j_i$ of the allele set $A_1, \ldots, A_m$ with no fixed points and let in (9.7.7) $j = j_i$:

$$\lambda(A_i B_k, A_i B_l) = \alpha'_{ij_i}(A_i A_i) + \beta'_{kl}(B_k B_l) + \lambda_{ij_i,k}. \qquad (9.7.8)$$

Subtracting (9.7.8) from (9.7.7) we have $\lambda_{ijk} = \epsilon_{ik} + \omega_{ij}$, where $\epsilon_{ik} = \alpha'_{ij_i}(A_i A_i) + \lambda_{ij_i,k}$ and $\omega_{ij} = -\alpha'_{ij}(A_i A_i)$. Letting $\tilde{\alpha}_{ij} = \alpha'_{ij} + \omega_{ij}$, $\tilde{\beta}_{kl} = \beta'_{kl}$ we rewrite (9.7.6) as $\lambda(A_s B_r, A_t B_q) = \tilde{\alpha}_{ij}(A_s A_t) + \tilde{\beta}_{kl}(B_r B_q) + \epsilon_{ik}$. Because $\tilde{\beta}_{kl}$ is symmetric with respect to k and l, ϵ_{ik} does not depend on k and can be included in $\tilde{\alpha}_{ij}$. So we can write $\lambda(A_s B_r, A_t B_q) = \tilde{\alpha}_{ij}(A_s A_t) + \tilde{\beta}_{kl}(B_r B_q)$. Then we see that $\tilde{\alpha}_{ij}(A_i A_i)$ does not depend on j. Define $\alpha(A_i A_i) = \tilde{\alpha}_{ij}(A_i A_i)$ for $1 \leq i \leq m_1$, $\alpha(A_i A_j) = \tilde{\alpha}_{ij}(A_i A_j)$ for $i \neq j$. Analogously define β. Then $\lambda(A_s B_r, A_t B_q) = \alpha(A_s A_t) + \beta(B_r B_q)$. $\square$

Finally we remark that recombination invariance means invariance under the action of the transformation group $T_{U|V}(g, h) = (g_U h_V, h_U g_V)$. It is natural to call it the *recombination group*. It is easy to see that it is a product of $l - 1$ cyclic groups of order 2.

Appendix A

Commentaries and References

Chapter 1

§1.1. The reader can become acquainted with the genetics using one of the standard textbooks, for example (SinE58) which includes, as an appendix, a translation of Mendel's classic paper (MenG66).

The evolution of populations is closely connected with the origin of species (CheS26; DobT70; MayE70; KimM71).

§1.2. The inheritance coefficients were first introduced by S.N. Bernstein (see (BerS24)), who used them to derive the evolutionary equations (1.2.12). The survival coefficients appeared in the classical works by Fisher, Haldane, Wright (see (HalJ24; FisR30; WriS46; WriS68)). According to Fisher, the mean fitness of the population determines the exponential growth or decrease of its size (*Malthusian parameter*). Ratner (RatV71) axiomatized the notion of generalized fitness of genotypes as numerical characteristics whose population average is the ratio of population sizes in subsequent generations; the evolutionary equations are then written in the standard form (cf. (1.2.9)). These requirements are satisfied in a number of concrete models which take into account the genetic and population mechanisms (see (RatV72; RatV73)).

§1.3. The study of evolution on two parallel levels, zygote and gamete, dates to the origin of population genetics in the works of Hardy (HarG08), Weinberg (WeiW08) (note Castle (CasW03) as their predecessor) and has been continuing until now. Usually preference is given to the more easily depicted level of gametes. The connection of levels in concrete situations can be easily traced. The general notion of the two level population was introduced in (LyuY71). The general theory described in §1.3 is first published here.

Chapter 2

A number of monographs (FisR30; HogL46; DahC48; MalG48; KemO57; MorP62; WebF67; WriS68; EweW69; WriS69; CroJ70; KojK70; LiC76; FriY77; NagT77) and a great number of journal articles (see the bibliography in the books mentioned above) have been devoted to the elementary models of evolutionary genetics. These models describe many interesting genetic situations from which we have chosen just a few basic ones. Because of the restrictions on space we are not considering such phenomena as *polyploidy, inbreeding* and other deviations from panmixia, *overlapping generations, frequency-dependent selection* etc., though the methods described in this book are applicable to many of these phenomena. Also we do not touch at all stochastic models connected with diffusion equations, such as questions of *gene drift* etc. All of these can be seen, for example, in (MorP62), where the main sources before 1960 are given. More recent works are reflected in, for example, Crow and Kimura's book (CroJ70) and Li's book (LiC76).

§2.1. The algebraic symbolism appeared in the founding work of Mendel (MenG66) in a form strikingly close to the contemporary one. Subsequently geneticists widely used elementary algebraic apparatus for the "Mendel" calculations (see (SerA70)). The first attempts to axiomatize this apparatus were undertaken by Serebrovsky (SerA34) and Glivenko (GliV36) (see also (KosV38)). The modern algebraic approach to the population genetics appeared in the pioneer work of Etherington (EthI39).

The Hardy-Weinberg Law was independently discovered by the authors in 1908 (HarG08; WeiW08). The equilibrium of a population in the first generation of descendants was noted for special cases) (for example, (1/4,1/4,1/2)) in prior works by Castle (Cas W03) and Pearson (PeaK04). The axiomatization of the Stationarity Principle is due to Bernstein (BerS22) who likened this principle to Newton's Law of inertia. The initial investigation of the Bernstein problem connected with the Stationarity Principle was conducted by himself (BerS23; BerS23a; BerS24). The importance of the problem was again noted by Bernstein in the report (BerS26).

The multiallele extension of the Hardy-Weinberg Law was first done by Weinberg (WeiW09).

§2.2. The equilibrium and dynamics of a free two-loci two- allele population with normal crossing-over was studied by Jennings (JenH17) (numerically) and Robbins (RobR18a) (analytically). The case of pseudolinkage when $r < 1/2$ has been considered in the work by Miké (MikV77). The general situation $r \neq \rho$ given in §2.2 is investigated here for the first time. A generalization to the case of three or more loci is a difficult problem.

§2.3. The case of X-linkage was investigated by Jennings (JenH16) (numerically) and Robbins (RobR18) (analytically). The results related to XY-linkage for $\rho_1 \neq \rho_2$ are first published here. The corresponding multiallele problem has not been solved.

§2.4. The study of the selection process, which was begun in Fisher (FisR22), was continued by him in (FisR30), by Haldane in the famous series of nine articles (HalJ24) and Wright (summarized in (WriS69)). Among the first works in this field there were also (WarH17; NorH28; KemW29). The Fundamental Theorem was discovered by Fisher (FisR30). Selection in the general case of Extended Hardy-Weinberg Law is investigated here for the first time.

§2.5. Wright (see (WriS69)) suggested the matrix description of the mutation process. The combinatorial condition which is necessary and sufficient for the convergence to the equilibrium with arbitrary initial condition is the primitivity of the indecomposable components of the matrix T.

The Haldane formula was obtained by him in (HalJ24,5)).

Chapter 3

§3.1. Theorem 3.1.1 was used in (LyuY77) as a lemma in proving of the nilpotence of Engel algebras of dimension $n \leq 4$. Suttles (SutD72) has shown that the latter is false when $n > 4$.

§3.2. Our definition of nilalgebra probably is somewhat different from the usual one, but coincides with it in the case of a power associative algebra (cf. (GerM60)). Theorem 3.2.1 is analogous to a theorem by Gerstenhaber (GerM60) who first applied categorical considerations.

§3.3. The concept of a baric algebra was introduced by Etherington (EthI39), and has been systematically used since. The theory stated in §3.3 was developed by the author beginning in (LyuY71) where, in particular, the notion of the invariant linear form was introduced (in the language of quadratic operators). Incidentally, the ideal $J^\perp$ appeared as the linear hull of the elements of the type $xy - 1/2(\sigma(y)x + \sigma(x)y)$ in (EthI40). A low estimate of $\dim J$ in Bernstein algebra is obtained in (KraA85).

A definition of conservativity ($\equiv$ correcteness $\equiv$ normality) was introduced in (LyuY71). These algebras are especially important for Bernstein problem.

§3.4. Stationary (Bernstein) algebras were prompted by the problem of Bernstein. They were used in the author's articles (LyuY73;LyuY74a) on the "working" level, and later were explicitly defined by Holgate (HolP75) as an algebraic equivalent of the notion of the stationary (Bernstein) quadratic operator. This class of operators, which actually appeared in Bernstein's works, was defined in modern terms by the author in (LyuY71), where the notion of type was introduced and Lemmas 3.4.1, 3.4.7, Theorem 3.4.3 and structure Theorem 3.4.8, together with a number of corollaries were obtained. It is interesting that the standard decomposition of Bernstein algebra is the same that classical Pierce decomposition (see (SchR66)) and, moreover, all the idempotents are indecomposable (WorA87).

The summary of most results stated in §3.4 was published in the note (LyuY77b). However, some of them were obtained earlier (for example, Theorems 3.4.27, 3.4.32 (LyuY75)). Theorem 3.4.34 is due to Bernstein (BerS24).

Holgate in (HolP75) initiated the classification of Bernstein algebras of dimension $n \le 4$. This problem for Bernstein algebras of the type (m, δ) where $m \le 3$ or $\delta \le 1$ (in particular, when $n \le 4$) was solved in (LyuY80; LyuY84). In the general case it may be "wild" because it contains as a subproblem the classification of linear bundles of operators and quadratic forms.

In work (AlcM89) the Lie algebra of derivations of Bernstein algebra is considered and the estimate of its dimension $\le m(m - 1) + \delta$ is obtained. It is achieved if and only if the algebra is generated by a projection (WorA89).

§3.5. The meaning of the term "genetic algebra" varies in the literature. This term first appeared as the title of the article of Etherington (EthI39), and in its context it means any algebra arising in genetics. Etherinton considered a number of concrete genetic situations (see (EthI39)). In an attempt to formalize them, he introduced the notion of the special train algebra and showed in (EthI40; EthI41b) that any special train algebra has a basis whose existence we took as the definition of the genetic algebra (in this we are following Gonshor (GonH60; GonH71)). Thus, Theorem 3.5.4

is due to Etherington. The failure of the converse was found by Schafer (SchR49) on an example of the duplication of the Mendel multiallele algebra of zygotes. This algebra is genetic (see Theorem 3.6.5), but B^2 is not an ideal in it. In Schafer's work the definition of genetic algebra is what we call Schafer's condition. Reiersöl (ReiO62) probably following the example of (EthI39) calls "genetic" some concrete algebras arising in genetics. At present genetic algebras (defined by Gonshor) cover a wide area of applications. Heuch (HeuI78) considered the genetic algebras as a manifold in the space of all algebras. Notice here that the first survey of algebraic aspects of the population genetics was written by Bertrand (BerM66). The systematic presentation was done by Wörz-Busekros (WorA80). The survey (LyuY78) is oriented rather "pragmatically".

The proof of the Theorem 3.5.1, is due to Shafer (SchR49), and to Holgate (HolP72). It is to refine Holgate's argument here that Theorem 3.2.3 was proved.

The notion of a train algebra was introduced by Etherington (EthI39). The train rank, which he also introduced, exceeds ours by 1.

The example of an Engel, non-nilpotent algebra is taken from (LyuY77), and a slight simplification of the Suttles' (SutD72) examples as the result of refusing power associativity. Abraham (AbrV76), using the Suttles' example, constructed a non-genetic train algebra; he simultaneously proved that any train algebra of rank $r \leq 2$ is special (in the example, $r = 3$).

Holgate (HolP75) showed that in a Bernstein algebra B^2 is always an ideal (Corollary 3.4.13). Theorem 3.5.6 strengthens this statement. Corollary 3.5.8 appeared in (LyuY77b) as "genetic" and in (WorA80) as "special train". Beginning from it and to the end of the section all the results except Theorem 3.5.10. are due to the author (a part of them have been published without proof in (LyuY77b; LyuY78)). Theorem 3.5.10 appeared as the conjecture in the Russian version of this book. The first proof was done by Grishkov (GriA86). Later Krapivin (KraA89) obtained an upper estimate of the degree of nilpotence of barideal.

In connection with Corollary 3.5.17, we will note that Bernstein algebra with associative powers (in particular, any conservative algebra) is Jordan, i.e. satisfies the identity $(x_1 x_2)x_1^2 = x_1(x_2 x_1^2)$. Because in any Jordan algebra the powers are associative, the identity (3.5.19) in the class of Bernstein algebras is equivalent to being Jordan. Wörz-Busekros showed in (WorA80) that a baric algebra with the identity (3.5.19) is Bernstein. Note that a Bernstein algebra (with the exception of the constant one) can be neither associative, nor even alternate.

§3.6. Etherington (EthI39; EthI41a) introduced the construction of the duplication. The structure constants figured in his definition. Gonshor (GonH71) introduced an invariant definition. Theorem 3.6.3 is due to Etherington, Theorem 3.6.5 to Schafer (SchR49) in the context of his definition of genetic algebra. Two-level algebras are first introduced here.

§3.7. The notion of the stochastic space and the subsequent theory are due to the author. The theorems about hyperplanes in the simplicial space played an important role in (LyuY77a). Theorem 3.7.8 was published in (LyuY79), Theorem 3.7.11 in (LyuY75). The latter gives the solution of Bernstein problem in linear case. The partition of weight appeared as early as in (LyuY71) together with Lemma 3.7.12 during the investigation of the so called elementary gene structure. The latter is an important case of Bernstein problem. The rest of the results are published for the first time.

§3.8. The contents of this paragraph is the development of the article (LyuY77c) and the geometrical part of the article (LyuY74a).

§3.9. The conditions of normality were formulated in (LyuY71), and the procedure of normalization in a special context is described in (LyuY74a). The notion of kinship and the corresponding theorems are published for the first time.

Chapter 4

This chapter is based on the works (LyuY71; LyuY73; LyuY73c;LyuY74a; LyuY77a) by the author. The problem of the explicit description of stationary gene structure, solved in these works, had been formulated in (LyuY71) as the narrowing of the general Bernstein problem to a genetically interpreted case. In the same work the elementary gene structure (i.e. the Extended Hardy - Weinberg Law) was described and interpreted. The Quadrille Law was discovered by Bernstein (BerS24) and its genetic interpretation was suggested in (LyuY71). The Extended Quadrille Law appeared in (LyuY73).

The combined description of the mechanisms of s.g.s is given in (LyuY73c). Of course, other models are not excluded. This question requires further discussion on the biological level.

Chapter 5

Bernstein in (BerS23; BerS23a; BerS24) solved the problem posed by him for dimensions $n \leq 3$. He showed the necessity of Mendel's First Law, established the theorem of the constant inheritance when $p_{ik,j} > 0$ and the theorem about the two non- splitting types, somewhat more generally than Theorem 5.4.1, i.e. without the normality condition. Though elementary, Bernstein's proofs are quite complicated. The proofs in Chapter 5 are given by the author in (LyuY73b; LyuY76a; LyuY79a). Theorem 5.2.1 was proved in (LyuY77c) by carrying over arguments from (LyuY73b). A non-negative projection associated with an isolated idempotent is contained in a very implicit form in (BerS24, page 112); but on the whole §5.3 is due to the author (LyuY79a). The solution of Bernstein's problem for all exceptional populations was obtained in (LyuY76b), (where exceptional populations are called *quasilinear* in extending of the terminology of (LyuY75)) and in (LyuY73a) even earlier for the populations of the type $(n-1,1)$; the result for the populations of the type $(2, n-2)$ was given in (LyuY75a). Thus it has been solved for $n = 4$. For $n \geq 5$ the problem in full generality remains unsolved.

Theorem 5.7.1 was proved in (LyuY77c) by the topological method described in §5.7. Theorem 5.7.13 was proved in (LyuY79b), and there the proof of the topological lemma was given.

There is a natural generalization of Bernstein problem: to describe explicitly all evolutionary operators V satisfying the condition $V^{l+1} = V^l$, i.e. such that every trajectory comes to equilibrium for at most l steps (generations). Krapivin extended the theorem on constant inheritance to this situation (KraA78). The adequate class of algebras is baric with the identity $x^{[l+1]} = \sigma^{2^l}(x)x^{[l]}$. They are called *quasistationary* or *Bernstein algebras* of l-th *order* (KraA76).

Chapter 6

§6.1. Geiringer, (GeiH44), introduced the notion of linkage distribution, which plays a fundamental role in Reiersöl's work (ReiO62). The theory stated in §6.1 is constructed on the whole in (LyuY71). Lemma 6.1.12 was found by Trow (TroA13) for the case of three loci. The general formula is given by Bailey (BaiN61, page 140), discussing in general the problems of the genetic linkage.

§6.2. Reiersöl algebra appeared in his work (ReiO62), where also the main Lemma 6.2.1 was established. The language of the projections is equivalent to the language of differential operators (by the cancellation of the loci which belong to $L\backslash K$). The latter was used by Reiersöl. The equation (6.2.19) is also due to Reiersöl, (ReiO62). This approach turned out to be rather fruitful (ReiO62; LyuY71; KirV73; KunL79; KunL80).

§6.3. The theorems on the structure of the set of equilibrium states and on the convergence to equilibrium are due to Geiringer, (GeiH44). She obtained analogous results for a number of other genetic situations, including the multilocus haploid system linked with the X-chromosome, (GeiH48). Her article (GeiH49) summarizes this series of works. Later on the discussed theorems were reproven and generalized in (ReiO62; EllB66; KesH70; LyuY71; KirV73; KirV74). The entropy approach at this point was proposed in (KunL80a). See also (AkiE79, page 138).

§6.4. Theorem 6.4.1 has been obtained in (LyuY71) by the described method. In reality, awareness of this theorem in the context of the genetic algebras had existed for a long time, (EthI41b), and its general proof in this context was noted by Bennett, (BenJ54), and composed in detail by Holgate in (HolP67). To include the case under consideration, it is necessary to know that a Reiersöl algebra is genetic. This result was obtained by Gonshor in (GonH65) (see also (HolP68)). The described approach is presented in chapter 7. The explicit formula (6.4.5) for the evolutionary spectrum had been found by the author in (LyuY71), who also is responsible for Theorems 6.4.4, (LyuY71), and 6.4.5. The latter corrects an inaccurate exposition of the problem of multiple evolutionary roots in (LyuY71).

§6.5. The explicit evolutionary formula was first derived by the author in (LyuY71). The derivation is given here without any changes. For $l \leq 3$ cf. (ReiO62) but even for $l = 4$ the formula was not known before (LyuY71).

§6.6. In this section, the corresponding results from (LyuY71) have been stated with some refinements (Lemma 6.6.2). The estimate of the delay of the convergence to equilibrium in the case of a negative interference for three or more loci seems interesting from the biological point of view. The phenomena of the negative interference has been in detail discussed in the book by Bailey, (BaiN61). It does occur, but apparently not very often.

§6.7. This section is based on Kirzner's note (KirV73).

Appendix. The convergence to equilibrium for the X-linkage was studied by Geiringer in (GeiH48) and by Reiersöl in (ReiO62). The general results for the system with X-linkage and XY-linkage are due to Kirzner, (KirV73). See the further developed in (KirV74), where the joint action of all the main genetic factors in a system without selection is taken into consideration.

Chapter 7

§7.1. The results of this section are due to Gonshor, (GonH65), and Holgate, (HolP68). The given proof of Theorem 7.1.3 was suggested by the author (cf. (HolP68)).

§7.2. Theorems 7.2.2 and 7.2.5 were proved by Gonshor (GonH60; GonH71), Theorem 7.2.4 by Wörz-Buzekros, (WorA80)(cf.(HolP68)) who also remarked that in the complex case the converse of Theorem 7.2.2 is true. Theorems 7.2.3 and 7.2.6 are due to the author (cf. (WorA80)).

§7.3. Theorem 7.3.1 was proved by Holgate who in (HolP67) used a different argument. Theorem 7.3.2 is a refinement (KurG76) of Holgate's description (HolP67). Lemma 7.3.3 is due to M. Lyubich. Theorem 7.3.4 is due to Heuch (HeuI77). A detail theory of the exact linearization is elaborated by Abraham (AbrV80); AbrV80a).

§7.4. The explicit evolutionary formula for genetic algebras was derived by Heuch (HeuI77) who used the author's work (LyuY71) but computed in eigenbasis of genetic algebra.

Chapter 8

§8.1. Theorem 8.1.3 was done by Kirzner together with the author. An analogous fact for autosomal populations with selection provided that all the coefficients are positive, was established by Kun (KunL77) who also remarked that in this situation the number of fixed points is odd.

The conjecture that the number of fixed points is finite if $p_{ik,j} > 0$ was formulated in (LyuY71). The example with the three fixed points was constructed by Krapivin. He also proved in (KraA75) Theorem 8.1.10 for $n = 4$. The "descent" construction comes back to Bernstein (BerS24). Lemma 8.1.9 and Theorem 8.1.12 were obtained in (KraA76a).

§8.2. The estimate of $l(V)$ per Theorem 8.2.1 was attained by Kesten (KesH70) who also considered fractional quadratic maps. Kesten's proof is complicated. The proof given here was proposed by the author (LyuY74b). The idea of the reduction to the set of extreme points was further developed in joint papers (KraA77; KraA77a). Lemma 8.2.3 was first proved by R.L. Dobrushin (DobR56), but later rediscovered several times, in particular by the author in (LyuY74b). It is closely connected to the so called *ergodic coefficient* (DobR56). The estimates of Lemma 8.2.5 and Corollary 8.2.6 are due to Kesten (KesH70). Incidentally, they imply the well known Hopf's inequality (HopE63) for the second eigenvalue of a positive stochastic matrix.

One-dimensional case see (KesH70; LyuY74b). If there are some negative coefficients of the given quadratic map on the segment then its asymptotic behavior can be exclusively complicated even chaotic (see, for instance, (ColP80). At present this subject is studying very intensively and deeply.

It is interesting that there is an explicit formula for the iterations of an arbitrary quadratic map on the segment (DorD84) but, unfortunately, it does not help solving the asymptotic problems in this case.

§8.3. Example (8.3.1) first occurs in Menzel, Stein, Ulam (MenM55) as a numerical experiment (see also (SteP64)). This paper notes that the trajectories diverge. The first rigorous study of this example is due to Kesten (KesH70) who proved by the argument given that the trajectories converge to the boundary. Vallander in (ValS72) established that the limit set is infinite. The ergodic conjecture of Theorem 8.3.1 is in Ulam's book (UlaS60). The counterexample (Theorem 8.3.1) was obtained by M.I. Zakharevich (ZakM78). The question posed at the end of the section was formulated by Vallander (ValS72).

In (ValS72) and (SarA81) the limit sets of some other quadratic maps $\Delta^2 \to \Delta^2$ were investigated.

Chapter 9

§9.1. Formulas (9.1.6) for $m = 3$ are due to Wright and Dobzhansky (WriS46). Theorem 9.1.2 is in Mandel (ManS59), but without the note that the matrix Λ has to be non-degenerate; this note is present in later papers, e.g. (WeiB70). Hughes and Seneta wrote a critical note, (HugP75), about inaccuracies and gaps in papers about multiallele one-locus equilibrium. (HugP75) gives a general solution for the system (9.1.5) for the degenerate case $\det\Lambda = 0$.

§9.2. Local stability via the linear approximation was considered in Mandel (ManS59; ManS70) and Kingman (KinJ61) mostly for the interior polymorphism case.

Fundamental Theorem 9.2.7 was proved almost simultaneously by a number of authors (SchR49; MulH59; AtkF60; KinJ61a). The more informative Theorem 9.2.8 was proved in (LyuY76) by refining Kingman's (KinJ61a) proof (see also(LyuY80)).

Theorem 9.2.10 is essentially due to Kingman (KinJ61). For the number of stable equilibria see (KarS80).

§9.3 and the first half of §9.4 up to Theorem 9.4.6 inclusive comprise the contents of the work by the author, Maĭstrovskiĭ and Olkhovsky (LyuY80). A short abstract was published earlier (LyuY76). Convergence to equilibrium for the case of finite set of equilibrium states was earlier proved by Blakley (BlaG64). The general case of three loci was studied by Heiden (HeiU75).

The general theory of relaxation processes was first developed to systematically study computational minimization processes. Beginning with author's note (LyuY65) it was developed in a number of works by the author and G.D. Maĭstrovskiĭ, whose review is given in (LyuY70). The next step is due to Yu.G. Olkhovsky (OlkY73), whose techniques were used in the proof of Theorem 9.4.4.

Beginning with Theorem 9.4.7, the results (except Corollary 9.4.17) are due to the author and M. Lyubich, in particular, the example at the end of §9.4 is due to M. Lyubich. We note that although Theorem 9.4.7, its corollary, and Theorem 9.4.9 were long assumed in the literature, they were considered almost obvious and were not quite rigorously proved. This point of view is contradicted by the sufficiency proof of Theorem 9.4.9, apparently adequate. But we note that Theorem 9.4.9 uses only the Theorem 9.4.7 and Corollary 9.4.8 (the criterion of asymptotic stability) can be proved independently using simple topological considerations. It should be said that all the proofs become considerably simpler if some genericy conditions are fullfilled, in particular, if every principal minors of the fitness matrix are nonzero. A systematic presentation of the results in the generic case see (KarS72; KarS78). But a degeneracy may appear as a consequence of some biological circumstances. For example, if the gene A_1 is dominant then all λ_{1k} are equal to λ_{11}. For this (and purely mathematical) reason we have developed the whole theory without of any assumption of genericy.

A new local Lyapunov function for the Fisher dynamics was discovered by Karlin and Lessard (KarS83; KarS84b). This is the spectral radius of the matrix $B(p) = (W_i(p)\delta_{ik} + p_i\lambda_{ik})_{i,k=1}^{m}$ closely related to the matrix of the differential of Fisher map at equilibria.

A generalization of Fisher system is such that the survival coefficients depend on sex, i.e. they are $\lambda_{ik}^{(f)}$ and $\lambda_{ik}^{(m)}$. It is important because an explanation for sex-ratio in populations can be based on it (see (KarS84a)). Karlin and Lessard (KarS84a; KarS84b) studied stability of equilibria in the case $\lambda_{ik}^{(f)} + \lambda_{ik}^{(m)} = 1$ and under some assumptions of genericy. They also proved the convergence of all trajectories when all homozygotes are females and all heterozygotes are males.

Dorninger and Länger (DorD83) suggested an explicit formula for the iterations of Fisher map.

It is easy to see that a sufficiently small mutation together with selection in hyperbolic case gives the same regular dynamics (due to structure stability). Recently, Hofbauer (HofJ85) has deduced cycling behavior in a single locus model due to the interaction of selection and mutation.

§9.5, §9.6 mostly repeat the paper (KunL80a) by Kun and the author, whose abstract is given in (KunL79). The system of equations (9.5.1) in other notation is given in (MorP62) (see (LewR60) for two loci). The generalization of the Fundamental Theorem to additive selection is due to Ewens (EweW69a). A more general condition (9.5.11) was found by the author. We note that even for two loci the Fundamental Theorem may not hold for non-additive selection (MorP63). This of course is the case in our final example. A number of counterexamples to the Fundamental Theorem are in (RatV72; RatV73). The Fundamental Theorem is the subject of the special study (AruV70).

For the case of two diallelic loci with additive selection and heterosis in each locus, the convergence of trajectories was proved by Karlin and Feldman (KarS70a). Under their conditions the set of equilibrium states is finite (if linkage is not total) and the interior equilibrium is asymptotically stable.

If the probability of crossing over between two loci is close to 1/2 (the required degree of closeness depends on zygote survival coefficients), the global convergence is present not only under additive but also under multiplicative selection (BodW67; MorP68).

A long paper by Bodmer and Parsons (BodW62) studies the interaction of selection and linkage. Local stability of equilibrium states for two loci was studied by Moran (MorP67), Karlin and Feldman (KarS70). We note Karlin's review (KarS74) of two-locus models.

The example of non-covergence at the end of §9.6 including Theorem 9.6.15 is added by E.Akin. Recently, new examples of non-convergence have appeared. Hopf bifurcation occurs in two-locus two-allele models, see (HasA80).

Appendix. The characterization of additive selection is first published here. Recombination group was introduced (in slightly different form) by Shikata (ShiM64).

Many results in this monograph can be transferred to a system with continuous time t $(0 \leq t < \infty)$. Usually this case is easier to study. For example the proof of Fundamental Theorem of continuous time reduces to computing the derivative of W because of the evolutionary equations which here have the form $p_i' = W^{-1}p_i(W_i - W)$. Systems with continuous time are widely described in the literature (e.g. (MorP62; GimA74; SviY82)). Lately this area has been penetrated by methods of general differential dynamics (see (AkiE79; ShaS79)). In particular, the monograph (AkiE83) represents Hopf bifurcations for two-locus two-allele system.

References

[AbrV76] V.M. Abraham. A note on train algebras. *Proc. Edinburgh Math. Soc., Ser II,* **20** (1976), 53-58.

[AbrV80] Lineariziring quadratic transformations in genetic algebras. *Proc. Lond. Math. Soc.* (3) **40** (1980), 346-363.

[AbrV80a] The induced linear transformations in a genetic algebras. *Proc. Lond. Math. Soc.* (3) **40** (1980), 364-384.

[AlcM89] M.T. Alcalde, C. Burgueno, A. Labra, A. Micali. Sur les algèbres de bernstein. *Proc. Lond. Math. Soc.* (3) **58** (1989), 51-68.

[AkiE79] E. Akin. *The geometry of population genetics.* Springer-Verlag, New York–Berlin–Heidelberg–Tokyo, 1979.

[AkiE83] E. Akin. Hopf bifurcation in the two locus genetic model. *Memoirs Amer. Math. Soc.* **284** (1983).

[AruV70] V. Arunachalam. Fundamental theorem of natural selection in two loci. *Ann. Hum. Genet.* **34** (1970), 195-199.

[AtkF60] F.V. Atkinson, G.A. Watterson, P.A.P. Moran, A matrix inequality. *Quart. J. Math.* **11** (1960), 41, 137- 140.

[BaiN61] N.T.J. Bailey. *Introduction to the mathematical theory of genetic linkage.* Clarendon Press, Oxford, 1961.

[BenJ54] J.H. Bennett. On the theory of random mating. *Ann. Eugen.* **18** (1954), 311-317.

[BerM66] M. Bertrand. Algèbres non associative et algèbres génétiques. *Mem. des Sci. Math.* 162, Gauthier-Villars, Paris 1966.

[BerS22] S.N. Bernstein. Mathematical problems in modern biology (in Russian). *Science in the Ukraine* 1 (1922), 14-19.

[BerS23] S.N. Bernstein. Demonstration mathematique de la loi d'hérédité de Mendel. *C. r. Acad. Sci. Paris* **177** (1923), 528-531.

[BerS23a] S.N. Bernstein, Principe de stationarité et généralisation de la loi de Mendel. *C. r. Acad. Sci. Paris* **177** (1923), 581-584.

[BerS24] S.N. Bernstein. Solution of a mathematical problem related to the theory of inheritance (in Russian). *Uch. Zap. n.-i. kaf. Ukrainy* 1 (1924), 83-115.

[BerS26] S.N. Bernstein. Present state of probability theory and its applications (in Russian). In *Trudy All-Russ. Math. Congr.* Moscow, 1926, 50-63.

[BlaG64] G.R. Blakley. The sequence of iterates of a nonnegative non-linear transformation. *Bulletin Amer. Math. Soc.* **70** (1964), 712-715.

[BodW62] W.F. Bodmer, P.A. Parsons. Linkage and recombination in evolution. *Adv. Genet.* **11** (1962), 1-100.

[BodW67] W.F. Bodmer, J. Felsenstein. Linkage and selection: theoretical analysis of the deterministic two locus random mating model. *Genetics* **57** (1967), 1, 237-265.

[CasW03] W.E. Castle. The laws of heredity of Galton and Mendel and some laws governing race improvement by selection. *Proc. Amer. Acad. Arts and Sci.* **39** (1903), 223-242.

[CheS26] S.S. Chetverikov. Certain moments in evolutionary process from the point of view of modern genetics (in Russian). *Zhurn. Eksperim, Biologii* **A-2** (1926), 1, 1-54.

[ColP80] P. Cole, J.P. Eckman. *Iterated maps on the interval as dynamical systems.* Birkhäuser, Basel-Boston-Stuttgart, 1980.

[CroJ70] J. Crow, M. Kimura, *An introduction to population genetics theory.* Harper and Row, New York, 1970.

[DahC48] C. Dahlberg. *Mathematical methods for populations genetics.* Interscience Publishers, New York, 1948.

[DobR56] R.L. Dobrushin. Central limit theorem for nonhomogeneous Markov chains (in Russian). *Probability Theory and its Applications* **1** (1956), 1, 72-79.

[DobT70] Th. Dobzhansky. *Genetics of the evolutionary process.* Columbia Univ. Press, New York, 1970.

[DorD83] D. Dorninger, H. Länger. An explicit formula for the solution of the Fisher-Wright selection model in population genetics. *Diskr. Appl. Math.* **6** (1983), 209-211.

[DorD84] D. Dorninger, H. Länger. A formula for the solution of the difference equation $x_{n+1} = ax_n^2 + bx_n = c$. *Acta Sci. Math.* **47** (1984), 487-489.

[EllB66] B.E. Ellison. Limit theorems for random mating in infinite populations. *J. Appl. Probab.* **3** (1966), 94-144.

[EthI39] I.M.H. Etherington. Genetic algebras. *Proc. Roy. Soc. Edinburgh A* **59** (1939), 242-258.

[EthI40] I.M.H. Etherington. Commutative train algebras of ranks 2 and 3. *J. London Math. Soc.* **15** (1940), 58, 136-148.

[EthI41] I.M.H. Etherington. Non-associative algebra and the symbolism of genetics. *Proc. Roy. Soc. Edinburgh A* **61** (1941), 24-42.

[EthI41a] I.M.H. Etherington. Duplication of linear algebras. *Proc. Edinburgh Math. Soc., Ser. 2,* **6** (1941), 222-230.

[EthI41b] I.M.H. Etherington. Special train algebras. *Quart. J. Math. Ser. 2,* **12** (1941), 1-8.

[EweW69] W.J. Ewens. *Population genetics.* Metheun and Co., London, 1969.

[EweW69a] W.J. Ewens. Mean fitness increases when fitnesses are additive. *Nature* **221** (1969), 5181, 1076.

[FisR22] R.A. Fisher. On the dominance ratio. *Proc. Roy. Soc. Edinburgh A* **42** (1922), 321-341.

[FisR30] R.A. Fisher. *The genetic theory of natural selection.* Clarendon Press, Oxford, 1930. Reprint: Dover Publications, New York, 1959.

[FriY77] Ye. Ya. Frisman, A.P. Shapiro. *Selected mathematical models of divergent evolution population* (in Russian). Nauka, Moscow, 1977.

[GalD60] D. Gale. *The Theory of Linear Economic Models.* McGraw Hill, New York, 1960.

[GeiH44] H. Geiringer. On the probability theory of linkage in Mendelian heredity. *Ann. Math. Stat.* **15** (1944), 25-37.

[GeiH48] H. Geiringer. On the mathematics of random mating in case of different recombination values for males and females. *Genetics* **33** (1948), 548-564.

[GeiH49] H. Geringer. On some mathematical problems arising in the development of Mendelian genetics. *J. Amer. Statis. Assoc.* **44** (1949), 526-547.

[GerM60] M. Gerstenhaber. On nilagebras and linear varieties of nilpotent matrices II. *Duke Math. J.* **27** (1960), 1, 21-31.

[GimA74] A.A. Gimelfarb, L.R. Ginsburg, R.A. Polyektov, Yu.A. Pych, V.A. Ratner.*Dinamical theory of biological populations* (in Russian). Nauka, Moscow, 1982.

[GliV36] V.I. Glivenko. Mendelian algebra (in Russian). *Doklady Akad. Nauk SSSR* **8** (1936), 4, 371-372.

[GonH60] H. Gonshor. Special train algebras arising in genetics I. *Proc. Edinburgh Math. Soc. Ser. 2,* **12** (1960), 41-63.

[GonH65] H. Gonshor. Special train algebras arising in genetics II. *Proc. Edinburgh Math. Soc. Ser. 2,* **14** (1965), 41-53.

[GonH71] H. Gonshor. Contributions to genetic algebras. *Proc. Edinburgh Math. Soc. Ser. 2,* **14** (1971), 4, 289-298.

[GriA86] A. Grishkov, On the geneticness of Bernstein algebras. *Soviet Math. Dokl.* **35** (1987), 3, 489-492.

[HalJ24] J.B.S. Haldane. A mathematical theory of natural and artificial selection. *Proc. Cambridge Phil. Soc.* (1924- 1932), **23**, 1) 19-41); 2) 158-163; 3) 363-372; 4) 607-615; 5) 838-844; **26**; 6) 220-230; **27**; 7) 131-136; 8) 137- 142 **28**; 9) 244-248.

[HarG08] G.H. Hardy. Mendelian proportions in a mixed population. *Science* **28** (1908), 706, 49-50.

[HasA80] A. Hastings. Stable cycling in discrete time genetic model. *Proc. Nat. Acad. Sci.* **78** (1980), 7224-7225.

[HeiU75] U. Heiden. On manifolds of equilibria in the selection model for multiple alleles. *J. Math. Biol.* **1** (1975), 321-330.

[HeuI77] I. Heuch. An explicit formula for frequency changes in genetic algebras. *J. Math. Biol.* **5** (1977), 43-53.

[HeuI78] I. Heuch. Genetic algebras considered as elements in a vector space. *SIAM J. Appl. Math.* **35** (1978), 695-703.

[HofJ85] J. Hofbauer. The selection mutation equation. *J. Math. Biol.* **23** (1985), 41-53.

[HogL46] L. Hogben. *An introduction to mathematical genetics.* Norton and Co. New York, 1946.

[HolP67] P. Holgate. Sequences of powers in genetic algebras. *J. London Math. Soc.* **42** (1967), 496-498.

[HolP68] P. Holgate. The genetic algebra of k linked loci. *Proc. London Math. Soc., Ser. 2,* **18** (1968), 315-327.

[HolP72] P. Holgate. Characterization of genetic algebras. *J. London Math. Soc.* **6** (1972), 169-174.

[HolP75] P. Holgate. Genetic algebras satisfying Bernstein stationarity principle. *J. London Math. Soc.* **9** (1975), 613-624.

[HopE63] E. Hopf. An inequality for positive linear integral operators. *J. Math. and Mech.* **12** (1963), 683-692.

[HugP75] P.J. Hughes, E. Seneta. Selection equilibria in a multi-allele single-locus setting. *Heredity* **35** (1975), 185-194.

[JacN79] N. Jacobson. *Lie Algebras.* Dover Publications, New York, 1979.

[JenH16] H.S. Jennings. The numerical results of diverse systems of breeding. *Genetics* **1** (1916), 53-89.

[JenH17] H.S. Jennings. The numerical results of diverse systems of breedings with respect to two pair of characters, linked or independent with special relation to the effects of linkage. *Genetics* **2** (1917), 97-154.

[KarS70] S. Karlin, M.W. Feldman. Linkage and selection: two-locus symmetric viability model. *Theor. Pop. Biol.* **1** (1970), 39-71.

[KarS70a] S. Karlin, M.W. Feldman. Convergence to equilibrium of the two-locus additive viability model. *J. Appl. Probab.* **7** (1970), 262-271.

[KarS72] S. Karlin. Some mathematical models of population genetics. *Amer. Math. Monthly,* **79** (1972), 699-739.

[KarS74] S. Karlin. General two-locus models: some objectives, results and interpretations. *Theor. Pop. Biol.* **7** (1975), 364-398.

[KarS78] S. Karlin. Theoretical aspects of multilocus selection balance. *I, Studies in Math. Biology, II:Populations and Communities* (S.A. Levin,ed.) *MAA Studies in Math.* **16**, MAA, Wash., D.C. (1978), 503-587.

[KarS80] S. Karlin. The number of stable equilibria for the classical one-locus multiple selection model. *J. Math. Biol.* **9** (1980), 189-192.

[KarS83] S. Karlin, S. Lessard. On the optimal sex-ratio. *Proc. Nat. Acad. Sci. U.S.A.* **180** (1983), 5931-5935.

[KarS84] S. Karlin. Mathematical models, problems and controversies of evolutionary theory. *Bull. of the Amer. Math. Soc.* **10** (1984), 221-273.

[KarS84a] S. Karlin, S. Lessard. On the optimal sex-ratio: A stability analysis based on a characterization for one-locus multiallele viability models. *J. Math. Biol.* **20** (1984), 15-38.

[KemO57] O. Kempthorne. *An introduction to genetic statistics.* J. Wiley and Sons. New York, 1957.

[KemW29] W.B.. Kemp. Genetic equilibrium and selection. *Genetics* **14** (1929), 85-127.

[KesH70] H. Kesten. Quadratic transformations: a model for population growth I. *Adv. Appl. Probab.* **2** (1970), 1-82.

[KimM71] M. Kimura, M. Ohta, *Theoretical aspects of population genetics.* Princeton Univ. Press, 1971.

[KinJ61] J.F.C. Kingman. A mathematical problem in population genetics. *Proc. Cambridge Phil. Soc.* **57** (1961), 574-582.

[KinJ61a] J.F.C. Kingman. A matrix inequality. *Quart. J. Math.* **12** (1961), 45, 78-80.

[KirV73] V.M. Kirzhner. Behavior of the trajectories of a class of genetic systems. *Soviet Math. Dokl.* **14** (1973), 2, 378-381.

[KirV74] V.M. Kirzhner, Yu.I. Lyubich. General evolutionary equation and the limit theorem for genetic systems without selection. *Soviet Math. Dokl.* **15** (1974), 2, 582-586.

[KojK70] K. Kojima, ed. *Mathematical topics in population genetics.* Springer-Verlag, New York–Berlin–Heidelberg–Tokyo, 1970.

[KosV38] V.A. Kostizin. Sur le coefficients mendéliens d'hérédité. *C.R. de l'Acad. des Sci.* **206** (1938), 883-885.

[KraA75] A.A. Krapivin. Fixed points of quadratic operators with positive coefficients (in Russian). *Teor. Func., Func. Anal. and Appl.* **25** (1975), 62-67.

[KraA76] A.A. Krapivin. Quasistationary quadratic maps (in Russian). *Vestnik Kharkov Univ., Ser. Mat. and Mech.* **134** (1976), 104-108.

[KraA76a] A.A. Krapivin, Yu.I. Lyubich. Estimate of the dimension of the set of fixed point of a class of quadratic operators. *Siberian Math. J.* **17** (1976), 4, 704-707.

[KraA77] A.A. Krapivin, Yu.I. Lyubich. Estimates of Lipschitz constants for polynomial operators in a simplex. *Soviet Math. Dokl.* **18** (1977), 3, 718-721.

[KraA78] A.A. Krapivin. Generalization of Bernstein theorem (in Russian). *Dokl. Akad. Nauk Uzbek SSR,* (1978), 2, 9-11.

[KraA85] A.A. Krapivin, Yu.I. Lyubich. Estimates of the number of the invariant linear forms in Bernstein algebras (in Russian) *Vestnick Kharkov Univ., Ser. Probl. of Control and Mech.* **277** (1985), 85-92.

[KraA89] A.A. Krapivin. On a characteristic of the geneticness of Bernstein algebra (in Russian). *Vestnik Kharkov Univ.,* **334** (1989),75-79.

[KunL77] L.A. Kun. Equilibrium set of a biological population (in Russian). *Autom. and Telemech.* (1977), 4, 136-138.

[KunL79] L.A. Kun, Yu.I. Lyubich. Convergence to equilibrium under the action of additive selection in multilocus multiallele population. *Soviet Math. Dokl.* **20** (1979), 6, 1380-1382.

[KunL80] L.A. Kun, Yu.I. Lyubich. The H-theorem and convergence to equilibrium for free multi-locus population (in Russian). *Kibernetika* (1980), 2, 150.

[KunL80a] L.A. Kun, Yu.I. Lyubich. Convergence to equilibirum in polylocus polyallele population with additive selection (in Russian). *Problemy Peredachi Inform.* **16** (1980), 2, 92-102. Transl. in *Probl. Inform. Transmis.* (1980) Oct., 152-161.

[KurG76] G.Ch. Kurinnoĭ. Note on evolutionary spectrum of genetic algebras (in Russian). *Vestnik Kharkov Univ., Ser. Mat. and Mech.* (1976), **134** 89-95.

[LewR60] R.C. Lewontin, K. Kojima. The evolutionary dynamics of complex polymorphisms. *Evolution* **14** (1960), 3, 458-472.

[LiC76] C.C. Li. *First course in population genetics.* Boxwood Press, 1976.

[LyuY65] Yu.I. Lyubich. General theorems of quadratic relaxation. *Soviet Math. Dokl.* **161** (1965), 6, 588-591.

[LyuY70] Yu.I. Lyubich, G.D. Maĭstrovskiĭ. The general theory of relaxation processes for convex functionals. *Russian Math. Surveys,* **25** (1970), 1, 57-117.

[LyuY71] Yu.I. Lyubich. Basic concepts and theorems of evolutionary genetics for free populations. *Russian Math. Surveys* **26** (1971), 5, 51-123.

[LyuY73] Yu.I. Lyubich. On the mathematical theory of heredity. *Soviet Math.Dokl.* **14** (1973), 2, 579-581.

[LyuY73a] Yu.I. Lyubich. Structure of Bernstein populations of the type $(n-1,1)$ (in Russian). *Uspekhi Mat. Nauk* **28** (1973), 5, 247-248.

[LyuY73b] Yu.I. Lyubich. About a theorem of S.N. Bernstein. *Siberian Math. J.* **14** (1973), 3, 474.

[LyuY73c] Yu.I. Lyubich. Analogues to the Hardy-Weinberg Law (in Russian). *Genetics* **9** (1973), 10, 139-144.

[LyuY74] Yu.I. Lyubich. A class of quadratic maps (in Russian). *Teor. Func., Func. Anal. and Appl.* (1974), 21, 36-42.

[LyuY74a] Yu.I. Lyubich. Two-level Bernstein populations. *Math. of the USSR Sbornik* **24** (1974), 1, 593-615.

[LyuY74b] Yu.I. Lyubich. Iterations of quadratic maps (in Russian), in *Mathematical economics and functional analysis.* Nauka, Moscow, 1974, 109-138.

[LyuY75] Yu.I. Lyubich. Linear Bernstein populations (in Russian). *Teor. Func., Func. Anal. and Appl.* **22** (1975), 107-111.

[LyuY75a] Yu.I. Lyubich. Structure of Bernstein populations of the type $(2, n-2)$ (in Russian). *Uspekhi Mat. Nauk* **30** (1975), 1, 247-248.

[LyuY76] Yu.I. Lyubich. G.D. Maĭstrovskiĭ, Yu.G. Ol'khovskiĭ. Convergence to equilibrium under the action of selection in a single locus population. *Soviet Math. Dokl.* **17** (1976), 1, 55-58.

[LyuY76a] Yu.I. Lyubich. Algebraic proof of S.N. Bernstein's theorem on necessity of Mendel's law (in Russian). *Vestnik Kharkov Univ., Ser. Mat. and Mech.* **134** (1976), 85-86.

[LyuY76b] Yu.I. Lyubich. Quasilinear Bernstein populations (in Russian). *Teor. Func., Func. Anal. and Appl.* (1976), 26, 79-84.

[LyuY77] Yu.I. Lyubich. Nilpotence of commutative Engel's algebras of dimension ≤ 4 over **R** (in Russian). *Uspekhi Mat. Nauk* **32** (1977), 1, 195-196.

[LyuY77a] Yu.I. Lyubich. Proper Bernstein populations. *Probl. Peredachi Inform.* **13** (1977), 3, 91-100. Transl. in *Probl. Inform. Transmiss.* (1978), Jan., 228-235.

[LyuY77b] Yu.I. Lyubich. Bernstein algebras (in Russian). *Uspekhi Mat. Nauk* **32** (1977), 6, 261-263.

[LyuY77c] Yu.I. Lyubich. Stochastic algebras and their applications to mathematical genetics (in Russian), In *Mathematical methods in biology.* Naukova Dumka, Kiev, (1977), 119-131.

[LyuY78] Yu.I. Lyubich. Algebraic methods in evolutionary genetics. *Biom. J.* **20** (1978), 5, 511-529.

[LyuY79] Yu.I. Lyubich. General form of nonnegative projectors in $\mathbf{R}^n$ (in Russian). *Teor. Func., Func. Anal. and Appl.* (1979), 31, 84-86.

[LyuY79a] Yu.I. Lyubich. Algebraic proof of S.N. Bernstein's theorem on two pure types (in Russian). *Vestnik Kharkov Univ., Ser. Mat. and Mech.* (1979), **177** 44, 86-94.

[LyuY79b] Yu.I. Lyubich. A topological approach to a problem in mathematical genetics. *Russian Math. Surveys* **34** (1979), 6, 60-66.

[LyuY80] Yu.I. Lyubich. Classification of certain types of Bernstein algebras (in Russian). *Vestnik Kharkov Univ., Ser. Mat. and Mech.* **205** (1980), 124-137. Transl. in *Selecta Math. Sov.* **6** (1987), 1, 1-14.

[LyuY80a] Yu.I. Lyubich, G.D. Maĭtrovskiĭ, Yu.G. Ol'khovskiĭ. Selection-induced convergence to equilibrium in a single locus autosomal population. *Problem Peredachi Inform.* **16** (1980), 1, 93-104. Transl. in *Probl. Inform. Transmis.*, (1980), July, 66-75.

[LyuY84] Yu.I. Lyubich. Classification of non-exceptional Bernstein algebras of the type $(3, n-3)$ (in Russian). *Vestnik Kharkov Univ., Ser. Mech and Mat.* **254** (1984), 36- 42.

[MalG48] G. Malecot. *Les mathématiques de l'hérédité.* Masson et Cie, Paris, 1948.

[ManS59] S.P.H. Mandel. The stability of a multiple allelic system. *Heredity* **13** (1959), 2, 289-302.

[ManS70] S.P.H. Mandel. The equivalence of different sets of stability conditions for multiple allelic systems. *Biometrics* **26** (1970), 840-845.

[MayE65] E. Mayr. *Animal speciel and evolution.* Belknap Press, Cambridge, Mass., 1965.

[MenG66] G. Mendel. Versuche über Pflanzen-Hydriden. *Verh. naturforsch. Verein in Brunn* (1866), 4, 3-47, (transl. in (SinE58)).

[MenM55] M.T. Menzel, P.R. Stein, S.M. Ulam, Quadratic transformations I, Los Alamos, 1955 (Rep. LA2305).

[MikV77] V. Mikè. Theories of quasilinkage and "affinity": some implications for population structures. *Proc. Nat. Acad. Sci.* **74** (1977), 3513-3517.

[MorP62] P.A.P. Moran. *The statistical processes of evolutionary theory.* Oxford Univ. Press, 1962.

[MorP63] P.A.P. Moran. Balanced polymorphisms with unlinked loci. *Austral. J. Exp. Biol. and Med. Sci.* **16** (1963), 1-5.

[MorP67] P.A.P. Moran. Unsolved problems in evolutionary theory, in *Proc. 5th Berkeley Symp. Math. Stat. Probl.*, California, 1967, 457-480.

[MorP68] P.A.P. Moran. On the theory of selection dependent on two loci. *Ann. Hum. Genet.* **32** (1968), 183.

[MulH59] H.P. Mulholland, C.A.B. Smith. An inequality arising in genetical theory. *Amer. Math. Monthly* **66** (1959), 673-783.

[NagT77] T. Nagylaki. *Selection in one and two locus systems.* Springer-Verlag, New York–Berlin–Heidelberg–Tokyo, 1977.

[NorH28] H.T.J. Norton. Natural selection and Mendelian variation. *Proc. London Math. Soc., Ser. 2,* **28** (1928), 1-45.

[OlkY73] Yu.G. Ol'khovskiĭ. The use of Loyasevich's inequality in the theory of relaxation processes. *Siberian Math. J.* **14** (1973), 5, 792-795.

[PeaK04] K. Pearson. On a generalized theory of alternative inheritance with special references to Mendel's laws. *Phil. Trans. Roy. Soc. London A* **203** (1904), 1, 53-86.

[RatV71] V.A. Ratner. Equations of dynamics of Mendelian populations and the concept of generalized fitness (in Russian). *Sbornik Trudov po Agr. fis.* (1971), 30, 131-141.

[RatV72] V.A. Ratner. Generalized fitness and Fisher's fundamental theorem on natural selection, in *Studies in Theoretical Genetics.* Novosibirsk, 1972, 10-26.

[RatV73] V.A. Ratner. Mathematical theory of evolution of Mendelian populations (in Russian). *Probl. Evol.* **3** (1973), 151-213.

[ReiO62] O. Reiersöl. Genetic algebras studied recursively and by means of differential operators. *Math. Scand.* (1962), 10, 25-44.

[RobR18] R.B. Robbins. Some applications of mathematics to breeding problems II. *Genetics* **3** (1918), 73-92.

[RobR18a] R.B. Robbins. Some applications of mathematics to breeding problems III. *Genetics* **3** (1918), 375-389.

[SarA81] A.T. Sarymsakov. On trajectories of some quadratic transformations of two-dimensional simplex (in Russian). *Izv. Akad. Nauk Uzbek. SSR* (1981), 34-37.

[SchP59] P.A.G. Scheuer, S.P.H. Mandel. An inequality in populations genetics. *Heredity* **31**, (1959), 519- 524.

[SchR49] R.D. Schafer. Structure of genetic algebras. *Amer. J. Math.* **71**, (1949), 121-135.

[SchR66] *An introduction to nonassociative algebras.* Acad. Press, New York - London,1966.

[SerA34] A.S. Serebrovskiĭ. Properties of Mendelian equalities (in Russian). *Doklady Akad. Nauk SSSR* **2** (1934), 1, 33-36.

[SerA70] A.S. Serebrovskiĭ. *Genetical analysis* (in Russian). Nauka, Moscow, 1970.

[ShaS79] S. Shahshahani. A new mathematical framework for the study of linkage and selection. *Memoirs Amer. Math. Soc.* **17** (1979), 211, 1-34.

[ShiM64] M. Shikata. Group representation of genetic recombination. *Bull. Math. Biophys.* **26** (1964), 91-100.

[SinE58] E.W. Sinnott, L.C. Dunn, Th. Dobzhansky. *Principles of Genetics.* Fifth edition, McGraw-Hill, New York, 1958.

[SteP64] P.R. Stein, S.M. Ulam. Nonlinear transformation studies on electronic computers. *Rozpz. Mat.* **39** (1964), 1, 1-15.

[SutD72] D. Suttles. A counterexample to a conjecture of Albert. *Notices Amer. Math. Soc.* A-566 **19** (1972), 2, 78.

[SviY82] Yu.M. Svirezhev, V.P. Pasekov *Foundations of Mathematical genetics* (in Russian). Nauka, Moscow, 1982.

[TroA13] A.H. Trow. Forms of reduplications: primary and secondary. *J. Genet.* **2** (1913), 313-324.

[UlaS60] S.M. Ulam. *A collection of Mathematical problems.* Interscience Publishers, New York, 1960.

[ValS72] S.S. Vallander. On the limit behavior of iteration sequences of certain quadratic transformations. *Soviet Math. Dokl.* **13** (1972), 1, 123-126.

[WarH17] H.C. Warren. Numerical effects of natural selection acting on Mendelian characters. *Genetics* **2** (1917), 3, 305-312.

[WebF67] F. Weber. *Mathematische Grundlagen der Genetik.* Fischer, Jena, 1967.

[WeiB70] B.S. Weir. Equilibria under inbreeding and selection. *Genetics* **65** (1970), 371-380.

[WeiW08] W. Weinberg. Über den Nachweis der Verebung beim Menschen. *Jahresber. Ver. vaterl. Naturk. in Würtemb.* **64** (1908), 368-372.

[WeiW09] W. Weinberg. Über Vererbungestze beim Menschen. *Zeitschr., Abst. und Vererb.* **1** (1909), 277-230.

[WorA80] A. Wörz-Busekros. *Algebras in genetics.* Springer-Verlag, New York–Berlin–Heidelberg–Tokyo, 1980.

[WorA87] A. Wörz-Busekros. Bernstein algebras. *Arch. Math.* **48** (1987), 388-398.

[WorA89] A. Wörz-Busekros. Further remarks on Bernstein algebras. Proc. Lond. Math. Soc. (3) **58** (1989), 69-73.

[WriS46] S. Wright, Th. Dobzhansky. Genetics of natural populations XII (Experimental reproduction of some of the changes caused by natural selection in certain populations of *Drosophila pseudoobscura*). *Genetics* **31** (1946), 121-156.

[WriS68] S. Wright. *Evolution and the genetics of populations I (Genetic and biometric foundations).* Chicago Univ. Press, 1968.

[WriS69] S. Wright. *Evolution and the genetic of populations II (The theory of gene frequencies).* Chicago Univ. Press, 1969.

[ZakM78] M.I. Zakharevich. On the behavior of trajectories and an ergodic hypothesis for quadratic mappings of a simplex. *Comm. of Moscow Math. Soc., Russian Math. Surveys* **33** (1978), 6, 265-266.

Index

MIX
Papier aus verantwortungsvollen Quellen
Paper from responsible sources
FSC® C105338

If you have any concerns about our products,
you can contact us on
ProductSafety@springernature.com

In case Publisher is established outside the EU,
the EU authorized representative is:
Springer Nature Customer Service Center GmbH
Europaplatz 3, 69115 Heidelberg, Germany

Printed by Libri Plureos GmbH
in Hamburg, Germany